Medicine and Biomedical Sciences in Modern History

The aim of this series is to illuminate the development and impact of medicine and the biomedical sciences in the modern era. The series was founded by the late Professor John Pickstone, and its ambitions reflect his commitment to the integrated study of medicine, science and technology in their contexts. He repeatedly commented that it was a pity that the foundation discipline of the field, for which he popularized the acronym 'HSTM' (History of Science, Technology and Medicine) had been the history of science rather than the history of medicine. His point was that historians of science had too often focused just on scientific ideas and institutions, while historians of medicine always had to consider the understanding, management and meanings of diseases in their socio-economic, cultural, technological and political contexts. In the event, most of the books in the series dealt with medicine and the biomedical sciences, and the changed series title reflects this. However, as the new editors we share Professor Pickstone's enthusiasm for the integrated study of medicine, science and technology, encouraging studies on biomedical science, translational medicine, clinical practice, disease histories, medical technologies, medical specialisms and health policies.

The books in this series will present medicine and biomedical science as crucial features of modern culture, analysing their economic, social and political aspects, while not neglecting their expert content and context. Our authors investigate the uses and consequences of technical knowledge, and how it shaped, and was shaped by, particular economic, social and political structures. In re-launching the series, we hope to build on its strengths but extend its geographical range beyond Western Europe and North America.

Medicine and Biomedical Sciences in Modern History is intended to supply analysis and stimulate debate. All books are based on searching historical study of topics which are important, not least because they cut across conventional academic boundaries. They should appeal not just to historians, nor just to medical practitioners, scientists and engineers, but to all who are interested in the place of medicine and biomedical sciences in modern history.

This series continues the Science, Technology and Medicine in Modern History series.

David Redvaldsen

A History of British Eugenics since 1865

From Francis Galton to Designer Babies

palgrave
macmillan

David Redvaldsen
Department of Sociology and Social
Work
University of Agder
Kristiansand, Norway

ISSN 2947-9142 ISSN 2947-9150 (electronic)
Medicine and Biomedical Sciences in Modern History
ISBN 978-3-031-72289-9 ISBN 978-3-031-72290-5 (eBook)
https://doi.org/10.1007/978-3-031-72290-5

Cover credit: Galton Archive at University College London

This Palgrave Macmillan imprint is published by the registered company Springer Nature Switzerland AG
The registered company address is: Gewerbestrasse 11, 6330 Cham, Switzerland

To Alexandra. And to Marius.

Preface

For about sixty years during the twentieth century, eugenics guided research about itself and interlinked topics at the University of London. Considered an academic discipline, its base was located at University College London, which in 2018 commissioned a study into what the connection had involved. When the Inquiry into the History of Eugenics at UCL reported two years later, it led to the de-naming of the Galton lecture theatre, the Pearson building and the Pearson lecture theatre. In 2021 the College made a formal apology for its role in developing eugenics, pledging to give additional support to disabled and minority ethnic students. In 2019 the London School of Economics, in an exhibition marking the centenary of appointing William Beveridge as its director, acknowledged that he had been a eugenist. The same year BBC television ran a mini-series called 'Eugenics. Science's Greatest Scandal' and in 2022 geneticist Adam Rutherford narrated a BBC radio documentary examining the doctrine called 'Bad Blood'.

My motivation for embarking on the project predated those events. I attended a conference at Oxford Brookes in 2016 and afterwards, seated at the venue, I pulled forward a table to begin writing what became Chapter 2. The long delay in completing the book was caused by the need to make my way in life. I am eternally grateful to Stefan Berger and the late John Rogister for helping me so assiduously in finding a permanent position. Their assistance went far beyond what might be expected of a former examiner and a former teacher, respectively.

During the writing, I have also benefited from the kindness of others. Mandy Wise from Special Collections at University College London and Holly Peel from the Wellcome Library sent me high-resolution digital copies of the illustrations. Adelphi Genetics Forum granted permission to reproduce the ones to which they own the rights. The rest form part of the Galton papers, Galton Laboratory papers, Pearson papers, Penrose papers and College photographs of UCL. I acknowledge their provenance here and elsewhere. Lucy Kidwell and Saranya Siva from Palgrave Macmillan answered all my queries proficiently. Chris Renwick sent me an offprint of an article which I needed at short notice. The two anonymous referees gave valuable suggestions on improving the manuscript. Amanda Machin took the time to read a chapter and provide feedback. I was pleased to act on it. My employer the University of Agder funded a trip to London in February 2023, when I finished the archival and library research. The University provides a friendly environment in which to teach and research. The same can be said for Oxford Brookes, where I worked 2016–2018. Its focus on the history of eugenics spurred me on and I enjoyed discussing the subject with Marius Turda and Paul Weindling. None of the above, however, bear any responsibility for the shortcomings of the book.

On a personal level, Alexandra has her own unique take on the world. My friends in Fitzmaurice Place make Fridays worth looking forward to. My parents allowed me to pursue a scholarly career even though they knew about its risks and drawbacks.

Kristiansand, Norway David Redvaldsen

Praise for *A History of British Eugenics since 1865*

"In this deeply researched and original book, Redvaldsen advances a political assessment of eugenics in Britain, offering an analysis of the political attitudes and efforts of its main protagonists and the institutions through which they operated, notably the Galton Laboratory of Eugenics at University College London and the Eugenics Society. Cosmopolitan in scope, it attends to the linkages between British eugenics and its sister movements in the United States and Germany. It probes the class and social biases that pervaded British eugenics, casting new light on familiar subjects, especially contraception, sterilization, and anxieties about the differential birth rate. And it maps the evolution of eugenic thinking in response to Nazism as well as to the development of the welfare state and its persistence in considerations of in vitro fertilization. In all, a thought-provoking work, valuable not least for its framing in a politically engaged eugenics of human reproductive technologies past, present, and possibly to come."

—Daniel J. Kevles, *Professor Emeritus of History, Yale University, and author of* "In the Name of Eugenics: Genetics and the Uses of Human Heredity"

"David Redvaldsen reconstructs the enduring history of British eugenics in vivid detail, outlining its transformations in response to evolutionary theories. Far from fading after 1945, he shows how eugenics has a capacity to update itself, enduring until the present day."

—Paul Weindling, *Wellcome Trust Research Professor in the History of Medicine, Oxford Brookes University, and author of* "Health, Race and German Politics between National Unification and Nazism, 1870–1945"

"A reappraisal and reassessment of the history of British eugenics was long overdue and David Redvaldsen has provided it. Drawing on years of research and attuned to the latest historiographic debates, his book invites us to rethink the social imaginaries, public policies and political visions of the world created by eugenics. This is an engaging work which will resonate for years to come."

—Professor Marius Turda, *Centre for Medical Humanities, Oxford Brookes University*

CONTENTS

About the Author

David Redvaldsen is Lecturer in the Department of Sociology and Social Work at the University of Agder. He is the author of one previous book, *The Labour Party in Britain and Norway. Elections and the Pursuit of Power between the World Wars* (London: I.B. Tauris, 2011). An historian of Britain and Norway since the Congress of Vienna (1814), he is a Fellow of the Royal Historical Society. In 2014 he jointly won the Emile Lousse Prize for the best article on an aspect of parliamentary history. He has also published on the teaching of history and is a Fellow of the Higher Education Academy.

List of Figures

List of Tables

CHAPTER 1

Introduction

Our understanding of the history of eugenics has widened and deepened considerably in recent decades. In the early days of the subject, the signpost cases of Britain, Germany and the United States made up the entirety of scholarship regarding what was seen as a regrettable and mistaken experiment in social control. A knowledgeable person is aware that eugenics originated in Britain under the tutelage of Francis Galton and was then implemented substantively in the United States though forced sterilization and immigration restrictions. The latter was justified through contentions that the early colonial settlers, and others who hailed from the northern part of Europe, possessed better heredity than the eastern and southern Europeans who started arriving in the late nineteenth century. This mentality has been seen as a staging post towards outright National Socialism, with Stefan Kühl, Peter Weingart and Egbert Klautke arguing that German racialism did not develop in a vacuum.[1]

[1] Stefan Kühl, *The Nazi Connection. Eugenics, American Racism, and German National Socialism* (New York: Oxford University Press, 1994). Peter Weingart, 'Science and Political Culture: Eugenics in Comparative Perspective', *Scandinavian Journal of History* 24:2 (1999), pp. 163–177. Egbert Klautke, '"The Germans are beating us at our own game": American Eugenics and the German Sterilization Law of 1933', *History of the Human Sciences* 29:3 (2016), pp. 25–43.

D. Redvaldsen, *A History of British Eugenics since 1865*, Medicine and Biomedical Sciences in Modern History,
https://doi.org/10.1007/978-3-031-72290-5_1

The original title of Kühl's major work on the International Federation of Eugenics Organizations was *Die Internationale der Rassisten*, which received acclaim among Anglophones under a different name.[2] The excesses of the Nazis in the field of eugenics are notorious. *Lebensborn* stud-breeding, attempts to create a master race through compulsory sterilization, euthanasia of the disabled followed by genocide were all motivated by eugenic ideology. It is a story which has been told often. That, of course, does not make it any less true or in need of revision.

When Alison Bashford and Philippa Levine edited the *Oxford Handbook of the History of Eugenics* (2010), they did not deny that the practice of eugenics, worldwide and not just in Germany, had led to horrendous consequences.[3] That forced sterilizations had also taken place in democratic countries, especially the United States, had been known since the 1980s. Reacting to an historiography which had accumulated over half a century, Bashford and Levine's volume also described the variability of the doctrine and charted its effects across countries. As the cases they included differed in cultural and historical contexts, chapters dealt with the doctrine in its myriad forms across the globe. The *Oxford Handbook* is probably the best full-length primer for serious students of the subject today. It incorporates many findings from scholars working in languages other than the Western European ones.

By elucidating the history of eugenics in Hungary, Romania and Eastern Europe generally, Marius Turda has shown how the ideology was both transferred to and reformatted in that region.[4] Eugenics is not a single entity. The more we learn about it in different places, the better we understand how it was an international movement which nevertheless emphasized local concerns. A relevant example is the book *Latin Eugenics in Comparative Perspective* by Turda and Aaron Gillette. It deals with the interconnectedness of eugenics across borders, especially in a part of the world sharing a similar culture and value-driven assumptions. By

[2] Stefan Kühl, *For the Betterment of the Race. The Rise and Fall of the International Movement for Eugenics and Racial Hygiene* (Basingstoke: Palgrave Macmillan, 2013).

[3] Philippa Levine and Alison Bashford, 'Introduction: Eugenics and the Modern World' in Alison Bashford and Philippa Levine (eds.), *The Oxford Handbook of the History of Eugenics* (Oxford: Oxford University Press, 2010), pp. 3, 6, 9.

[4] Marius Turda, *Eugenics and Nation in Early 20th Century Hungary* (Basingstoke: Palgrave Macmillan, 2014). *Idem*, *Science and Ethnicity II. Biopolitics and Eugenics in Romania, 1920–1944* (Bucharest: Municipal Museum of Bucharest, 2019).

considering France, Italy, Spain, Portugal, Romania and settler states in Latin America, they demonstrate how, because the Catholic faith could not brook sterilization, eugenics had to find other ways of making its mark.[5] Combined with and immersed in natalism, puericulture and public health measures, eugenics could be transformed into a relatively benign and inclusive ideology.

These innocuous cases do not detract from the darker side of the doctrine. Sweden is thought to have sterilized 63, 000 of its citizens before 1975, many of whom had not consented.[6] Depriving men and women of the right to reproduce is an egregious violation of human rights. For non-specialists eugenics largely features as the set of ideas which provided the rationale for sterilization. In most countries such procedures often targeted especially women, vulnerable people and ethnic minorities. Assertions stemming from supposed heredity are still used to justify unfair treatment of the least powerful and to rationalize why life-chances differ profoundly according to social background. As will be argued here, eugenics was a modern bourgeois doctrine. While other aspects varied, that was true for every version of it, across time and space. Eugenics has never or hardly ever been utilized to give support to ordinary people over specialists or indigenous tribes in preference to European settlers. Eugenics could *sound* radical if the intelligentsia berated aristocrats for being divorced from science and modernity, which only confirms its middle-class nature. As such it offered a conception of the world attractive to social reformers, who tended to be thinkers guided by a theoretical understanding of contemporary problems. That eugenics was able to muster support from every political persuasion is by now so well established that little effort will be spent on proving it in what follows.

The book trails British eugenics from its inception in the second half of the nineteenth century and virtually to the present day. Along the way the United States, a country of which British intellectuals often kept abreast, is referenced. The two nations form a community of language and, to a lesser extent, of ideas. Germany, another paradigm in the history of eugenics, will sometimes appear too, as a contrast. Eugenics in Britain has previously been analysed in works by Geoffrey Searle,

[5] Marius Turda and Aaron Gillette, *Latin Eugenics in Comparative Perspective* (London: Bloomsbury Academic, 2014).

[6] Paul Weindling, 'International Eugenics. Swedish Sterilization in Context', *Scandinavian Journal of History* 24:2 (1999), p. 187.

Pauline Mazumdar and Clare Hanson. Searle's *Eugenics and Politics in Britain 1900–1914* was the first study and introduced the term 'reform' eugenist. A conceptualized dichotomy between 'mainline' and 'reform' eugenics was originated by Daniel Kevles. His seminal monograph *In the Name of Eugenics* considered the ideology in both Britain and the United States. Its findings became instrumental in the field. These included the early conflicts in the movement, the shift towards 'reform' eugenics in the 1930s and how 'positive' eugenics reappeared through artificial insemination. Mazumdar's *Eugenics, Human Genetics and Human Failings* followed the Eugenics Society from its founding and into the 1950s. Hanson's *Eugenics, Literature and Culture in Post-war Britain* assessed the cultural and social effects of eugenics after 1945. Specialized studies include Dan Stone *Breeding Superman*, Debbie Challis *The Archaeology of Race,* Bradley Hart *George Pitt-Rivers and the Nazis* as well as Patrick Merricks *Religion and Racial Progress in Twentieth-century Britain.* Two of Greta Jones's books, examining social Darwinism and social hygiene, had significant coverage of eugenics. There have also been several biographies of Galton and one of Karl Pearson.

This contribution is an institutional history of the Galton Laboratory, the scientific unit researching eugenics at University College in London, and the Eugenics Society. In addition, it reflects on how eugenics might still be with us through the new reproductive technologies. The emphasis is on eugenics as a political movement, rather than as a science. The doctrine emerged from biology and until the 1930s almost all geneticists agreed with its main tenets. The vocational debate instead concerned how to interpret and apply eugenics meaningfully. The scientific discourse which led to eugenics being challenged is not narrated here.[7] It is proposed instead that political developments, and internal manoeuvres within the Eugenics Society, caused a change of tack. The explanation for how the doctrine accommodated itself to post-war realities is contextual.

A global ideology like eugenics lends itself admirably to being investigated through transnational and comparative research. Eugenists attended international conferences, kept up to date on the work of colleagues abroad, corresponded across borders and followed developments elsewhere with interest. Many were physicians or academics who belonged

[7] See Pauline M. H. Mazumdar, *Eugenics, Human Genetics and Human Failings. The Eugenics Society, its Sources and its Critics in Britain* (London: Routledge, 1992), chapters 3 and 4.

to transnational networks within their professions, which affected their eugenic endeavours too. But since eugenics accommodated itself to each country in which it had a presence, furthering knowledge of its presence in a single society has its advantages. Britain was the homeland of the doctrine and its eugenists enjoyed a certain prestige on account of that. They were nevertheless nowhere near as successful as their counterparts in other nations. Crucially, they were unable to pass a law on 'voluntary' sterilization, seen as the apex of the doctrine. They were surpassed in this regard by North American, Swiss, German, Scandinavian, Finnish, Icelandic and Baltic eugenists.

The fact that the campaign for 'voluntary' sterilization stalled later spared Britain the opprobrium which accompanied revelations of human rights abuses by other states. Undoubtedly, it was traumatic for nationals of those countries to learn that their own polities had practised surgical procedures depriving citizens of a fundamental natural right. Financial compensation cannot properly make amends for such a grievous loss. Whole lives changed their course through being childless. The psychological damage includes feeling belittled and rejected. However, for the purposes of scholarly discourse, it makes a difference that sterilization remained illegal in Britain, except if the operation was regarded as life-saving.[8]

A few eugenic ideas were of value. The institutionalization of 'moral imbeciles' and, occasionally, mental defectives may have averted the odd tragedy. Contraception, though initially advanced from motives of snobbery, helped especially women to live better and more fulfilling lives. An excessively large family is a classic cause of poverty, while living standards are raised through having fewer children. Mental testing proved that pupils not belonging to the highest social strata could be talented too. Eugenics thus played a part in moving Britain towards meritocracy. Left-wing eugenists argued that the redistribution of resources improved social efficiency. Though individuals varied in their capabilities, only genuine equality of opportunity would allow such distinctions to be confirmed and acted upon. And though lack of respect for the global poor surely fuelled concerns over overpopulation, the concomitant risks of conflict and harm to the planet are nonetheless real.

[8] 'Voluntary Sterilization. Report of the Conference at the Caxton Hall on May 23rd 1932', *Eugenics Review* 24:2 (1932), pp. 133, 134.

This examination of eugenics in Britain consists of five full-length chapters. Chapter 2 considers the early days of eugenics. It finds its origins in Darwinism and sees Galton, in the context of his time, as initially seeking to address problems thrown up by the theory of evolution. Not incidentally, he rooted for his own social class, believing it to better than the masses on the one hand and the idle aristocracy on the other. The ideal person for Galton was intelligent, healthy and virtuous. Where outsiders shared these characteristics, they should be recruited into the ranks of the educated. Although theoretically sympathetic to socialism, he did not think that redistribution was a cure for social problems. Eugenics became entangled in concerns about 'degeneracy' and the residuum no later than the first decade of the twentieth century. That was when the Eugenics Society emerged and its members were usually social Darwinists. They had a profound influence on how the doctrine developed.

Chapter 3 narrates the story of the Galton Laboratory, which arose from Galton's financial support to the University of London. Starting out as the Eugenics Record Office in 1904, it came under the supervision of Karl Pearson in 1907 when it was renamed the Eugenics Laboratory. Upon Galton's death in 1911, it was merged with the Biometric Laboratory at University College and his bequest founded a chair in eugenics with Pearson as the first holder. Although the Galton Laboratory carried out scientific research at a high level, it had little impact in government circles, for which Pearson implicitly aimed. Its publications were suited for physicians and geneticists, with the *British Medical Journal* and Establishment newspapers supporting its efforts. In the 1930s its research on race dismissed the theory that miscegenation might be harmful and one employee published a book opposing German anthropological claims. Live discoveries in genetics were attempted through the introduction of a serological laboratory. The departure of Pearson's successor Ronald Fisher in 1943 changed the nature of the research.

Chapter 4 charts the Eugenics Society from its founding in 1907 to 1945. It represented the popular strand of the doctrine and had the greatest say in how British eugenics would develop. The Society was a pressure group seeking to implement eugenic policies. These were inherently right-wing, such as tax cuts for wealthy parents and greater control of the residuum. The sterilization which became its key demand after 1931 was intended for the lowest classes. In the 1930s the organization increasingly sided with 'reform' eugenics. This was partly motivated by trying to bolster support for sterilization, but also occurred due to changing attitudes and personnel. Hence there was a leftward shift through the 1930s and during the Second World War.

Chapter 5 covers post-war eugenics until 1990. The Galton Laboratory started researching value-free human genetics under Lionel Penrose, who was appointed in 1944. The changeover to ordinary medical science was symbolized by removing the eugenic name tag in 1963. The Society continued to promote eugenic policies, concentrating on 'problem families' in the late 1940s and early 1950s. It still supported 'voluntary' sterilization and reversed its previous progressive stance on William Beveridge's notion of flat-rate child benefit. Social class continued to be shorthand for eugenic 'fitness', although now supplemented by measured intelligence. In the 1960s it pivoted towards research and becoming a learned association. This did not hinder eugenic activity on a reduced scale, facilitating birth control for what it considered unsuitable parents and defending abortion rights vigorously.

Chapter 6 is about the new technologies of reproduction. In the 1930s the Society discussed artificial insemination for eugenic purposes, returning to the subject in the 1950s. Apart from contributing in 1959 to an inquiry conducted by the Home Office, it researched techniques of low-temperature semen preservation in the 1960s. In the 1970s Society member Robert Edwards succeeded in developing in vitro fertilization in collaboration with the gynaecologist Patrick Steptoe. This led to a whole range of technical reproductive innovations gradually becoming available. Prenatal screening, preimplantation genetic diagnosis, sex selection, cloning and parthenogenesis are examined for their eugenic effects. Although justified by reproductive choice, such medical procedures enable what is called 'new eugenics'. Additionally, the chapter argues that the substance of both 'mainline' and 'reform' eugenics continues to exist.

CHAPTER 2

Francis Galton and the Early Days of Eugenics

The field in which eugenics flowered was the world view inaugurated by Charles Darwin in *On the Origin of Species by Means of Natural Selection*, published in 1859. Its alternative title was *The Preservation of Favoured Races in the Struggle for Life*. Darwin described how variants of species changed characteristics over long periods of time, according to what benefited individual plants or animals in their struggle to survive and reproduce. The theory negated the traditional view of the fixity of species and explained how each geological age produced a different set of flora and fauna. Thus it incorporated recent scientific discoveries of how the earth was millions, not thousands, of years old, as shown by Sir Charles Lyell in his *Principles of Geology* (1830–1833).[1] Geologists had uncovered a trail of fossil evidence which made it obvious that sets of animal populations had succeeded each other.[2] Darwin had gestated the theory of evolution over two decades and was in the process of writing a longer volume when he was roused to publish speedily by the receipt of

[1] Charles Darwin, *On the Origin of Species by Means of Natural Selection or The Preservation of Favoured Races in the Struggle for Life* (London: Folio Society, 2006), p. 381.

[2] Peter J. Bowler, *Monkey Trials and Gorilla Sermons. Evolution and Christianity from Darwin to Intelligent Design* (Cambridge MA: Harvard University Press, 2007), pp. 30–31.

D. Redvaldsen, *A History of British Eugenics since 1865*, Medicine and Biomedical Sciences in Modern History,
https://doi.org/10.1007/978-3-031-72290-5_2

a manuscript in 1858 from another naturalist, Alfred Russel Wallace, who asked for comments on what were essentially the same findings. Because Darwin had been working on evolution for so long, the resulting book was able to give an impressively wide defence of the theory from every scientific angle, through his familiarity with the work of other researchers.

On the Origin of Species dealt with humans only very perfunctorily. It was, however, entirely clear to its readers that our species too had developed by means of the same process. At the famous Oxford debate in 1860, the biologist Thomas Henry Huxley and Bishop Samuel Wilberforce clashed on whether humans were descended from an ape-like creature. It led assorted thinkers to reflect on what characteristics had been serviceable to humans in various stages of their evolution. Darwin dealt with the social imperative as providing increased chances of survival in *The Descent of Man and Selection in Relation to Sex*, published in 1871. In that book, he also answered his critics on questions such as whether morality is innate or the consciousness of a divinity universal.[3] Darwin rejected both these views, but he accepted that the idea of God had been ennobling. He implicitly made normative judgements on the quality of human beings, by noting how superior people and civilizations had greater empathy with those who differed from themselves. The highest form of morality was exemplified by Marcus Aurelius, who refused even to *think* badly of others.[4]

Darwin as a scientist had no blueprint for how to improve society or the human condition. He described himself as 'a Liberal or Radical' and gave a public subscription of £ 10 to the Jamaica Committee, which included, among others, the MPs Charles Buxton, John Bright and John Stuart Mill. It had been established to prosecute Edward Eyre, the island's governor, who had put down an uprising in Morant Bay in 1865 with real savagery.[5] Darwin considered slavery 'a great crime' and yet noted that this institution had only recently been abolished (as late as 1833 in the colonies).[6] At home, he supported the care of the weak and vulnerable,

[3] Charles Darwin, *The Descent of Man and Selection in Relation to Sex* (London: John Murray, 1909), pp. 191, 194, 933, 936.

[4] Darwin, *The Descent of Man*, pp. 187–189.

[5] University College London (UCL), Galton Archive, Galton/1/1/9/5/7/15. Darwin to Galton 28 May 1873. Adrian Desmond and James Moore, *Darwin's Sacred Cause. Race, Slavery and the Quest for Human Origins* (London: Allen Lane, 2009), p. 356.

[6] Darwin, *The Descent of Man*, pp. 180, 205–206.

in spite of those who were using his theory to argue that such charity was misplaced. Darwin nevertheless thought it would be better if those who were 'weak in body or mind' did not reproduce. In *The Descent of Man,* he quoted Alfred Russel Wallace, his cousin Francis Galton and the writer William Rathbone Greg on how modern civilization was allowing individuals with poor hereditary endowment to survive and breed. It meant that one of the mechanisms underlying evolution was being eroded.

Ultimately, none of these intellectuals could justify denying the masses medical aid to prevent them from impeding the continual improvement of the species. But Galton was interested in how it might be possible for mankind to take control of its own destiny and evolution, instead of blindly letting nature do its work over uncounted years. His reading of *On the Origin of Species* upon publication had made him think about how heredity worked (although how the environment shaped species was the main message of the book).[7] Gregor Mendel's work on peas and flowers, published in 1865, was only made known to British scientists in 1900, when William Bateson championed it, becoming the first biologist to use the word 'genetics' six years later.[8] Galton studied heredity independently, through his own measurements and experiments. One of his methods was to postulate the existence of a hypothetical 'mid-parent' with characteristics which were the mean of the actual parents. He then investigated how closely the offspring matched this 'mid-parent' and discovered that they approached it in a predictable manner. Most observed cases were close to the 'mid-parent' and the farther one departed from it, the fewer the number of individuals (Fig. 2.1).

Galton's Personal Background

Francis Galton was born in Birmingham in 1822 into a prosperous family of manufacturers and bankers. He was the youngest son of Samuel Tertius Galton and Violetta Galton, née Darwin. The Galtons were traditionally Quakers, but Francis's father had converted to Anglicanism in 1816

[7] Francis Galton, *Memories of My Life* (London: Methuen, 1908), pp. 287–288.

[8] Nils Roll-Hansen, 'Eugenics and the Science of Genetics' in Alison Bashford and Philippa Levine (eds.), *The Oxford Handbook of the History of Eugenics* (Oxford: Oxford University Press, 2010), p. 83.

Fig. 2.1 Charles Darwin (reproduced from the Galton Archive, UCL Special Collections)

probably under the encouragement of his wife.[9] Quakerism nevertheless continued to affect Francis, who did not attend a public school for adolescents of his social background.[10] His third sister, Adèle, twelve

[9] Karl Pearson, *The Life, Letters and Labours of Francis Galton*. Volume 1 (Cambridge: Cambridge University Press, 1914), pp. 25, 62.

[10] Galton, *Memories of My Life*, p. 20.

years his senior, suffered from curvature of the spine, which necessitated frequent rest. Not one to waste all the time spent lying down, she had educated herself to a remarkable standard. When Francis was born, she asked if she could take charge of his pre-school education.[11] Their parents agreeing, Francis was trained to be a scholar from infancy. Adèle taught him Latin and some French before he was five years old, also arithmetic and the clock.[12] Before the age of seven, he was familiar with the work of many great English writers, such as Pope and Shakespeare.[13] His proud mother, the daughter of Erasmus Darwin the naturalist, kept records of his accomplishments in childhood. Thus, Galton grew up valuing intelligence over any other characteristic. His later conception of eugenics celebrated the milieu into which he had been born. He himself believed that the taste for studies and furthering knowledge were inborn. And although his beloved sister Adèle was disabled, his eugenic schemes were not welcoming towards people with physical impairments.

Galton's interests turned in a scientific direction during adolescence. He did well in his studies of medicine at Birmingham General Hospital and King's College London. Then, in 1840, he matriculated at Trinity College, Cambridge, where his academic record dipped. He gained a fourth class in the May 1843 examinations and graduated in 1844 with a poll degree in mathematics.[14] The death of his father later that year released him from his proposed vocation as a medical doctor, as he came into a considerable inheritance. He took up travelling and exploring, attaining the fellowship of the Royal Geographical Society and winning its gold medal in 1853 for mapping uncharted parts of Africa. In 1855 he wrote a book called *The Art of Travel*. In those days, he might be described as a geographer with a bent towards measurement and quantification. One of his axioms was 'when in doubt, count'. In 1856 he was elected a fellow of the Royal Society, the most prestigious scientific body in Britain. In a career that spanned six decades, Galton wrote hundreds of scientific or popular articles and more than twenty books. In the nineteenth century it was possible to make contributions to a

[11] Galton/1/2/1/6/2, Biographical Sketch of Francis Galton by his Mother, folio 1r.

[12] Galton/1/2/1/3, folder 2 of 2, Francis Galton to Adèle Galton 15 Feb. 1827, folios 8r–9v.

[13] Galton/1/2/1/4, folio 1r.

[14] Pearson, *Life, Letters and Labours*, volume 1, pp. 179, 181.

wide range of disciplines. He discovered the anti-cyclone in meteorology and the uniqueness of fingerprints, originated the correlation-coefficient and regression analysis in statistics as well as inventing various machines and devices such as the ultra-sonic whistle. He was among the first to conduct blood transfusions, inspired by a desire to test Darwin's theory of pangenesis, and laid much of the groundwork for individual differences in psychology.

Galton married Louisa Butler, who belonged to another intellectually distinguished family, in 1853. Her late father had been a clergyman and headmaster of Harrow School 1805–1829. There were no children of this marriage, which may have been on account of Louisa's delicate health, the after-effect of Francis's contracting a venereal disease as a young man abroad or lack of passion between them.[15] Neither did Francis's brothers nor Louisa's sisters have any issue.[16] Galton thus saw for himself how the English intelligentsia was declining in numbers. He and Louisa had both belonged to large families. It is probably not coincidental that Galton germinated eugenics as an idea when he and Louisa, who was the same age as him, were in their late forties. By extension, he was aware of how the number of children born to a couple depended on their age at marriage. Eugenics initially advocated earlier marriage among the able classes in the community. It sounded innocuous enough, but the explicit ordering of humanity on a scale of worth and the hereditarian assumptions underlying it would in time make eugenics a controversial ideology and thoroughly despised by many.

It was intimately connected to Galton's views. What qualified someone as being valuable to humanity was primarily the capacity for intellectual activities. The test of that ability came through one's output or results in scholastic examinations. Not having children, he may have seen the doctrine as a surrogate contribution to society. Galton's knowledge of statistics helped to make his conclusions on heredity plausible after he became interested in the topic in 1859. Darwin had held the environmentalist assumption that men (to the Victorians it was always 'men')

[15] Daniel J. Kevles, *In the Name of Eugenics. Genetics and the Uses of Human Heredity* (Berkeley CA: University of California Press, 1986) p. 12.

[16] Richard A. Soloway, *Demography and Degeneration. Eugenics and the Declining Birthrate in Twentieth-century Britain* (Chapel Hill NC: University of North Carolina Press, 1995), pp. 19–20.

did not differ much in intellect, only in zeal and hard work.[17] Galton at least superficially convinced him that the penchant for study was mostly inherited.

Heredity and the Early Conception of Eugenics

The first steps towards eugenics were taken in a two-part article published in *Macmillan's Magazine* in June and August 1865. Galton argued that mental faculties were hereditary because a high proportion of people who had been successful in art, music, literature and scholarship or were otherwise celebrated for their great minds had kinsmen who were themselves distinguished.[18] Taking 85 men listed under the letter A from Walford's *Men of Today*, 25 were found to have relations also included in the volume, of which twelve were brothers and eleven sons or fathers. The incidence of accomplished individuals being interrelated was much higher than chance alone would allow. Galton did not consider that since education in the nineteenth century was the preserve of the few and only a limited number had the financial means to obtain it as well as the leisure to pursue intellectual interests, most work qualifying for distinction would indeed be conducted by a restricted range of families. To him, the findings were proof that most talent was to be found among those who had already been successful, with a few individuals from the masses also occasionally rising to that level. He recognized that high birth was a major advantage in politics and the army, but thought that science and literature were open to all. It is not surprising that distinguished people should have family connections, because they were a subgroup of the educated and marriage usually occurs between individuals of similar station in life.

Galton had made a beginning towards understanding heredity. In 1865 it had not been systematized how children resembled their parents physically or mentally.[19] It would take decades before it became clear that his discoveries did not amount to a full proof. He had found that abilities ran in families among those who were *active* in the fields concerned, but this had no bearing on the *inactive* general population. One could

[17] Galton, *Memories of My Life*, p. 290.

[18] Galton, 'Hereditary Talent and Character. Part I', *Macmillan's Magazine* 12 (1865), pp. 160–161, 163.

[19] Adam Rutherford, *Control. The Dark History and Troubling Present of Eugenics* (London: Weidenfeld & Nicolson, 2022), p. 55.

not say whether it had talents too and, if so, how that would affect his scheme of inheritance. Galton either ignored or did not realize that he had only obtained a partial proof of heredity. He continued arguing along the same lines in *Hereditary Genius* (1869), adding copious data to his findings. He believed that only a little talent had been squandered due to outsiders not receiving the training required to compete with his pantheon of heroes. The United States, which valued the education of the masses more, hardly surpassed his own country in literature, philosophy or art.[20] Galton characterized differing levels of attainment by letters of the alphabet, the highest being X, which stood for achievements occurring at the rate of one in a million among the population.[21] He determined each individual's level by what was said about him (or occasionally her) by others in the same field and general reputation in society. Alternatively, it could be found by performance in scholastic examinations. Because he had discovered from the Belgian astronomer Adolphe Quetelet's calculations that characteristics varied above and below the mean in the same manner, he used an identical system of letters but in lower case, to label people who were below average. Thus someone rated 'B' was as much above average as someone rated 'b' was below it.

Galton believed that the civilization which Darwin lauded as improving all its members' chances of survival was primarily the work of gifted individuals. He thought that mankind's progress would accelerate if more such people were born. His taxonomy of the higher ranges of ability included the 'eminent' (one in 4, 300 individuals or class F) and the 'illustrious' (rarer than one in 79, 000 individuals or above class G).Everyone varied from the central tendency of his or her family. Therefore accomplished families must be encouraged to raise more children, which had to come about through earlier marriage. Selective breeding had already been recommended in other settings. In the medical context, eugenics existed in Britain in earlier decades of the nineteenth century.[22] Plato outlined schemes of selective breeding in *The Republic*, so it had also surfaced in ancient Greece. Galton regarded the Athenians between 530 and 430 BC

[20] Francis Galton, *Hereditary Genius. An Inquiry into its Laws and Consequences* (London: Macmillan, 1892), p. 40.

[21] Galton, *Hereditary Genius*, pp. 34, xii–xiii.

[22] J. C. Waller, 'Ideas of Heredity, Reproduction and Eugenics in Britain, 1800–1875', *Studies in the History and Philosophy of the Biological and Biomedical Sciences* 32:3 (2001), p. 458.

Table 2.1 Classification of intellectual gifts

Ability level	*Rarity*	*Percentile*	*Galton's description*
A	1 in 4	75	Within mediocrity
B	1 in 6	84	Intelligent provincial
C	1 in 16	93.75	Better foreman of jury
D	1 in 64	98.4375	Wins ordinary prizes
E	1 in 413	99.757	High ability
F	1 in 4,300	99.9767	Eminent
G	1 in 79, 000	99.9987	Borderline illustrious
X	1 in 1, 000,000	99.9999	Deserves state funeral

as the most gifted race.[23] Both he and Plato took it for granted that it was men who would bring about the progress they sought. Galton later defined a large family as one in which there were three male children.[24] The import of *Hereditary Genius* was not just proposing eugenics, but the construction of a nomenclature whereby inheritance became the main explanation for intellectual differences. They had never been studied so systematically before (Table 2.1).

The book contained other novel ideas. Although Galton later mused that he should perhaps have called the volume *Hereditary Ability* instead, it was the first scientific study of 'genius' as a concept, in other words exceptional intelligence or creativity.[25] When the book was reissued in 1892, Galton took the opportunity to comment on a rival notion of genius promulgated by the Italian criminal anthropologist Cesare Lombroso since the appearance of the first edition. Lombroso thought the genius was subject to 'degeneration' and psychoses.[26] Galton, conversely, noted how capable and vigorous most intellectually distinguished men were. He suggested that if an 'eleven' were chosen from any of his mentally talented subgroups, it would outmatch a similar selection of mediocre men of the same class, even if the latter were two or three

[23] Galton, *Hereditary Genius,* pp. 340–341.

[24] Francis Galton, 'Eugenics: Its Definition, Scope and Aims' in *idem*, *Essays in Eugenics* (London: Eugenics Education Society, 1909), p. 40.

[25] Galton, *Hereditary Genius*, pp. viii–ix.

[26] Cesare Lombroso, *The Man of Genius* (New York: Garland Publishing, 1984), p. vi.

times more numerous.[27] Galton therefore postulated that intelligence was correlated with both strength and health.

This was important in making eugenics a reasonable scheme. If Lombroso was right that the most intelligent were weak and eccentric, perhaps mad, breeding more of these would not take society to a higher level. Galton gave a little ground to Lombroso by admitting that some of his recent studies had shown a connection between genius and insanity. An overpowering idea could take hold of an individual, making him either a genius or a lunatic. Galton also conceded that it was inevitable that a few of the kinsmen of the genius would be mentally subnormal or chronically ill (such as Darwin, who suffered from recurrent spells of dizziness). Nevertheless, it did not alter his main conclusion: that it would be worthwhile breeding as many of the target group as possible. That genius was associated with undesirable traits was in any case part of the theory of compensation, which Galton had dismissed in *Inquiries into Human Faculty and its Development* (1883). There was no basis for the belief that sensory deprivation in one area (e.g. sight) led to improvement in the others.[28]

In later years, Galton sought to get around the problem identified by Lombroso through controlling for three variables among those encouraged to have more children: intelligence, health and morality. Deficiency in just one of the characteristics would outweigh superiority in the other two.[29] In *Hereditary Genius* he included strength as a desideratum through the chapters 'Oarsmen' and 'Wrestlers of the North Country'. Galton tended to value whatever led to success in schools and universities, including sporting honours.

Hereditary Genius should be seen as an inspired reaction to *On the Origin of Species*, which preceded it by ten years. Darwin wrote about artificial selection in breeding animals for certain qualities and had implicitly broken down the dichotomy between animal and human. Artificial selection existed alongside natural selection since various animal species had been domesticated. Galton simply transferred this to humans, wishing to improve mankind by a sprinkling of artificial selection through the

[27] Galton, *Hereditary Genius*, pp. 331–332, ix–x.

[28] Francis Galton, *Inquiries into Human Faculty and its Development* (New York: Macmillan, 1883), p. 30.

[29] Francis Galton, 'Local Associations for Promoting Eugenics' in *idem*, *Essays in Eugenics*, p. 104.

identification of valuable strains. His social interpretation was that race improvement could be facilitated through a greater degree of meritocracy. In the article 'Hereditary Improvement' of 1873, eugenics was seen as ushering in a better and more compassionate society much faster than would otherwise happen. Ideally, he wanted eugenics to be the ruling ideology, as it would be in his unpublished utopian novel 'The Eugenic College of Kantsaywhere', written the year before his death in 1911. The common outlook with Darwinism was clearer in the early days of eugenics, as Galton wrote about how it could remove some of the cruelty of natural selection.[30]

Darwin was very complimentary about *Hereditary Genius* and welcomed ways of overcoming the severe challenge to the doctrine of a beneficent world order, occasioned by his own theory. Because he died in 1882, when the ideology was still taking shape, Darwin never became a convinced eugenist.[31] He was interested in heredity though, having formulated his own theory of pangenesis. It involved somatic cells forming hereditary copies of themselves, which were then transmitted to the reproductive cells through the blood. Although allowance should be made for family ties, after finishing just fifty pages of his cousin's book, he wrote to him saying, 'I do not think I ever in all my life read anything more interesting and original'.[32] This letter delighted Galton and his reply illustrates how his personal world view derived from his cousin's. *On the Origin of Species* had eliminated the unsatisfactory arguments from design and given him freedom of thought.[33] (Fig. 2.2).

The Ideology of Eugenics

The goal of improving our species became a replacement religion for Galton. It should primarily occur through hereditary measures, which he believed would conserve the gains made for future generations. Environmental reforms were useful too, but had far lower potency than those

[30] Galton, *Memories of My Life*, p. 323.

[31] Daniel Pick, 'The Politics of Nature. Science and Religion in the Age of Darwin' in Gareth Stedman Jones and Gregory Claeys (eds.), *The Cambridge History of Nineteenth-century Political Thought* (Cambridge: Cambridge University Press, 2001), p. 664.

[32] Galton, *Memories of My Life*, p. 290.

[33] Galton/3/3/7/7 Francis Galton to Charles Darwin 24 Dec. 1869.

Fig. 2.2 Francis Galton (reproduced from the Eugenics Society Archive, Wellcome Library, London)

transmitted through heredity.[34] The theory contrasted with folk belief at the time concerning the value of the surroundings. Galton did not think that the lifestyle of pregnant women had noteworthy impact on the personality of the unborn child. It was the heredity they transmitted which mattered. Artistic talent ran in families and was not affected, for instance, by the mother visiting picture galleries before the child was born.[35]

While eugenics was a forward-looking ideology, it came to incorporate anxieties prevalent at the time. Along with other scientists, Galton was concerned about what had been named 'degeneration' by the French physician Benedict-Augustin Morel in 1857.[36] Darwin used the term two years later to signify evolution going backwards, thus an individual reverting to an earlier form of life less in tune with current conditions. Agreeing with an established discourse, Galton thought that life in the industrial cities was less salubrious than in the countryside. Urban life encompassed physical and mental disease, crime and prostitution.[37] He was contemplating creating a gifted caste and desired that it should be brought up in healthy rural conditions. In the 1873 article, he suggested bringing such people together and, without disturbing existing markers of social status, giving them honorific advantages such as diplomas and entry into a social register.[38] He continued to follow through the identification of valuable strains, writing another book similar to *Hereditary Genius*, this time solely about scientists. It was published in 1874 as *English Men of Science. Their Nature and Nurture.*

Galton did not think that everyone was created equal. He was opposed to female suffrage and his theoretically sympathetic view of socialism was tempered by not believing that it could be brought about in the present.[39] Society was getting so complex that the average citizen was unable to play a full part in it. The solution was the improvement of

[34] Francis Galton, 'Hereditary Improvement', *Fraser's Magazine* 7:37 (1873), p. 116.

[35] Martin Brookes, *Extreme Measures. The Dark Visions and Bright Ideas of Francis Galton* (London: Bloomsbury, 2004), p. 172.

[36] Pick, 'The Politics of Nature', p. 676.

[37] Mike Hawkins, *Social Darwinism in European and American Thought, 1860–1945. Nature as Model and Nature as Threat* (Cambridge: Cambridge University Press, 1997), p. 219.

[38] Galton, 'Hereditary Improvement', pp. 127–128.

[39] Soloway, *Demography and Degeneration*, p. 121.

human stocks, which involved the institutionalization of people with hereditary afflictions or who were of low ability. It was a draconian policy, more understandable in the context of the Victorian era in which a fifth of the population lived in abject poverty. The inmates' lives would in practice be closed down, in preference to the cut-throat competition and suffering which a laissez-faire economy visited upon them. It also represented an alternative to the redistribution of wealth and the creation of a welfare state.

Galton was thus at odds with both the social Darwinism of Herbert Spencer and, less manifestly, the contemporary labour movement. Spencer stood for non-intervention in the economy, which he believed was justified by the theory of evolution. Galton did not think mankind's improvement would come about through natural competition. It would not produce good results and only lead to suffering, just as the fir trees which grow in the upper limits of vegetation do not become particularly tall.[40] He was disappointed that morality did not play a part in natural selection, which was one of the reasons he wanted eugenics to supersede the workings of nature. He saw his new ideology as an *alternative* to natural selection, not identical to or an exacerbation of it. He wrote that eugenics could be 'more merciful and not less effective' than nature in selecting individuals for survival and reproduction.[41] Equally, social reform aimed at creating better living conditions would fail unless it were accompanied by the improvement of the race. Working-class men were sadly incapable of furthering their own interests, so the ameliorations they sought needed to be implemented from above.

In 1883, in *Inquiries into Human Faculty and its Development*, Galton finally coined the term 'eugenics' to signify raising the hereditary standard of the population. It came from *eugenes* in Greek, meaning 'good in stock' or 'hereditarily endowed with noble qualities'.[42] He had previously used the words 'viriculture' and 'stirpiculture' for the same. Although a decade and a half had passed since Galton first mooted the idea, it had not yet made waves. He had merely suggested diplomas for those with valuable heredity and the opportunity to marry in Westminster Abbey, with

[40] Francis Galton, 'The Part of Religion in Human Evolution', *The National Review* 23 (1894), p. 757.

[41] Galton, *Memories of My Life*, p. 323.

[42] Galton, *Inquiries into Human Faculty*, pp. 24–25.

an added cash prize, for the winners of all-round scholastic and athletic competitions.[43] Competitions of this type would later be at the heart of Kantsaywhere, the ultimate statement of his eugenic creed. The novel's contents were remarkably consistent with what he had been promoting throughout the genesis of eugenics, a process lasting more than forty years. Those were his convictions, articulated towards the end of his life.

In 1891 Galton published an article in which he disagreed with the social Darwinist Benjamin Kidd. It concerned the meaning of religion from an evolutionary point of view. If current religious thought was erroneous and based on false revelations, why had it not been weeded out in the same way that a badly adjusted species becomes extinct? Galton answered that religion created a tribal community, which aided the individual's chances of survival.[44] It gave believers a sense of purpose and direction. But as to whether there is a God, Galton argued that we should assume the negative case since it is more probable than the positive. Human effort was thus crucial. Galton took a proposal from John Stuart Mill to extend the meaning of religion to include any grand collective purpose. The hereditary improvement of the race was one such aim. This introduced the notion of eugenics as a secular religion, which Galton promoted openly in the first decade of the twentieth century.

To the disillusioned mind in search of a new creed, eugenics constituted an alternative to religion. Adèle had taught Francis to be a pious Christian and he went along with this for a while.[45] The Galton household was quietly religious. Francis, however, was by nature a freethinker and had an inquiring mind. As an undergraduate, he joined a lodge of Freemasons concerned especially with science.[46] In adulthood, he was increasingly dissatisfied with the lack of proof inherent in religion. After *On the Origin of Species* was published in 1859, Galton considered himself an agnostic. Darwin too had been an agnostic since his return from the voyage of the *Beagle*. Galton in fact practised Christianity at home with his wife, but his scientific mindset involved publishing fearless investigations which seemed inimical to religion. In a paper of 1872 entitled 'Statistical Inquiries into the Efficacy of Prayer', Galton showed that public prayers

[43] Galton, 'Hereditary Talent and Character. Part I', p. 165.

[44] Galton, 'The Part of Religion', pp. 755, 760, 763.

[45] Galton/1/2/1/3, folder 1 of 2. Booklet by Francis Galton (4 years), folios 3r- 4v.

[46] Pearson, *Life, Letters and Labours*, Volume 1, p. 187.

for members of the royal family, bishops and other clergy had no effect.[47] They did not live longer than members of professions, such as lawyers or physicians, who were not prayed for in church. There was a high correlation between poor health and piety.[48] One example was his very devout wife Louisa, who had suffered from various ailments since childhood. Given that Galton valued health greatly in his eugenic schemes, he probably did not see piety as particularly desirable. Crucially, however, he did not abandon belief in the moral standards which Christianity had inculcated.

Eugenics and Politics

Apart from his religious opinions, Galton also held political views. He was not a supporter of capitalist accumulation because it was better that everyone made their own way in life. The possession of riches distorted sexual selection, as the choice tended to go in favour of the potential spouse with money, rather than the one with laudable characteristics. These attitudes were noted by progressive thinkers and partly explain why many of them found eugenics congenial. Galton's approach to socialism was that it provided another argument for why eugenics was needed. This is how he put the case:

> All this gives hopes for the future of our race, especially if 'viriculture' be possible, notwithstanding that our present moral nature is as unfitted for a high-toned civilisation as our intellectual nature is unfitted to deal with a complex one. […] But the case would be very different in those higher forms of civilization, vainly tried as yet, of which the notion of personal property is not the foundation, but which are, in honest truth, republican and co-operative, the good of the community being literally a more vivid desire than that of self-aggrandisement or any other motive whatever. This is a stage which the human race is undoubtedly destined sooner or later to reach, but which the deficient moral gifts of existing races render them incapable of attaining. It is the obvious course of intelligent men—and I venture to say it should be their religious duty—to advance in

[47] Francis Galton, 'Statistical Inquiries into the Efficacy of Prayer', *Fortnightly Review* 12 (1872), pp. 128–129.

[48] Brookes, *Extreme Measures*, pp. 183–184.

the direction whither Nature is determined they shall go; that is, towards the improvement of their race.[49]

A society in which 'personal property is not the foundation' and which is 'republican and cooperative' could be interpreted as socialism. As his biographer Karl Pearson noted, it was not generally known that Galton wanted race improvement to aid socialism, yet that is exactly what he was arguing for in this article.[50] In Kantsaywhere the state owned almost all the land.[51] This fostered a mentality in which citizens considered the commonwealth above their own interests. But without eugenics, socialism would be a long time in coming. In the article reacting to Benjamin Kidd from 1891, he noted that socialist experiments tried in the present condition of mankind were bound to fail.[52]

Eugenics was nothing if not a utopian ideology. There was a literary trend towards the perfect society which incorporated elements of it. Robert Desborough published *State Contentment: An Allegory* in 1870. It was an ideal socialist society with a eugenic element, though not inspired by Galton, as weak children were culled at birth.[53] Jane Hume Clapperton wrote *Scientific Meliorism and the Evolution of Happiness*, published in 1885. In it she advocated that the state should segregate (and sterilize) those who were not fit for parenthood.[54] Besides Galton, Clapperton admired Malthus, Spencer and Darwin. She thought evolution would lead to a communist polity where even households were shared. Clapperton's collectivist view regarded the individual as of little consequence. Her novel *Margaret Dunmore, or, A Socialist Home* (1888) warned against indiscriminate charity because it harmed the race.[55] There

[49] Galton, 'Hereditary Improvement', p. 119.

[50] Karl Pearson, *The Life, Letters and Labours of Francis Galton*. Volume 2 (Cambridge: Cambridge University Press, 1924), pp. 119–120.

[51] Galton/2/4/19/6/1 folio 4.

[52] Galton, 'The Part of Religion', p. 759.

[53] Gregory Claeys (ed.), *Late Victorian Utopias: A Prospectus* (London: Pickering & Chatto, 2009), Volume 1, p. xviii.

[54] Stephen Trombley, *The Right to Reproduce. A History of Coercive Sterilization* (London: Weidenfeld & Nicolson, 1988), pp. 20–21.

[55] Angelique Richardson, *Love and Eugenics in the Late Nineteenth Century. Rational Reproduction and the New Woman* (Oxford: Oxford University Press, 2003), p. 67.

were even some who practised what they preached, especially a community called the Perfectionists Oneida in New York state. They were sexual radicals who disdained monogamy and instituted selective breeding in 1869.[56]

Intellectual middle-class socialists tended to be supporters of eugenics before the turn of the century. Amongst these were Beatrice Webb, Havelock Ellis, Graham Wallas, Edward Aveling and Eden Paul.[57] They saw the breeding of better people as not just an end in itself, but a heuristic for creating an equitable society. Eugenics was regarded as scientific with a real potential for making life better. Since socialism and eugenics were both new-fangled, untested ideas gathering momentum in the 1870s and 1880s, they both seemed to have the future on their side. Galton preferred professional men to any other section of the community and thought that they in particular were harbingers of amelioration. Socialists correctly reasoned that this did not rule out labour movement action.

Biometry

Though eugenics was beginning to be recognized, Galton was involved in many other scientific pursuits. It was only towards the end of his life that he returned to the issue in a sustained way. He was aware that eugenics was controversial because it involved outside expertise in personal decisions about how large families should be. Taken to extremes, it might lead to the breakup of the family unit. Although Galton was not always promoting eugenics, his continued statistical inquiries had relevance for the doctrine and could be used in honing it. The correlation-coefficient which he invented showed how closely two variables are interrelated. And he devised a laboratory of physical tests and measurements for the International Health Exhibition in London in 1884, afterwards demonstrated for six years at the Science Museum (Fig. 2.3).

For a fee of 3d visitors had their height, weight, arm-span, sense of touch, hearing, eyesight, etc., measured and received a report on where they stood compared to the average.[58] Fingerprints were added after

[56] Kevles, *In the Name of Eugenics*, p. 21.

[57] Diane Paul, 'Eugenics and the Left', *Journal of the History of Ideas* 45:4 (1984), pp. 567–568.

[58] Francis Galton, 'On the Anthropometric Laboratory at the International Health Exhibition', *Journal of the Anthropological Institute* 14 (1885), p. 213.

Fig. 2.3 The Anthropometric Laboratory at the International Health Exhibition 1884 (reproduced from the Galton Laboratory, UCL Special Collections)

the laboratory moved to the museum. More than 13, 000 people went through the rigour of these tests at either the exhibition or museum. It gave Galton data on the physical dimensions of human beings in numerous categories. He wrote up some of the results in the *Journal of the Anthropological Institute* in 1884, using the concept of percentiles which he had invented in 1875.[59] Although the tests were physical and practical, the principles underlying them were applied to the emerging field of mental testing in the early twentieth century. The psychologist

[59] Nicholas Wright Gillham, *A Life of Sir Francis Galton. From African Exploration to the Birth of Eugenics* (Oxford: Oxford University Press, 2001), pp. 213–214.

and eugenist Cyril Burt used similar tests of dexterity to gauge differences of intelligence between children in a fee-paying preparatory school and a state elementary school in 1909.[60]

The combination of statistics with biology and anthropology later became the scientific rationale for eugenics as promulgated by Galton and his followers. In 1901 Karl Pearson, professor of applied mathematics at University College in London, and W. F. R. Weldon, a biologist at the same institution, founded the journal *Biometrika*, with Galton superficially lending his name to the venture.[61] The aim was to publish studies of biological phenomena using statistics, which the three men had christened 'biometry'. It did not become as influential as Mendelian genetics, which was introduced in Britain at this time. Biometry studied inheritance empirically instead of taking account of dominant and recessive characteristics. Galton never adopted Mendelism , probably because he was too old when acquainted with it. Mendelism nevertheless became the main genetic system behind eugenics internationally and would play a major role in British eugenics too.

Galton's Efforts to Promulgate Eugenics

In 1901 Galton was invited to give the Huxley Lecture at the Anthropological Institute in London and chose as his topic 'The Possible Improvement of the Human Breed under Existing Conditions of Law and Sentiment'. The lecture was evidently about eugenics and marked the onset of Galton's final stage promoting the ideology. The years ahead would mark his most serious effort to hold a candle for eugenics. He paid heed to contemporary realities and as in 'Heredity Improvement', it was clear that girls, and not just boys, would have the opportunity to show their talents.Galton continued to value ability to do intellectual work above all else, to which he added other favoured traits of character, disposition, energy and strength.[62]

[60] Cyril Burt, 'Experimental Tests of General Intelligence', *British Journal of Psychology* 3 (1909), pp. 94–177.

[61] Francis Galton, 'Probability, the Foundation of Eugenics' in *idem*, *Essays in Eugenics*, p. 78.

[62] Francis Galton, 'The Possible Improvement of the Human Breed under Existing Conditions of Law and Sentiment' in *idem, Essays in Eugenics*, pp. 3, 22.

As a new ideology, eugenics was permeated by existing debates and challenges coming to the fore. Both eugenics as conceived by Galton and left-wing progressive opinion were geared towards solving social problems. In 1904 he spoke to the nascent Sociological Society at a meeting convened at the School of Economics, the brainchild of the Fabians. The Sociological Society itself contained many Fabians and other progressives. Galton, who had made a small donation towards the foundation of the Sociological Society, chose as his topic 'Eugenics: Its Definition, Scope and Aims'.[63] He gave a social reformist, rather than social Darwinist, definition of what eugenics involved.[64] Instead of the old definition that it allowed the more valuable strains of humanity to prevail more speedily over the others, he proposed two new formulations. Eugenics was all influences, thus hereditary and environmental, which improved the inborn qualities of a race.[65] It was also whatever sought to take talents and abilities to their highest level.

It was through this presentation that Galton cemented the support of many Fabians for the doctrine. George Bernard Shaw displayed notable enthusiasm and went so far as to say that only a eugenic religion could save civilization as we know it. He praised Galton's idea of bringing into existence individuals whose tasks would be to repair society, as opposed to getting rid of unworthy people on the battlefield and the gallows. Shaw's only difference with Galton concerned whether gifted people needed to be restricted to monogamous marriage or not. If one had to conform to society's norms and prejudices, eugenics might well end up as mere speculation.[66] Galton had an interest in artificial insemination and did not rule out reproduction outside families.[67] A socialist society populated by supermen happened to be Shaw's goal and was compatible with what Galton believed. Yet Galton was far less willing to ruffle feathers than Shaw, as he feared that if eugenics developed into the plaything of cranks, it would be doomed.

[63] Galton/2/4/19/9/1 folio 3v.

[64] Michael Freeden, 'Eugenics and Progressive Thought: A Study in Ideological Affinity', *Historical Journal* 22:3 (1979), pp. 645–646.

[65] Francis Galton, 'Eugenics: Its Definition, Scope and Aims', p. 35.

[66] Galton/2/4/19/9/2. Sociological Society, folio 14.

[67] Galton/2/5/2/1. A plea to experiment on the artificial fertilisation of mammalia by F.G, folio 1.

H. G. Wells also became a supporter of eugenics. His reaction to the paper was that in society as it existed at the time, people of the very highest ability were unlikely to survive and reproduce as easily as those below them on the scale of intellect. He was also unsure whether health and energy were monolithic qualities, as he felt there must be several types. In a sign that Galton was not among the most racist of eugenists, Wells suggested that two parents both in good health might nevertheless have sickly offspring if they belonged to different races.[68] Galton agreed that races varied by average ability, writing that eugenics sought to create beings who were as superior to modern Europeans as they in turn were to the least advanced black people.[69] Yet he never gave race as a criterion for parenthood, usually concentrating on intelligence, health and morality. Galton was disappointed by the reactions garnered in the Sociological Society, irritably suggesting that this new venture risked becoming irrelevant. Despite this disparagement, he gave a second paper to the Society, on restrictions in marriage, on 14 February 1905.

The Eugenics Record Office and the Eugenics Laboratory

Galton, by now in his early 80s, was in a hurry to make eugenics count. In 1904 he contacted the University of London about funding a research fellow in 'national eugenics'. The job description entailed studying talented families. Galton would provide £ 1, 500 over three years to cover the fellow's salary and, if the scheme proved successful, would make it permanent at about £ 500 a year.[70] The scientist chosen for this role was Edgar Schuster, a zoologist. Galton published *Noteworthy Families* alongside him in 1906. The scheme was based in Gower Street near University College's quadrangle. Gradually, Karl Pearson became involved as an employee of the College. He had chaired Galton's first presentation to the Sociological Society on 16 May 1904 at the latter's request, despite not being a member.[71] Schuster, on the other hand, was half-hearted about his work as the research fellow. He did not wish to

[68] Galton/2/4/19/9/2. Sociological Society, folio 17.

[69] Galton, *Hereditary Genius*, p. x.

[70] Galton/2/4/12/1/1 folio 2r.

[71] Galton/3/3/7/20 folio 350r. Galton to Pearson, 12 Apr. 1904.

continue beyond the three years to which he had committed. Pearson ended up rescuing the scheme upon Schuster's resignation. In 1907 the Francis Galton Laboratory was set up at University College. Created as a continuation of the earlier Eugenics Record Office, it was under Pearson's control. When Galton died in 1911, Pearson was appointed to the newly endowed chair in eugenics, funded through Galton's last will and testament. The two men had been in close touch for more than a decade. Galton had chosen Pearson as his successor through a codicil stating that Pearson should be the first holder of the professorship.[72] (Fig. 2.4).

The fact that Galton made Pearson the torchbearer for eugenics suggests that he did not see the doctrine as reactionary. Pearson had been a socialist since the 1880s, as a result of coming into contact with a scholarly type of socialism as a postgraduate student in Germany.[73] Although Pearson was not active in the labour movement, his views were generally known and may have cost him a chair at Oxford, where his applications had been turned down.[74] Galton's aim was that eugenics should first be made into a subject for academic study until its importance had been recognized, then as a practical notion for development and, finally, turned into a secular religion which could guide society.[75] He thought it had potential as a religion because it accorded with nature, which sought the improvement of the various species within it.

Pearson began the process of making eugenics a science, although he was slow in converting his team's research findings into anything practical. In the Galton Laboratory, he was aided by David Heron as research fellow, Ethel Elderton as research scholar and Amy Barrington as 'computer', i.e. someone handling routine calculations.[76] Unlike Pearson, after some hesitation, Galton chose to engage with the Eugenics Education Society, set up in 1907. He was made honorary president upon joining

[72] Galton/2/4/12/1/1 folios 6–10, folios 10–13.

[73] D. MacKenzie, 'Karl Pearson and the Professional Middle Class', *Annals of Science* 36:2 (1979), p. 131.

[74] UCL, Pearson Papers, Pearson/1/8/1.

[75] Galton, 'Eugenics: Its Definition, Scope and Aims', p. 42.

[76] Pearson/4/8/1 page 1.

Fig. 2.4 Karl Pearson (reproduced from the Pearson papers, UCL Special Collections)

in 1908. Among Galton's ideas for the Society was to use it as a certification institute for good heredity.[77] He gave a lecture to it the year he joined, suggesting that local eugenics associations could act as a kind of social register of hereditary qualities.[78] The new Society would in time lend a popular edge to eugenics which was at odds with Pearson's scientific conception of the doctrine. Galton gave priority to the laboratory he was funding over the Society, which he saw as subsidiary. In 1909 he opined that enough was known about eugenics to begin promoting it to the public.[79] Galton was not completely ready for eugenics to become a secular religion yet, as that stage needed to be preceded by scientific research on a variety of issues. Whether eugenics really could be a science is debatable since it is clearly teleological, based on a number of value-judgements and sought to find methods to 'improve' the human race, rather than studying it dispassionately.

Popular Eugenics

Although it is true that the Galton Laboratory investigated eugenic themes using the scientific method, it was another aspect of its founder's thinking which roused many recent converts. Galton had warned against Malthusianism, noting that if the prudent and insightful restricted their breeding to counter a future crisis of resources while the feckless did not, the average level of morality and competence in humanity would decline.[80] This was the kernel of a classic eugenic discourse which was not always scientific: the differential birth rate. William Rathbone Greg had argued that it was a menace to civilization as early as 1872.[81] Galton did not directly use social class as a criterion of eugenic worth, but it became a shorthand for it among many eugenists. Galton's Huxley Lecture transposed his graph of 'civic worth' onto Charles Booth's classification of

[77] Galton//2/4/19/1/1. Draft of a Diploma to be issued by the Council of the Eugenics Education Society, folio 3.

[78] Francis Galton, 'Local Associations', pp. 103–104.

[79] Francis Galton, 'Preface' in *idem*, *Essays in Eugenics*, [p. 2] not paginated.

[80] Galton, *Hereditary Genius*, pp. 356–357.

[81] Angelique Richardson, '"I differ widely from you": Darwin, Galton and the Culture of Eugenics' in Eckart Voigts, Barbara Schaff and Monika Pietrzak-Franger (eds.), *Reflecting on Darwin* (Farnham: Ashgate, 2014), pp. 32–33.

the London poor.[82] It is nevertheless clear that Galton recognized that talented individuals were to be found in all walks of life. When considering local eugenics associations, he noted that tact would be required since these gatherings would involve people of various social strata.[83] Another idea was creating a settlement for artisans, selected for heredity, where their children might be brought up in healthy suburban conditions.[84]

To many eugenists who followed Galton, the individual person was forgotten and efforts to make professionals have more children, while reducing the number of working-class births, became virtually the entirety of the doctrine. They saw eugenics as counteracting social trends in the early twentieth century whereby the upper and middle classes had fewer children than manual workers. The discourse relating to the differential birth rate, which incorporated into it the earlier apprehension about less vigorous individuals surviving and breeding through the blessings of civilization, led to another form of the doctrine, 'negative' eugenics, as defined by Caleb Saleeby.[85] Saleeby, an Edinburgh gynaecologist and medical journalist, also claimed to have coined the word 'eugenist'.

'Negative' eugenics focused on reducing the birth rate among people with poor hereditary qualities or, as interpreted by many, the working class. Galton's version concentrating on breeding intellectually able, healthy and virtuous individuals consequently became known as 'positive' eugenics. As outsiders started making contributions to the doctrine, Galton too began thinking about restricting the birth rate of the least 'fit'. Being of an older generation, he never recommended sterilization or the use of contraception (which could cause 'race suicide').[86] He formulated two 'negative' eugenic policies, which indicated where the doctrine stood in the early years of the twentieth century. The first was to suggest

[82] Galton, 'The Possible Improvement of the Human Breed', pp. 8–11.

[83] Galton, 'Local Associations', p. 103.

[84] Galton/2/4/12/1/3 folio 10r. Eugenic Enactments.

[85] G. R. Searle, *Eugenics and Politics in Britain, 1900–1914* (Leyden: Noordhoff International Publishing, 1976), p. 73.

[86] Donald K. Pickens, *Eugenics and the Progressives* (Nashville: Vanderbilt University Press, 1968), p. 28. There is evidence that Galton adumbrated the eugenic drive for sterilization in the 1930s. In a letter to Pearson of 6 January 1907 he wrote, 'except by sterilization I cannot yet see any way of checking the produce of the unfit who are allowed their liberty and are below the reach of moral control.' Kevles, *In the Name of Eugenics*, pp. 94, 325.

that the right to marry could be removed from people with hereditary deficiencies.[87] In an essay on the topic, he pointed to marriage customs being socially based, rather than deriving from natural instincts.[88] Therefore, it might be possible to take society with him on this issue. The second 'negative' eugenic policy he advocated was withdrawing charity from recipients where there was no prospect of self-induced improvement, instead spending the money on people who would benefit in the long run.[89]

Social Hygiene and Social Darwinism

Eugenics may be posited as interacting with two other ideologies in particular: social hygiene and social Darwinism. Social hygiene focuses on improving living conditions and health without changing the existing distribution of resources in society.[90] Social Darwinism specifically applies natural selection to humans in the present, thus to some extent breaking down the barrier between civilization and a state of nature. Pearson was both a socialist and a social Darwinist. For him, however, the struggle for existence occurred *between* societies, not *within* them.[91] Every citizen needed to be equipped with material resources and an education to enable the country as a unit to resist other states. This tallied with socialism too.[92] An important consideration for Pearson was not only that the British population was smaller than Germany's, but that the gap was widening.[93] To counteract the larger nation's advantage in numbers, there needed to be genuine investment in individuals. Britain could not afford to allow a proportion of its people to languish in poverty and unproductive passivity. The Second Boer War (1899–1902), entailing the rejection of 37 per cent of recruits for military service, is often seen as

[87] Galton, 'Eugenics: Its Definition, Scope and Aims', p. 42.

[88] Francis Galton, 'Restrictions in Marriage' in *idem, Essays in Eugenics*, pp. 57–58.

[89] Galton, *Memories of My Life*, p. 322.

[90] See Greta Jones, *Social Hygiene in Twentieth-century Britain* (London: Croom Helm, 1986).

[91] Pearson/1/8/1. Article by Lawrence Small in *Liverpool Labour Chronicle*.

[92] Karl Pearson, *The Grammar of Science* (London: Walter Scott, 1892), p. 433.

[93] G. R. Searle, *The Quest for National Efficiency. A Study in British Politics and Political Thought, 1899–1914* (Oxford: Blackwell, 1971), p. 9.

alerting policy makers to a danger.[94] By 1900 these two types of social Darwinism and a third, struggle between species, had become ubiquitous.[95] The addition of the word 'national' to eugenics at the time when it became a subject for academic research was clearly informed by such realities.

In Galton's days an informal hierarchy of civilizations existed, based on races and classes differing in their innate capabilities. He did not dissent from this notion, believing that intelligent lads from humble backgrounds would rise in society provided they had the right hereditary endowment. On race, Galton shared the views of his contemporaries. He assumed that Europeans stood at the top of the great chain of being and that colonized people stood lower down it. He had respect for descendants of old civilizations, such as the Chinese and Jews. He urged the study of Eurasians, partly to investigate Mendelian inheritance, and wrote some uncomplimentary things about black people.[96] There was, however, more to these views than might at first be assumed.

During his travels Galton had lived in Syria and adopted an Oriental lifestyle. He was one of those Englishmen who blended in with the 'natives', wearing robes and learning to speak common Arabic fluently. Galton's opinions were likely to be more nuanced than the prevalent contemporary racial hierarchy. He did believe that white people were, on average, superior to black people. When these differing capabilities were added up, they explained why Europe was more advanced than Africa. Galton suggested that the 'worth' of a race could be determined from its median member. Yet there was a fair amount of overlap between ethnic groups: 'the Negro race is by no means wholly deficient in men capable of becoming good factors, thriving merchants, and otherwise considerably raised above the average of whites'.[97] Although the median black person stood below the median white, it did not follow that there were no black people of distinction. He gave the example of Toussaint L'Ouverture, a

[94] Lucy Bland and Lesley A. Hall, 'Eugenics in Britain: The View from the Metropole' in Bashford and Levine (eds.), *The Oxford Handbook*, p. 213.

[95] Gregory Claeys, 'The "Survival of the Fittest" and the Origins of Social Darwinism', *Journal of the History of Ideas* 61:2 (2000), p. 226.

[96] It was believed at the time by some biometricians that Mendelism only applied to hybrids. *The New Statesman* Vol III No. 65 Saturday, July 4, 1914 (London: The Statesman Publishing Company, 1914), p. 411.

[97] Galton, *Hereditary Genius*, p. 338.

commander who had led a slave rebellion against the French in Haiti in the 1820s. He also acknowledged that since black people in the United States suffered legal and social impediments, the comparison was not an entirely fair one.

Galton would assuredly have been counted as a racist by today's standards. His views were typical for the age in which he lived. Many whose treatment in history has been more favourable shared such attitudes. Abraham Lincoln wrote that if two races lived side by side, they could not be on an equal footing, and that he, as much as anyone else, wished the dominant and privileged position to be accorded to white people.[98] In addition, Galton implicitly did not believe that the ranking of a particular nation was set in stone. He regarded the ancient Athenians as the most gifted civilization hitherto, but did not feel that Greeks in the Victorian era matched the achievements of their distinguished forbears.[99]

Since Galton's eugenics was about improving society, he agreed with state intervention in order to make this come about. It was not the creation of wealth but scientific and cultural achievements which he saw as particularly valuable. The other side of the coin was segregation for those whose heredity was thought to be poor. His illiberal views illustrate how the doctrine, though ostensibly based on noble motives, easily slid into repression. Average working men, he wrote, were 'essentially the mediocre class, standing as far below the highest in civic worth as it stands above the lowest class with its criminals and semi-criminals'.[100] The latter group, amounting to roughly 9 per cent of the population, should be eliminated over time through not being allowed to breed. As an hereditarian, Galton did not envisage that they could become decent citizens through the redistribution of wealth and the provision of an education for them. It was, however, possible to raise the standard of ordinary labourers. They were among the 91 per cent who should benefit from both environmental and eugenic reforms.

Pearson regarded himself as a progressive. This was partly because laissez-faire was so dominant, whereas he supported state intervention. His understanding of eugenics was moderate because he recommended little action and much study. Pearson's first wife Maria Sharples had

98 Stephen Jay Gould, *The Mismeasure of Man* (London: Penguin, 1992), p. 35.

99 Galton, *Hereditary Genius*, pp. 342–343.

100 Galton, 'The Possible Improvement of the Human Breed', p. 10.

been a fellow member of the Men and Women's Club, established to discuss gender differences and feminism. Their marriage was companionate. Pearson supported women pursuing careers and chided Galton for disagreeing with female suffrage. But he did hold some odd views for a socialist. Pearson was war-like in his attitudes, while almost all labour movement adherents were pacifists.[101] He supported imperialism, as did a majority in the Fabian Society, but which mainstream opinion in the Labour party saw as unjust.[102] Most bizarrely, he was not actually against child labour. He argued that its abolition would reduce the economic value of the child to the parents and thus decrease the birth rate.[103] (This incidentally proves that he wanted families in the lower socio-economic brackets to have more rather than fewer children, which went against the grain of eugenic opinion.) In the 1880s he wrote that the value of all sorts of work was the same.[104] After the First World War, he came to appreciate professionals over workers in his eugenic schemes.[105] Even so, Pearson publicly continued to believe in socialism and, unlike Galton who was knighted in 1909, declined honours because he could not square them with his principles.

Neither Galton nor Pearson were politically active. Pearson had last interacted with the labour movement in the 1880s, when he gave some speeches to working men. After becoming a eugenist though, Pearson set great store on gaining influence in government circles. In pursuit of this goal, he would postpone his department's work in order to take on statistical assignments that might ease eugenics into the consciousness of the authorities.[106] Pearson also occasionally convened public lectures on

[101] Bernard Semmel, 'Karl Pearson: Socialist and Darwinist', *British Journal of Sociology* 9:2 (1958), p. 115.

[102] Bernard Shaw (ed.), *Fabianism and the Empire. A Manifesto by the Fabian Society* (London: Grant Richards, 1900), pp. 15–16. Rhiannon Vickers, *The Labour Party and the World*. Volume 1 (Manchester: Manchester University Press, 2003), p. 38.

[103] Galton/2/4/12/1/2 folio 2r-v. *The Lancet* Oct 9, 1909.

[104] Karl Pearson, *The Ethic of Freethought. A Selection of Essays and Lectures* (London: T. Fisher Unwin, 1888), p. 356.

[105] Theodore M. Porter, *Karl Pearson. The Scientific Life in a Statistical Age* (Princeton: Princeton University Press, 2004), p. 283. Kevles, *In the Name of Eugenics*, p. 23.

[106] Pearson/4/4/1. Drapers Company Grant, p. 1. Pearson/4/4/2. Report by Karl Pearson, p. 5.

eugenic topics in the Galton Laboratory. These were primarily designed to showcase his team at work rather than opening the subject up to outsiders.[107] Pearson was more cautious than Galton in stating eugenic ambitions. William Ralph Inge, a Cambridge theologian who later became dean of St Paul's Cathedral, accused Pearson of wanting to wait fifty years before doing anything practical about the doctrine in which they both believed.[108]

Pearson was a better mathematician than Galton, meaning that he supplied formal proofs to buttress eugenics. Galton was pleased when Pearson managed to demonstrate that mental inheritance functioned in the same way as physical. Conversely, he was puzzled when Pearson could find no correlation between the size of the head and intelligence.[109] Galton tended to take the side of Pearson when there was an altercation with the Eugenics Education Society. He covered up disagreements between the Galton Laboratory and the Society. Pearson's vision for eugenics was a limited academic endeavour with little public support. He seldom took any steps without conducting a major study of the question first. Galton agreed with this theoretically, though he was also moving towards implementing eugenics.

Galton's approach to social amelioration was hereditarian and typical of the Victorian and Edwardian eras. It was only after the Labour Party began to have influence that redistribution became a political concern and slum clearance moved onto the agenda.[110] These turned out to be better solutions to the problem of extreme poverty than eugenics. At the time when eugenics was launched, a number of other possible reactions were still in contention. The focus on innate qualities which Galton and Pearson pursued was conducive to social hygiene only, and created taxonomies from the snobbery and racism which was affecting society anyway. The fact that Galton made attempts to quantify individual differences (using an intellectual scale based on achievements or 'civic worth', a composite of ability and morality) has raised the suspicion that the purpose of the exercise was to hold the non-privileged down by means

[107] Pearson/4/17. Journal of the Galton Laboratory, p. 10.

[108] Soloway, *Demography and Degeneration*, p. 33.

[109] Galton/3/3/7/20. Galton to Pearson 17 Nov. 1901, folio 260r, Galton to Pearson 31 Dec. 1901, folio 262r.

[110] Cf. Greta Jones, *Social Darwinism and English Thought. The Interaction between Biological and Social Theory* (Sussex: The Harvester Press, 1980), p. 169.

of science. Science does not exist in a vacuum and prevalent opinions do have an effect on its practitioners. Eugenics, however, had more in common with social hygiene than with scientific racism or seeking to justify economic disparities. Why should Galton go about creating an ideology for the express purpose of proving something that seemed self-evident to most of his contemporaries? If that was all the doctrine offered, why did he leave £ 45, 000, the residue of his estate, to the Galton Laboratory to found a professorship?[111]

Hereditarian notions were soon used as arguments in opposing social reform, though Galton himself had written that he was in favour of it. Informed by Darwinism, the most powerful scientific rationale in his days, Galton initially sought to replace natural selection with something kinder, while accelerating progress towards better human beings. Towards the end of his life, eugenics received new support through formal meetings and, especially, the establishment of the Eugenics Education Society in 1907. Galton did not foresee that the Society would have a greater footprint than the Laboratory which he was funding. This had the result of changing the emphasis of the doctrine. Galton too was affected by current concerns, becoming anxious about the consequences of 'degeneration'. He began to concoct 'negative' eugenic schemes, suggesting in 1901 that they were important and in 1908 that they must take precedence. His recent allies, comprising reactionary professional men and their female relations, were turning eugenics into applied social Darwinism. There was still a socialist and progressive minority within the movement, many of whom were Fabians hardly free from prejudice themselves. The differential birth rate became key. Galton's not particularly offensive doctrine of breeding additional star performers in scholastic, aesthetic and athletic pursuits gave way to an interpretation whereby society risked dissolution from the multiplication of the feckless and hereditarily afflicted. Eugenics began transforming itself into a fundamentally conservative creed.

A prime example is the 'better dead' school of social Darwinists. They were extremists who wished to speed up the rate at which the 'unfit' died, supposedly because hereditary characteristics would thereby reach a higher standard. This attitude came to colour the mainstream of British eugenics and held sway at least until the 1920s, when many eugenists started reflecting on tactics. As 'reform' eugenics extended its

[111] Galton/2/4/12/1/1 folio 4.

reach, Galton's original attitudes and aims were revisited. Even his 'negative' eugenic policies of institutionalization and restricting marriage might be worth a second look. The next phase for the movement would be contraception and sterilization, which raised misgivings internally and externally.[112] While Galton's proposals would affect fewer people, they might prevent such opposition growing. If combined with 'positive' eugenics, it would still enable the doctrine to advance. Both Galton and the renamed Eugenics Society wanted to create a eugenic conscience. Even if the doctrine became less manifest, as long as it remained an influence, that might be enough for cautious minds among its supporters.

Eugenics was malleable and could indeed be moulded into a late version of social Darwinism. Galton had drawn attention to how natural selection had ceased functioning in modern societies, leading to the survival of weak human stocks. Eugenics might therefore be construed as accelerated social Darwinism. From this point of view, natural selection would be revived, rather than replaced, by eugenic policies. Eugenics had emerged from the same field as social Darwinism. Victorian Britain and Edwardian Britain were almost exclusively free-market economies. The differential birth rate could be defined as a market failure. Towards the end of his life, Galton wrote with consternation about the number of 'feebleminded' people in society. He estimated their numbers at just under 250, 000 in England and Wales in 1909, the year after the publication of the report of the Royal Commission on the Care and Control of the Feebleminded.[113] He recommended their institutionalization and was joined by three other eugenists in reacting to the report. Segregation was part of his design for eugenics. That can be seen when Galton allowed himself the luxury of surveying where the doctrine might ultimately lead.

Galton's Ideal Society

His unpublished novel 'The Eugenic College of Kantsaywhere' belonged to the same genre as Plato's *The Republic*. Written around 1910, it was based on an earlier idea, a 'report' by 'Professor I. Donoghue of Dunno

[112] *The New Statesman. A Weekly Review of Politics and Literature* Vol II No 39 Saturday, January 3, 1914, p. 410.

[113] *The Problem of the Feeble-Minded. An Abstract of the Report of the Royal Commission on the Care and Control of the Feeble-Minded* (London: P. S. King, 1909), p. 81.

Weir' on a family of artisans.[114] The manuscript of 'Kantsaywhere' was purged after Galton's death by his niece, Millicent Lethbridge, worried that its love scenes would reflect badly on her uncle. Writing primarily for himself allowed Galton to express his ultimate ideas freely. Although he is on record as not taking the novel too seriously, the explanation is that he realized that the utopia which he was sketching would not be brought about anytime soon.[115]

The novel described an island state which sought to improve mankind and had come a long way. Eugenics acted as its ideology and ruled supreme. The political system was a pure meritocracy, where both men and women achieved their status in life through an ordinary and an honours examination taken in early adulthood. These resembled aptitude tests more than university papers, with the ordinary exam being an intrusive medical check-up. The results made it possible to sort citizens into categories according to 'worth'. Failing the ordinary examination in effect meant being ejected from society. Such people were encouraged to emigrate and were only tolerated on the island if they did not procreate. If they did so, 'kindness was changed into sharp severity'.[116] Those who passed the ordinary examination became full citizens. In the honours examination seventy marks out of 120 conferred upon both sexes the right to vote, while higher scores might entail additional privileges. The state, known as the Eugenic College, gave successful candidates stipends for the purpose of rearing large families.[117] Really high scorers were allowed to join the government.[118] Under this system, mankind's improvement would happen much faster than through natural selection.

This type of civilization seems a nightmare in view of the totalitarian regimes mankind had to endure in the twentieth century. Like Sir Thomas More's original Utopia, with its slavery, it was a purported ideal society which in fact contained disturbing features. Because the novel was written very late in Galton's life, it attracted little attention after being rejected by a single publisher. In public, Galton was scathing about

[114] Galton/2/4/19/5 folios 5–7.

[115] Karl Pearson, *The Life, Letters and Labours of Francis Galton*. Volume 3a (Cambridge: Cambridge University Press, 1930), p. 412.

[116] Galton/2/4/19/6/1 folio 12.

[117] Galton/2/4/19/6/1 folio 27.

[118] Galton/2/4/19/6/1 folio 9.

George Bernard Shaw's extravagant claims on behalf of eugenics. Yet the only scientific difference between Galton and Shaw may have been that the latter believed in environmental influences on future offspring (neo-Lamarckism).[119] And *Man and Superman*, Shaw's play from 1903, did much to popularize 'positive' eugenics.[120]

Progressive and Reactionary Eugenics

Since Galton was cagey about his ultimate aims, the bare facts of eugenics tended to be combined with pre-existing political and social views. He accepted this, stating in a late article that eugenics sought to create a high-toned civilization, leaving the future political direction to its citizens.[121] When it started organizing itself, eugenics usually aligned with right-wing tendencies. Its adherents regarded the doctrine as a justification for inequality and believed that the 'solution' to starkly contrasting living standards was to prevent those at the bottom from being born. Redistribution would cause additional degeneration, as 'unfit' beneficiaries would thereby survive and breed for longer. Daniel Kevles named those who held such attitudes 'mainline' eugenists and David Stack called them 'hard hereditarians'.[122] 'Mainline' eugenists utilized the schemas Galton had provided for ranking people and echoed his conviction that such differences were hereditary. The form of eugenics they arrived at was based on contentions by Galton, but their social Darwinist assertion that nothing need be done by way of social reform and creating a more meritocratic society jarred with what he had recommended. Probably their goals were to cut taxes when less expenditure was required on poor or disabled people and to reduce social problems at what they saw as the human source. Eugenics for them consisted of measures to speed up the process whereby the least 'fit' lost out in the struggle for survival. They conflated the Darwinian notion of 'fitness' with social superiority. Left-wing eugenists pointed out that, if social conditions were awful, the 'fit' person would be a survivor with few scruples.

119 Gareth Griffith, *Socialism and Superior Brains. The Political Thought of Bernard Shaw* (London: Routledge, 1993), p. 126.

120 Claeys (ed.), *Late Victorian Utopias*, Volume 1, p. xvii.

121 Galton, 'Eugenics: Its Definition, Scope and Aims', pp. 36–37.

122 Kevles, *In the Name of Eugenics*, pp. 88–89. David Stack, *The First Darwinian Left. Socialism and Darwinism 1859–1914* (Cheltenham: New Clarion Press, 2003), p. 87.

Progressive liberals and socialists focused on another aspect of Galton's eugenics. He wanted to achieve social amelioration and breed better human beings, capable of contributing to those new societies. For them eugenics was an optimistic and forward-looking doctrine. They were only a minority within the movement taking shape towards the end of Galton's life, but their interpretation was as valid as that of the right-wing supporters. Although they were progressive, it did not mean that they necessarily focused on 'positive' eugenics.[123] Some Fabians argued for sterilization even before eugenists made this their key demand in the 1930s. Virtually all these thinkers came from educated backgrounds and believed in science. They did not have close contact with the masses. Their assumption may have been that eugenic techniques could be used for identifying the respectable and worthy among the working class. The 'feckless' could then be encouraged to breed less. Meanwhile, as the national citizenry improved in quality, it would become easier to promote liberal or socialist values. They saw eugenics and their political views as complementary, both having high-minded motives according to themselves. Among socialists with working-class backgrounds and a practical knowledge of poverty, such as George Lansbury, support for eugenics was not widespread. As early as 1912, the *Daily Herald* warned in an editorial that a section of the working class would not be permitted to have children under the influence of eugenics and biotechnology.[124] All socialist eugenists and most of the liberals believed in environmental improvements too. This was also true for some others. The naturalist Frederick Webb Headley supported eugenics, but agreed that what was called degeneration was in fact largely due to a poor living condition.[125]

The 'mainliners' used arguments derived from heredity to promote laissez-faire economics and to denigrate the option of social reform. They concurred with Spencer's view of competition as continuing to yield benefits to society. There was also a far-right fringe inspired by eugenics. English Misteri was active in the 1930s. The group believed in a rural

[123] Daniel Becquemont, 'Eugénisme et socialisme en Grande Bretagne 1890–1900', *Mil neuf cent* 18:1 (2000), p. 64.

[124] Angus McLaren, *Reproduction by Design. Sex, Robots, Trees, and Test-tube Babies in Interwar Britain* (Chicago: University of Chicago Press, 2012), p. 75.

[125] F. W. Headley, *Darwinism and Modern Socialism* (London: Methuen, 1909), pp. 135, 336.

utopia which would see a return to a medieval social order.[126] Even this vision could be related to facets of Galton's thought. He had been sceptical about industrialization and what he believed to be its concomitant effects, including slums, stunted growth and the spread of diseases. Eugenics was so plastic that it could be packaged to oppose modernity, finding its *raison d'être* in the breeding of an aristocracy with healthy bodies and minds. Lord Lymington and the writer Anthony Ludovici belonged to English Misteri. Before it had materialized, the far-right fringe consisted of Nietzsche supporters and the type of social Darwinists who believed in war as a purifier of civilization.[127] They saw eugenics as perhaps the closest one could get to war in a time of peace. Conversely, in 1919 the budding anthropologist George Pitt-Rivers saw the survival of the 'unfit' as a cause of the outbreak of the First World War.[128] He later became a fascist and a eugenist. In the 1930s, it was from this tendency that support for Mussolini and Hitler could be found.

As noted above, it has often been assumed that eugenics existed as a scientific justification or medicalization of disparities in wealth and power or for why life-chances were so unequal. Galton was in favour of quantifying all things and may have seen this as a useful application of his doctrine. He was far from being an egalitarian. Yet eugenics matched the world view of the middle classes, rather than the traditional rulers. It was brainwork which raised someone above the fray, while there were parts of the nobility which did not deserve to be so prominent. Galton complained how dukes in some cases had been created merely on account of being the illegitimate sons of Charles II, yet there was always one duke, oftener two, in every cabinet, wielding power equivalent to hundreds of thousands of ordinary Englishmen.[129] In *Hereditary Genius* he stated that the best polity for eugenic purposes was one in which incomes were chiefly derived from salaries rather than inheritance. These views suggest that justifying inequality had not been his prime objective. Supporting professional families, with an added element of meritocracy, was closer

[126] Dan Stone, *Breeding Superman. Nietzsche, Race and Eugenics in Edwardian and Interwar Britain* (Liverpool: Liverpool University Press, 2002), p. 35.

[127] Maximilian A. Mügge, 'Eugenics and the Superman: A Racial Science, and a Racial Religion', *Eugenics Review* 1:3 (1909), p. 193.

[128] Bradley W. Hart, *George Pitt-Rivers and the Nazis* (London: Bloomsbury Academic, 2015), pp. 32,34.

[129] Galton, 'Statistical Inquiries into the Efficacy of Prayer', p. 131.

to Galton's heart. Of course, no one person owns an ideology. Eugenics originated as a reaction to Darwinism and perhaps the knowledge that humans in the future would have to rely on their own powers of amelioration. Why it gained a hold on minds is a separate issue. The Eugenics Education Society developed independently of Galton and had a different emphasis to what was motivating him.

There were psychological reasons of a baser sort than improving humanity at play here. The enormous differences in living standards in Victorian and Edwardian Britain caused anxiety and guilt on the part of those who were well off. The report of the Royal Commission on the Care and Control of the Feebleminded defined the subjects as consisting of people only capable of making their living through wages when general economic conditions were good.[130] Below those classified as 'feebleminded' were 'imbeciles' and the lowest tier of mental defectives, 'idiots'. Imbeciles were only able to work at the easiest of tasks under supervision, while idiots could not work at all and needed to be institutionalized. However, the economy did not provide jobs for all. Lack of employment served as a convenient excuse for scapegoating the 'feebleminded', shifting the onus to the individual who allegedly had a medical condition which made gaining a living impossible. There were too many manual workers, the majority tier in society, to blame all of them, but the focus on the 'feebleminded' among the unemployed allowed the privileged to argue that heredity was the underlying cause of unequal living standards. This helped to assuage a restless conscience without there being a price to pay in terms of higher taxes or other forms of redistribution. The 'survival of the fittest' had placed the bourgeoisie in comfortable circumstances and those lower down the social scale were there owing to lack of talents, energy or self-control.[131] The Victorians liked to think of men as independent and dynamic. *Self-Help* by Samuel Smiles became a bestseller when it was published in 1859.

The 'feebleminded' of course existed in all social classes, but greater numbers were poor than rich. There were more of the former than the latter in general and the aetiology of mental deficiency makes social deprivation a risk factor. The 'feebleminded' roaming the streets were also

[130] *The Problem of the Feeble-Minded*, p. 5.

[131] Ruth Clifford Engs, *The Eugenics Movement. An Encyclopedia* (Westport CT: Greenwood Press, 2005), p. 23.

more visible than inmates in homes or those living under the care of their resourceful families. Those at large were assuredly a part of the residuum and survived by whatever means they could. Galton believed that they should be eliminated over time, rather than helped to live decent lives, which as an hereditarian he did not think possible. He gave credence to 'negative' eugenic policies going beyond his own by writing about 'criminals and semi-criminals' as a tier of society, using Charles Booth's classification of the London poor. He undoubtedly thought of them as 'unfit'.

Intellectual socialists had their own reaction to Darwinism. For them it was not competition between human stocks which represented its essence. Rather they took note of how Darwin explained the function of society in *The Descent of Man*. It boosted survival rates and made life safer for everyone. The law of the jungle could be superseded by proper social arrangements. Competition could become peaceful and cognitive, instead of being based on brute force.[132] The pre-eminent defence of such ideas was the book *Mutual Aid. A Factor in Evolution* by the Russian anarchist Peter Kropotkin, published in London in 1902. Ramsay MacDonald, an enthusiast for science, wrote that 'socialism is nought but Darwinism' in 1905.[133] His *Socialism and Government* (1909) accepted eugenics in part but not practices such as segregation.[134]

Eugenics was an interpretation of Darwinism, perhaps the reading which is the most profoundly inegalitarian. This has caused some to question whether it was ever compatible with socialism.[135] Those socialists who supported the doctrine are either regarded as not being genuine or their support for eugenics is either not known or ignored. Early socialists did have a belief in environmentalism, the notion that people can become more moral, intelligent, healthy and content through social reform. Darwin too believed that forming societies advanced both the

[132] David G. Ritchie, *Darwinism and Politics. With Two Additional Essays on Human Evolution* (London: Swan Sonnenschein, 1895), pp. 36,48.

[133] Stack, *The First Darwinian Left*, p. 53.

[134] Anna Davin, 'Imperialism and Motherhood', *History Workshop* 5 (1978), p. 18.

[135] Examples include Stack, *The First Darwinian Left*, pp. 87, 89, 107. John Macnicol, 'The Voluntary Sterilization Campaign in Britain, 1918–1939', *Journal of the History of Sexuality* 2:3 (1992), p. 432. Clyde Chitty, *Eugenics, Race and Intelligence in Education* (London: Continuum, 2007), pp. 26, 48–49.

intellect and moral feelings.[136] Galton was arguing virtually the opposite: that social reform was useful, but did not in itself have the power to elevate mankind. Additionally, degeneration was caused primarily by innate factors. Later eugenists were even more dismissive of improving the surroundings in which people lived and worked.

In the late nineteenth and early twentieth centuries various forms of biological thinking were powerful. Ordinary socialists, political activists and trade unionists could simply ignore these or, if pressed, say that they represented a bogus understanding of Darwinism. This was Keir Hardie's response in his opening speech to the Independent Labour Party conference in 1914, denying that his party's rightful demands were 'against the laws of nature' or similar.[137] Such tactics functioned well enough, but labourite intellectuals needed to address whatever ideas were circulating around them. They gave a little ground to the hereditarians, agreeing that the working class contained *some* people who were not so virtuous and that it might not be possible to create a perfect society. In the face of impossible odds, caused by a market economy, a few paupers took an easy way out and became criminals or prostitutes. Once conditions had been made more liveable, their numbers would diminish sharply. While socialism was a much better system than laissez-faire, it could not eliminate every single social problem.[138] Examples of such arguments may be found in Enrico Ferri's *Socialism and Positive Science* (1905) and Laurence Small's *Darwinism and Socialism* (1907), both published by the Independent Labour Party.[139]

Intellectual socialists generally recognized that individuals were not equal in ability, morality and health, with innate factors making up part of the disparities. It hardly negated their emphasis on improving the environment, providing clean dwellings, nutritious food, better education and safer working conditions. In *The Prevention of Destitution* (1911) Sidney

[136] D. Stack, 'Charles Darwin's Liberalism in "Natural Selection as Affecting Civilised Nations"', *History of Political Thought* 33:3 (2012), p. 541.

[137] Independent Labour Party, *Report of the Coming-of-Age Conference. Held at Bradford, April 1914* (London: Independent Labour Party, 1914), p. 49.

[138] Eden Paul, *Socialism and Eugenics* (Manchester: National Labour Press, 1911), p. 13.

[139] Enrico Ferri, *Socialism and Positive Science (Darwin-Spencer-Marx)* (London: Independent Labour Party, 1905), pp. 14, 33. Laurence Small, *Darwinism and Socialism* (London: Independent Labour Party, 1907), p. 6.

and Beatrice Webb quantified the underclass at four million people. In the face of the intractable problem of a large residuum, eking out a miserable existence, they and other left-wing thinkers were willing to consider a sprinkling of hereditary measures too, provided that social reform was not neglected.[140] This is as understandable as why some confirmed supporters of free-market economics became social imperialists, willing to introduce limited welfare provisions. It would help the nation compete against rising great powers, such as Germany or Japan. To argue for a climate amenable to eugenics owing to Britain's imperial position slipping vis-à-vis its rivals has become commonplace.[141] And while racism forms no part of contemporary socialism, in the early twentieth century almost everyone, socialists and workers included, assumed that Europeans were of higher calibre than natives of the colonies. Galton argued that 'national eugenics' would enable Britain to perform imperial tasks more efficaciously.[142] This might be attractive to Fabians, who often believed that the Empire was a pathway to the World State which they desired.[143]

In his first official lecture on the topic, 'Eugenics: Its Definition, Scope and Aims', Galton was willing to take on board modern opinions regarding environmental improvement. Taking human potential to its highest level by bettering the race through all sorts of methods was both a social reformist and an hereditarian goal. Galton thought that mankind's progress was largely the responsibility of the talented. By raising the mettle of the median citizen, a consequence would be more individuals of the highest order being born. The lecture makes clear that he valued a variety of attributes. Everyone's ideals and way of life would be respected in choosing the traits to be fostered, except that criminals would not be allowed to contribute their idea of the perfect person. He argued that morality was relative rather than absolute and that life would become very dull if everyone shared the otherwise exemplary attitudes of Marcus Aurelius and Adam Bede.[144]

140 Leo Lucassen, 'A Brave New World: The Left, Social Engineering, and Eugenics in Twentieth-century Europe', *International Review of Social History* 55:2 (2010), p. 284.

141 An early contribution which argued this was Searle, *The Quest for National Efficiency*, p. 60.

142 Galton, 'Eugenics: Its Definition, Scope and Aims', p. 38.

143 H. G. Wells, *A Modern Utopia* (London: Chapman & Hall, 1905), pp. 88–89.

144 Galton, 'Eugenics: Its Definition, Scope and Aims', p. 36.

Grounded on the latest revolutionary scientific discovery, evolution, eugenics was designed to fill the gap left by an increasingly unconvincing Christian orthodoxy. As a call to action, eugenics had the capacity to correct part of the problems with evolution. Darwin had described suffering and short lives without meaning as fundamental to the existence of species, greatly extending what was known about this before.[145] Eugenics might help to minimize the suffering and held out hopes that the refinement of mankind could happen faster than would otherwise occur. The Eugenics Education Society from the outset would also officially believe in environmental reform, although most of the membership probably did not.[146] Its president between 1911 and 1928, Leonard Darwin, a son of Charles Darwin, needed to remind eugenists that the environment was indeed of importance and to request that they did not argue against social reform.[147] But neither Leonard Darwin nor Galton suggested that life could be made better through the redistribution of resources. That was a facet of socialist eugenics alone.

German and American Eugenics

British eugenics would not necessarily develop along the same lines as the versions obtaining in other countries. That does not preclude the whole movement being international with cross-fertilization and the exchange of ideas across borders. Because their doctrine was an insurgent tendency, eugenists looked abroad with interest and congratulated each other on successes achieved within their national polities. Galton was hailed as the originator of German racial hygiene by that country's pioneering eugenist Alfred Ploetz, though the biologists Ernst Haeckel and August Weismann had laid the underpinning. Ploetz offered Galton the honorary presidency of the International Racial Hygiene Society in 1909.[148] Galton, who had recently been knighted, accepted. Eugenics was nonetheless

[145] Alastair E. McGrath, *Darwinism and the Divine. Evolutionary Thought and Natural Theology* (Chichester: Wiley-Blackwell, 2011), p. 171.

[146] The editorial in the first ever issue of its *Eugenics Review* noted the importance of the environment too and advocated social reform. *Eugenics Review* 1:1 (1909), p. 4.

[147] Leonard Darwin, 'Heredity and Environment. A Warning to Eugenists', *Eugenics Review* 8:2 (1916), p. 94.

[148] Paul Weindling, *Health, Race and German Politics between National Unification and Nazism, 1870–1945* (Cambridge: Cambridge University Press, 1993), p. 151.

such a variegated and malleable doctrine that it is scarcely possible to extrapolate from one setting to another. In France the doctrine was neo-Lamarckian and thus asserted that acquired characteristics could be handed down genetically to the next generation.[149] In Germany social class never played a preponderant part in the doctrine.[150] In the United States significant effort was expended on showing that northern Europeans were better citizens than southern and eastern Europeans. These contexts help explain their divergence from British eugenics. An implication is that each version should be related to its own setting. Only a small minority of British eugenists disdained democracy. Few were particularly bothered about race, at least until the beginning of large-scale immigration in the aftermath of the Second World War. The dominant position accepted laissez-faire economics and saw adjustments to it as boosting the number of 'degenerates', which would create problems in the future. The national variants nevertheless existed on a spectrum which makes up the entirety of the doctrine.

Eugenics was more influential in the United States than anywhere else prior to the accession to power of the Nazis in Germany in 1933. America presented a paradigm to other English-speaking countries in this regard. Especially the early stateside embrace of sterilization had a profound effect on how eugenics developed in Britain. It radicalized a doctrine which initially had differed little from Victorian values. After 1926 British eugenists too began clamouring for 'voluntary' sterilization. Had this been implemented, there is no doubt that many people would have lost the capacity to reproduce on spurious grounds. Even if the relevant authorities had stuck to the spirit of the law, mutilating citizens is a contradiction of why society exists in the first place: to keep individuals safe and protect their bodies from the transgressions of others.[151] Salpingectomy and vasectomy were taken all the way to euthanasia in Germany.

[149] William Schneider, 'Towards the Improvement of the Human Race: The History of Eugenics in France', *Journal of Modern History* 54 (1982), p. 269. Marius Turda and Aaron Gillette, *Latin Eugenics in Comparative Perspective* (London: Bloomsbury Academic, 2014), p. 29.

[150] Pauline M. H. Mazumdar, *Eugenics, Human Genetics and Human Failings. The Eugenics Society, its Sources and its Critics in Britain*, (London: Routledge, 1992), p. 4.

[151] See 'The Sterilization Bill', *Eugenics Review* 23:2 (1931), pp. 153–154.

In London the Battersea Dogs' Home was unable to rehome every animal brought to it. In 1884 it began using physician Benjamin Ward Richardson's patented carbonic acid gas method to humanely put down the surplus dogs. Knowledge of this practice gave a few people ideas. A handful of British thinkers began to argue that human lives of no value should be extinguished in what became known as the 'lethal chamber'.[152] H. G. Wells said specifically that he did not want this, but nevertheless supported the culling of deformed newborns.[153] Thus there was a minuscule strand in the doctrine which paved the way for the Holocaust. Apart from the notorious excesses Germany carried out during the Second World War, mentally ill patients were put to death in late 1939, backdated to the outbreak of the war on 1 September.[154]

The Case for Sterilization and Segregation

The fact that a sterilization law was not passed in Britain avoided damage to the country's reputation. In Germany a horrendous experiment in social control and persecution by the state took place. It is thought that 360, 000 Germans lost the capacity to reproduce under a surgeon's knife.[155] The Sterilization Law of 1933 was compulsory, wide-ranging and introduced by the Nazis after democracy had come to an end. It was not for lack of trying that British eugenists failed to implement a weaker equivalent. There was a difference in political culture between the two states. Proposed sterilization in Britain needed to be passed by Parliament. The type recommended by the Eugenics Society was touted as 'voluntary' and initially applicable only to mental defectives. Other hereditary afflictions only entered the picture after its first attempt failed in 1931. It might perhaps have been extended to carriers who were close relatives of the same, as the Departmental Committee on Sterilization chaired by Laurence Brock advised in 1934.

This form of sterilization would not in fact have been 'voluntary'. Applications could have been made by the Board of Control or the

[152] McLaren, *Reproduction by Design*, p. 73.

[153] Wells, *A Modern Utopia*, p. 143.

[154] Anne Waldschmidt 'Genetic Screening and German-based Eugenics—Old and New' in Annette Burfoot (ed.), *Encyclopedia of Reproductive Techniques* (London: Routledge, 2018), pp. 374–375.

[155] Weindling, *Health, Race and German Politics*, p. 533.

guardian of a patient (including the director of an asylum) and a signature, where required, obtained under pressure. One formulation by the Eugenics Society even admitted so. It was for reasons of persuasion that it sounded better to call it 'voluntary'. Yet even if it did not deserve the label 'voluntary', the number of sterilizations would not have been as high as under the German law. Compulsory sterilization for nine different reasons in Germany permitted mass sterilizations, not just of individuals already known to the medical authorities.[156] Both the 'mainliners' and 'reform' eugenists agreed with sterilization in the 1930s and sometimes before. Therefore, there was a risk that Britain would join those countries which came to regret what they had allowed to happen to their own citizens.

Sterilization was not a question of left or right, but of generations within the eugenics movement. The Liverpool physician Robert Rentoul advocated sterilization from 1903 and the procedure was occasionally mooted in the Eugenics Education Society before 1914. However, the early eugenics movement supported segregation, not sterilization.[157] This included Galton to whom surgical intervention would have been too radical. Contraception also was taboo to him. Institutionalization of the hereditarily afflicted was the preferred policy. Eugenists were proud of having succeeded with the Mental Deficiency Act of 1913. The reasons for the shift towards sterilization were partly because eugenics had achieved momentum, becoming a factor in social policy, and partly because the replacement of dysgenic with eugenic types was occurring too slowly. Starting with Indiana in 1907, American eugenists were securing sterilization legislation in state after state. It provided inspiration to their British counterparts. There were nonetheless dissenters from this policy within the Society in the 1920s.[158] Not everyone believed in the necessity of adopting such extreme measures. It is difficult to gauge how many supporters took it on face value as being 'voluntary'. These ingenues saw it as a permanent means of contraception for people who did not wish to have (additional) children.

[156] For, mostly hereditary, 'feeblemindedness', schizophrenia, manic-depression, epilepsy, Huntington's chorea, blindness, deafness, physical deformity and severe alcoholism. Weindling, *Health, Race and German Politics*, p. 525.

[157] Searle, *Eugenics and Politics*, p. 93.

[158] Wellcome Library, London. Eugenics Society Archive. SA/EUG/A18. Annual Report 1925–1926, 'An Outline of a Practical Eugenic Policy', p. 1.

With a variety of perspectives buttressing eugenics, sterilization too rested on a myriad of grounds and received support from outside the movement. In the 1930s several women's organizations recommended it, mainly due to the misunderstanding that it would counteract the scourge of rape and sexual abuse. In 1936 the Association for the Blind passed a motion to approve 'voluntary' sterilization for hereditary sightlessness.[159] A survey of 108 organizations in 1937 who supported the conclusions by the Departmental Committee included a few mutual help societies and associations for the unemployed.[160] On a theoretical level, an attempt could conceivably be made to justify it from the nineteenth-century philosophy of utilitarianism. Its focus on the greatest happiness principle and its collectivist approach had been popular half a century earlier. Galton agreed with the utilitarian notion of science as helping to achieve social reform rather than as a purely academic endeavour.[161] In the Edwardian age the argument was put forward that societies had become so advanced that they could override the rights of individuals.[162] Utilitarians might claim that the community had the right to compel an individual who suffered from an hereditary illness to refrain from procreating. If the illness was incurable and painful or stigmatizing to such an extent that ordinary life would be well-nigh impossible, the morally correct decision would probably be to avoid passing it on to others. If the individual disagreed, the wrongdoing involved in preventing him or her from having children might be lower than the harm arising from the individual reproducing. This would make it permissible according to utilitarianism. It seriously jars with post-Second World War human rights or notions of equity existing today. Since even hereditary afflictions are affected by social conditions, it is thought fairer to improve the environment in which individuals are raised first.

If the ethics of utilitarianism are accepted though, denying matrimony to such an individual, as Galton proposed, would be a less serious infraction of his or her rights than sterilization. If the diagnosis turned out to be

[159] Modern Records Centre, University of Warwick. MSS 229/NA/4/1/9. Mental Hospital and Institutional Workers' Journal 25:2 (1936), p. 4.

[160] British Library of Political and Economic Science, London. Carr-Saunders papers B/3/10. Evidence for Royal Commission Appendix II.

[161] Waller, 'Ideas of Heredity, Reproduction and Eugenics', p. 477.

[162] Marouf Arif Hasian, jr., *The Rhetoric of Eugenics in Anglo-American Thought* (Athens GA: University of Georgia Press, 1996), p. 26.

faulty, it would be easier to make restitution. According to utilitarianism, segregation in institutions where reproduction is ruled out could also be justified if the quality of life the inmate experienced there was higher than that from which he or she was taken. In the case of destitute people, the provision of regular meals and the safety of being supervised could conceivably make an institutional life better than one spent in natural surroundings. Even if the individual in question disagreed, if crime and disease were reduced in society through the institutionalization of many of these unfortunates, that too would be the morally correct decision from a utilitarian point of view.

Utilitarianism as a fount of classical liberalism suited the Victorian age better with its laissez-faire economy and lack of sympathy for individuals, neglected in a system which could not bring prosperity to everyone. As awareness of residual social problems rose, it bred pessimism among the public. Social Darwinism argued that nothing could or should be done and eugenics as an hereditarian ideology served to justify that the poor were struggling. British popular eugenists were generally reactionary and retained a Victorian mentality. Exceptions existed, like William Ralph Inge and Ernest Barnes, a clergyman with scientific training, but they had generally abandoned belief in Christianity and were looking for other moral justifications. Barnes was able to square sterilization with his Christian faith anyway.[163] For modern thinkers, it was easier to contemplate a sexual reform like sterilization to add to their own concoction of fear and politics. They perceived eugenics as leading the way to a rational and scientific society. Utilitarianism had lost its influence by the 1930s when sterilization was foregrounded, but lack of respect for the poor and downtrodden was endemic in British society.

British eugenics was class-based. It was only after the Second World War that race began to have effects upon it, and when it did, the interaction was not as negative as might be imagined. Many supporters of eugenics by the 1960s or 1970s had overcome the backward tendencies of their doctrine. Only a few wished to be involved in anti-immigration rhetoric or similar. Class is the main distinction in Britain, perhaps only surpassed by gender in moulding someone's life. While John Stuart Mill and Galton disagreed on hereditarianism, the former stating that the human mind could achieve anything with the correct training, they

[163] Patrick T. Merricks, *Religion and Racial Progress in Twentieth-century Britain. Bishop Barnes of Birmingham* (Cham: Palgrave Macmillan, 2017), p. 174.

were as one regarding how mental ability was distributed in the population. Mill saw professionals as more intelligent than small tradesmen or supervisors, who again were more capable than ordinary workers.[164] This matched the social status of each group. Mill's empiricist epistemology supported the idea, which was held before Galton, that the mind was a *tabula rasa* at birth.[165] Mill did not completely rule out innate powers, but his theory of associations involved new learning being accommodated through what had already been assimilated.[166] In other words, learning got easier through experience. Environmentalism was a hardy theory and even underwent a renaissance from the 1880s to 1900, when it was incorporated into Darwinism as the doctrine of neo-Lamarckism. [167] Gradually, as August Weismann's experiments with mice from 1883 became known and Mendelian genetics made progress in Britain after 1900, environmentalism lost credibility. *Essays upon Heredity* (1889–1892) and *The Germ Plasm* (1893) by Weismann were available in English.[168] Although his experiments were crude, cutting the tails of mice did not cause their offspring to have shorter tails, it seemed to prove that traits acquired during a life were not transmitted further. It meant that there were few scientific alternatives to hereditarianism, which buttressed eugenics and made it seem credible on a popular level until around 1930.[169]

[164] John Stuart Mill, *Considerations on Representative Government* (New York: Prometheus Books, 1991), pp. 182–183.

[165] Peter Dickens, *Social Darwinism. Linking Evolutionary Thought to Social Theory* (Buckingham: Open University Press, 2000), pp. 7–8.

[166] Gillham, *A Life of Sir Francis Galton*, p. 228.

[167] Denis R. Alexander, 'Creation and Evolution' in J. B. Stump and Alan G. Padgett, *The Blackwell Companion to Science and Christianity* (Chichester: Wiley-Blackwell, 2012), p. 234.

[168] Robert C. Bannister, *Social Darwinism. Science and Myth in Anglo-American Thought* (Philadelphia: Temple University Press, 1979), p. 138.

[169] André Pichot, *The Pure Society. From Darwin to Hitler* (London: Verso, 2009), p. 129.

Interaction with Americans

American society took to eugenics easily, one of the reasons for which was a sense of community with people of the same or similar national background. Social class probably did not provide a common identity to the same extent in Britain. As a relatively ethnically homogeneous state in the late nineteenth and early twentieth centuries, race did not matter much domestically. Eugenics in the United States consequently interacted with racial thinking earlier. Famous American eugenists like Madison Grant, Henry Fairfield Osborn and David Star Jordan are perhaps better known for scientific racism today than for eugenics. Grant's *The Passing of the Great Race* (1916) was a homage to Anglo-Saxons at a time when they began to decline within the population. New immigration from southern and eastern Europe was initially viewed as a threat, an attitude which found its way into eugenics. Between 1898 and 1910, two million Italians, one million Jews and one million subjects of the Habsburg Empire had come to the United States.[170] Galton and Pearson held racist assumptions too, but it would be well into the 1920s before the question of race was even mooted by British eugenists. It was in the context of miscegenation, on which tentative and somewhat contradictory results were obtained from research in that decade and the next.

An early American eugenist who would become a leading name in the international movement was the biologist Charles Davenport. He was inspired by Pearson's work in biometry, which he had come across while acting as an instructor in zoology at Harvard University in the 1890s.[171] At the turn of the century, he visited London and had dinner at Galton's home in Knightsbridge. This inspired him further, and Davenport soon originated plans for an American laboratory of eugenics which would make the doctrine count in the New World. It was duly founded in 1904 in Cold Spring Harbor on Long Island, becoming the Eugenics Record Office in 1910, six years after its namesake was founded by Galton at the University of London.

The Americans generally considered eugenics to be a continuation of social Darwinism. Herbert Spencer's thinking had been lauded in the United States and suited conditions where Native Americans had

[170] Marius Turda and Maria Sophia Quine, *Historicizing Race* (London: Bloomsbury Academic, 2018), p. 19.

[171] Kevles, *In the Name of Eugenics*, p. 45.

been conquered and seen their traditional way of life curtailed, where robber capitalists sought to justify their actions, where every man could make his fortune if hard-working enough and where freedom to live under a minimum of government interference was highly prized. Some of these notions parallelled British popular eugenics too, particularly the respect for wealth. The main way in which American eugenics stood apart from British or European was its success on sterilization. The 1907 law in Indiana mandated the procedure for mentally subnormal people and those found guilty of sex crimes. Eugenics furnished the reasoning behind this, with previous legislation from 1905 having already banned the mentally deficient, sufferers from hereditary conditions and inebriates from marrying.[172] California followed in 1909 and by 1918 a total of eighteen states had introduced such legislation (although six had subsequently repealed the laws).[173] Eugenists everywhere were inspired by the laws relating to hereditary conditions, with the American example serving as a potent model when British eugenics too began agitating for sterilization in the late 1920s.

Although Davenport had originally been enthused by biometry, he was among the first to take up Mendelian inheritance in the 1900s. It offered greater insight than biometry, though its early formulations did not account for the complexity of heredity. It was thought that feeblemindedness was a recessive trait among just two genes. By studying a family pedigree and noting whether particular persons were mentally subnormal or not, it therefore seemed possible to eliminate feeblemindedness relatively quickly. Books on hopelessly inadequate families (Jukes, Nams, the Tribe of Ishmael, etc.) had been written in the United States in the 1870s and 1880s, including calculations on what their crimes and misdemeanours were costing society.[174] Galton was aware of these investigations, which influenced British eugenics. The Eugenics Education Society began to compile pauper pedigrees in 1912.

[172] Trombley, *The Right to Reproduce*, pp. 50–51.

[173] Engs, *The Eugenics Movement*, p. 55.

[174] See Harry H. Laughlin, *The Second International Exhibition of Eugenics* (Baltimore: Williams & Wilkins, 1923), pp. 146–151.

The Effects of Galton's Eugenics

Galton's multifarious work encompassed individual differences in psychology. The research field eventually led to the understanding that able students from humble backgrounds should be supported. It accorded with what Galton had written in *Hereditary Genius* and which he continued to advocate. Back in 1869, he reckoned that the best sort of society for the improvement of the race was 'where every lad had a chance of showing his abilities, and, if highly gifted, was enabled to achieve a first-class education and entrance into professional life, by the liberal help of the exhibitions and scholarships which he had gained in his early youth'.[175] Within a few years he had included girls in his schemes. Women were tested for all sorts of abilities in Kantsaywhere, including gender-specific household tasks. The examinations would aid an evolutionary process leading to the continual betterment of individuals and society. Until they were implemented, he urged wealthy patrons to take an interest in promising lads and help them along the way.[176] He was struck by how often future luminaries had received such assistance. This again proved that he accepted that talents were to be found in other classes than those which were comfortably off.

Galton's eugenics could be summarized as (1) a concern with heredity as being more important than the environment in human affairs, (2) controlling evolution so that our species would improve faster than nature could manage and with less associated suffering, (3) support for eugenics as a guiding principle of society, (4) discovering able and talented youth through specially convened examinations and encouraging them to marry early and (5) segregating hereditarily afflicted people and, if he could take society with him on the issue, restricting marriage to normal men and women free of hereditary taints such as congenital mental deficiency.

Those principles gained only a little traction over the course of the twentieth century. Concurrent with early eugenics, the labour movement was growing and its emphasis on environmental amelioration represented an immediate attempt to counter social misery. The early eugenists believed in hereditarian measures almost to the exclusion of everything else. It is not uncommon for adherents of a new movement, apparently making progress and providing a novel perspective, to regard it as a

[175] Galton, *Hereditary Genius*, p. 362.

[176] Galton, 'The Possible Improvement of the Human Breed', pp. 30–32.

panacea. Caleb Saleeby thought that eugenics would create a paradise on earth compared with which the theologian's was a foolish dream.[177] Left-wingers like him (he joined the Fabian Society in 1910 and wrote in the *New Statesman* under the pseudonym 'Lens') were particularly prone to utopian thinking. The socialist writer George Whitehead dreamt of 'the evolution of a race of beings physically beautiful, morally pure, soaring to an intellectual platform hitherto unconceived, living a full, happy, human life, with liberty to develop to the sublimest heights that which we call soul'.[178] Eugenics advanced steadily in the 1910s and 1920s, including finding favour in government circles. Then eugenists overreached themselves through launching their campaign for sterilization. It crystallized opposition to the ideology and brought much of the progress which had hitherto accompanied its trajectory to a halt.

By the standards of its own time, eugenics as conceived by Francis Galton contained a mixture of ameliorative and reactionary features. It did not improve the conditions of life of the least fortunate in society, but that was no change from their situation in the Victorian and Edwardian eras anyway. The founding of the asylums occurred in the nineteenth century and provided a model for how dangerous or deficient individuals could be kept away from the community. In England the number of 'lunatics' in asylums doubled between 1844 and 1860.[179] There were no known cures for various debilitating hereditary conditions in the early days of eugenics. Contemporaries believed that being mentally deficient often resulted in a predilection towards crime or prostitution. Galton had recommended as early as in *Hereditary Genius* that society should encourage the weak to enter celibate monasteries or nunneries which would care for them, but not allow them to breed. Institutionalization on medical grounds served the same purpose. Eugenics presented a cogent argument for segregation and social considerations too played a part. Due to the dire living standards of the residuum in the early twentieth century, it might even be in their own interests to be confined.

The idea of prohibiting marriage for sufferers from hereditary afflictions would only have had a piecemeal effect. The residuum was

[177] Trombley, *The Right to Reproduce*, p. 72.

[178] George Whitehead, *Socialism and Eugenics* (London: Twentieth Century Press, 1911), p. 2.

[179] Keir Waddington, *An Introduction to the Social History of Medicine: Europe since 1500* (Basingstoke: Palgrave Macmillan, 2011), p. 319.

submerged below respectable society anyway and did not pay especial attention to legislation standing in the way of natural instincts. Galton also suggested that charity be directed away from such people towards genetically desirable stock who might thereby have more children. This would have caused hardship to some paupers and reduced the life expectancy of a fraction of them. The collective consequences would have been limited as they were eking out a wretched existence from virtually nothing as things stood. Charity seldom resulted in the transformation of lives. Galton's attitude was still reactionary because even if charity was a palliative, it amounted to a foreshadowing of a durable solution being found. Denying certain people the right to marry or institutionalizing them in places where no breeding would occur was justified by the well-being of their potential offspring and the social costs of the crime and diseases which it was thought that they were spreading. Today it is appreciated that inheritance is less straightforward than Edwardians imagined.

Galton's eugenics was not laissez-faire, as he supported state interference in the lives of its citizens. His initial 'positive' eugenics would probably have accelerated social stratification, while his 'negative' eugenics was a wholly backward project. The environmental reforms meant to accompany his main hereditary measures might have done some good. Greater reliance on exams, rather than patronage, along with scholarships and practical help for poor people with talents, was a modernization project designed to improve outcomes. He formulated plans which would enable artisans of pedigree to have larger families through a social housing scheme. Galton's support for meritocracy was nevertheless somehow a supplement to maintenance of the class system. The recording of heredity in his aptitude tests would give an advantage to scions of Galton's notable families. It smacks a little of Cyril Burt who would revise marks upwards or downwards on his mental tests after consulting the teacher of the pupil in question.[180] The exams in Kantsaywhere were divided by social class, with one type for the children of farmers and manual workers and another for the offspring of cultured types.[181]

When snobbish and racist assumptions to a greater degree than occurring in Galton's original conception are added, the doctrine becomes sinister in itself. His reforms would particularly benefit the professional

[180] Gould, *The Mismeasure of Man*, p. 278.

[181] Galton/2/4/19/6/1 folios 21–22.

classes and, at times, those who stood in line to join them. They were not of much use in tackling the vast amounts of poverty and crime in Victorian or Edwardian Britain. Although his views on race and class were typical of the time, he may have legitimized prejudice by lending a scientific gloss to it. His understanding of history and what causes a nation to flourish or decline were naïve. He believed that the Dark Ages had been brought on by clerical celibacy because sensitive minds were guided towards the church in medieval times. Galton seemed to have earnest intentions and, in the context of the society of which he was a part, eugenics was meant to be ameliorative and forward-looking. Anchored in the theory of evolution, it sought to correct some of the moral difficulties which emerged from this understanding of the world. It was a related though slightly different interpretation to social Darwinism. The distinction involved whether the state should be utilized to secure improvement to human life. Should the weak, vulnerable and least fortunate be left to die? That was the social Darwinist contention. Or should they be cared for but not be allowed to breed? That was Galton's opinion.

Conclusion

The mere fact that he would discourage or prevent the above individuals from reproducing and thus satisfying their natural instincts did not bode well for the future. The abject and brutal poverty which affected so many people in his time presented too great a challenge for the accepted solutions. As charity, philanthropy or preaching were not convincing either, Galton's position seemed more reasonable than it does today. And because studies of heredity were only incipient, important objections had not yet been raised. The environment in which someone grows up is surely a great determinant of how that person's personality is formed. Galton himself was a product of his family's values and interests, especially through being trained from childhood to become a scholar. If he or his cousin Darwin had needed to earn their keep from an early age and had not received an education, they probably would not have had the opportunity to become scientists. It was not true, as Galton believed, that science and literature were open to all. But what was meant by environmentalism in the Victorian era consisted of prenatal influences, the value of self-help or the notion that everyone begins life at the same starting point. A little later, a work like Frank Challice Constable's *Poverty and Hereditary Genius*, published in 1905, stood out precisely because it argued that opportunity was more important than heredity in charting success.[182] Other socialists understood the reasoning, but it took a few decades before it began to affect genetics. In 1925 American biologist Thomas Hunt Morgan cautioned against blithe assumptions that feeble-mindedness was due to a recessive gene. He said it was too nebulous exactly what it involved and what its causes were.[183] The discourse was nevertheless of long standing. Charles Dickens's novel *Bleak House*, first published as a serial from 1852, featured a character called Harold Skimpole. Skimpole had almost a dozen children although being childlike himself.[184]

It does not make sense to argue that Galton was a 'reform' eugenist because he died in 1911, before such interpretations had coalesced into a set of principles. However, in many ways 'reform' eugenics was a revival of what Galton had put forward before eugenics became a political movement. It extricated eugenics from all but the most general observations of social Darwinism. Galton's main contention that heredity counted for more than the environment had been stressed by eugenists repeatedly prior to the 1930s. 'Reform' eugenists could therefore allow themselves to consider the environment too. Their version counterposed Galton's limited social reforms with the eugenics he had preached concurrently. His name still carried weight among eugenists, as the founder of the creed. Because 'mainline' eugenics had originated as social Darwinism and right-wing attitudes, it was often possible for the 'reform' eugenists to find passages Galton had written which were at odds with 'mainline' axioms.

Eugenics alternated between optimism and pessimism. Galton had held great hopes for the future until the 1900s. He lived through marvels of engineering, such as railways and man-made canals being constructed, telegraphy permitting instant communication across continents, anaesthetics and improved hospital care, the second industrial revolution and the suffrage being extended to the majority of adult men.[185] Those developments nourished hope, but there was also an awareness of the problems

[182] F. C. Constable, *Poverty and Hereditary Genius. A Criticism of Mr Francis Galton's Theory of Hereditary Genius* (London: Arthur C. Fifield, 1905), pp. 19, 104–105.

[183] Steven A. Farber, 'US Scientists' Role in the Eugenics Movement' (1907–1939). A Contemporary Biologist's Perspective', *Zebrafish* 5:4 (2008), p. 245.

[184] Charles Dickens, *Bleak House* (London: Penguin, 2012), p. 81.

[185] Cf. C. P. Blacker, *Eugenics. Galton and After* (London: Gerald Duckworth, 1952), p. 129.

thrown up by nature and the ending of belief in a beneficent world order. Darwin had told his readers to console themselves with the thought that for animals losing out under conditions of natural selection, death generally came quickly, not much fear was felt and it enabled the happy and well-adjusted to live better lives. Even so, his theory drew attention to the precarious position incumbent upon all species. The majority of life did not grow up, there was not enough sustenance for everyone, to be preyed on by stronger species was merely a fact of life and any purpose behind this suffering was nullified when a new geological age began and almost all the previous species became extinct.[186] These conditions affected humans too. Infant mortality rates were high in Victorian Britain and being the victim of crime was not uncommon.

Eugenics mirrored the society in which it arose in having both a benign and a dark side. Using science and thus immersed in empirical reality, it attempted to find solutions to these existential problems. Given the same environmental conditions, some infants survived while others did not. To the eugenist, the objective was to study the heredity that was successful under conditions of natural selection. As a political and only ostensibly scientific ideology, it had already decided that such heredity was to be found primarily among the middle classes and above. The Malthusian problem of the lack of sustenance would be solved through controlling human breeding. Eugenics would reduce the personal suffering to which natural selection gave rise. But on a different level these solutions were cruel in themselves, especially when they involved the replacement of natural selection by restrictions in marriage or confining the 'feebleminded' to institutions where their lives would in effect be closed down. It was an unfeeling method of stopping the residuum's 'unusually prolific breeding'. Later developments would show that Galton's ideas in this respect were simplistic, besides offering no sympathy for the poor and downtrodden. Instead of thinking about how crime and poverty could be removed in a society which had social justice, as Aristotle had done, Galton's proposals were to eliminate, over a few generations, those who suffered from poverty and committed many of the crimes.

[186] Darwin, *On the Origin of Species*, pp. 48, 53.

CHAPTER 3

The Galton Laboratory Until 1945

The Eugenics Society promoted eugenics as a social movement, whereas the Galton Laboratory sought to make it a science.[1] Unlike what its founder had intended, the latter's publications were usually dense scientific literature with only limited appeal beyond a narrow readership.[2] They tended to generate debate only when they touched upon controversial issues or if the findings contradicted received opinion.[3] For the general public, it was the Eugenics Society which epitomized the ideology. As the carrier, it shaped how British eugenics is perceived. Prominent members Caleb Saleeby, Ernest MacBride and Leonard Darwin presented it in book series such as Harmondsworth Popular Science, Home University or the magazine *Armchair Science*.[4]

They regarded eugenics as scientific and were concerned about the dimming of natural selection in civilized societies. 'Degeneration' might

[1] Lyndsay Andrew Farrall, *The Origins and Growth of the English Eugenics Movement 1865–1925* (New York: Garland Publishing, 1985), pp. 103–104.

[2] Karl Pearson, *The Life, Letters and Labours of Francis Galton*. Volume 1 (Cambridge: Cambridge University Press, 1914), p. 57.

[3] *The New Statesman. A Weekly Review of Politics and Literature*. Vol III No. 70 Saturday August 8, 1914 (London: The Statesman Publishing Company, 1914), p. 668.

[4] Peter J. Bowler, *Science for All. The Popularization of Science in Early Twentieth-century Britain* (Chicago: University of Chicago Press, 2009), p. 28.

D. Redvaldsen, *A History of British Eugenics since 1865*, Medicine and Biomedical Sciences in Modern History,
https://doi.org/10.1007/978-3-031-72290-5_3

result from this, an endemic process of stunted growth, rising crime and contagion. Francis Galton pointed to the proliferation of 'feeblemindedness' as an example. By seeking to discover through an actuary what a child of each of his statistical classes' net contribution to society might be, he foreshadowed the differential birth rate argument. The statistician William Farr had already calculated it for a typical child of Essex labourers.[5] A common opinion in educated circles was that the poor were having too many children and well-situated parents too few. Correcting this became the primary task of eugenics.

As universities were professionalized in the twentieth century, they began to take on additional roles. Science had been seen as important to the nation at least since the second industrial revolution in the 1870s. The subject 'national eugenics' was considered a science and came to be housed within the federal University of London. England's third research university thereby came to be associated with an innovative field of inquiry, which seemed destined for great things. Even if the hopes were not fulfilled, it might still solve isolated medical and social problems. The university's principal, Sir Arthur Rücker, believed his institution to be a suitable venue for Galton's project due to its past connection with statistics. Karl Pearson, whom Galton later designated as the first professor of eugenics, had been involved in this.[6] When making an appeal for funds in 1911, the university stated that social legislation needed a statistical basis.[7] The data had to be analysed dispassionately without bias or ulterior motives. In view of the reputation eugenics subsequently gained, these assertions seem ironic today.

The Eugenics Record Office

On 10 October 1904 Galton wrote to the University of London offering to fund a fellowship in 'national eugenics'. He defined it as socially controllable influences which affected the population and its health,

[5] Francis Galton, 'The Possible Improvement of the Human Breed under Existing Conditions of Law and Sentiment' in *idem, Essays in Eugenics* (London: Eugenics Education Society, 1909), p. 11.

[6] University College London (UCL), Galton Laboratory. Galton Laboratory/1/1 Rücker to Galton 19 Dec 1906, p. 1.

[7] UCL, Pearson papers. Pearson/4/9/1 Appeal for Funds to Build the Francis Galton Laboratory, p. 5.

although he was less interested in health on this occasion. It was a wide definition, stretching beyond heredity and resembling the one Galton had used in his talk to the Sociological Society in May that year. The relevant task was applying statistics to determine the characteristics of the populace. Galton had underlined the word 'exact' when describing how he wanted the subject to be studied.[8] The fellowship would be tenable for three years, attracting a total grant of £ 1, 500, and if it yielded positive benefits might be made permanent at about £ 500 a year.

Just four days later a meeting was held at which Galton, Pearson, Sir Edward Busk, chairman of convocation of the university, Halford Mackinder, recently appointed as director of the School of Economics, and Sir Arthur Rücker attended.[9] It established the principles by which the scheme would be governed. A special committee of Galton, Pearson, Busk and Mackinder would be convened to review the applications received and recommend a candidate for the position. Galton, anxious that the right person be appointed, secured a veto on this matter. There were also enough funds for an assistant. Expected to devote all his time to eugenics, the fellow appointed was Edgar Schuster. A nephew of the physicist Arthur Schuster, he had held a biology scholarship at New College, Oxford, under W. F. R. Weldon. As Weldon was one of the founders of biometry and had been a professor of zoology at University College in London 1889–1899, the fellowship was kept within a close intellectual coterie. Schuster's assistant would be Ethel M. Elderton, a former teacher who had studied at Bedford College. They were based in two rooms in 50 Gower Street, later gravitating to no. 88, in accommodation rented by University College near its quadrangle. This was the kernel of the Galton Laboratory, the main research institution for eugenics in Britain.

Known initially as the Eugenics Record Office, its work began along the lines Galton had pioneered. Forming a register of able families, through investigating whether the kinsfolk of distinguished people also displayed superior ability, was the main assignment. Although *Noteworthy Families* by Galton and Schuster published in 1906 dealt only with scientists, it was envisaged as the first of a series. The book was an extension of Galton's 1904 article in *Nature*, which analysed responses to his survey

[8] UCL, Galton Archive. Galton/2/4/12/1/1 folio 2r.

[9] Galton/2/4/12/1/1 folio 3r.

of fellows of the Royal Society. It might be seen as a continuation of *Hereditary Genius* and *English Men of Science*. In other work, Schuster collated performance in the Oxford B. A. examination with school records at Harrow and Charterhouse, and professional activities since undertaken. The rationale was to discover whether the best performers could be identified as early as their school days. As Schuster had held scholarships at Charterhouse and Oxford, there was an autobiographical element to his research. These two studies became the first and third volumes respectively of the Eugenics Laboratory Memoirs. Schuster additionally studied 'idiocy' tentatively using the records of Dr John Langdon Down and the National Association for Promoting the Welfare of the Feebleminded.[10] Langdon Down's research gave rise to what is now known as 'Down's syndrome'.

This excepted, the Eugenics Record Office dealt with the 'positive' side of the doctrine. It was in line with Galton's usual procedure before he too became worried about degeneration. His idea of segregation received support in 1904 when the Interdepartmental Committee on Physical Deterioration, led by Almeric Fitzroy, reported. It had been set up as a response to the anxiety over the physical condition of the people, uncovered by enlistment to the army during the 1899–1902 Boer War.[11] The Committee suggested that people unable to meet ordinary standards of decency could be held in labour colonies.[12] It used the word 'degeneration' in connection with alcohol, but was not persuaded that genetic causes lay behind it.[13] No known eugenists served on the committee. Its recommendations included foster care for children, better surveys, improving education on practical matters including gymnastics and enforcing already existing laws. It did not think that the deterioration was severe.

Galton decided to hold meetings to discuss which topics his new institute and the doctrine's supporters should investigate. These took place at the Eugenics Record Office after hours or in Galton's home.[14]

[10] Pearson/4/4/13 Report of the Francis Galton Research Fellow, folio 2r.

[11] Anna Davin, 'Imperialism and Motherhood', *History Workshop* 5 (1978), pp. 15–16.

[12] *Report of the Inter-departmental Committee on Physical Deterioration* (London: H. M. Stationery Office ,1904), Volume 1, p. 85.

[13] *Report of the Inter-departmental Committee*, Volume 1, p. 87.

[14] Galton/2/4/12/1/3 folios 1–2.

At the meeting held on 30 October 1905, the pathologist Frederick Mott suggested compiling a register of the 'feebleminded' based on the knowledge of local medical officers. The psychologist William McDougall proposed that the mental capacities of various nationalities and of 'cross-breeds' might be studied. He agreed to confer with Schuster about this and to contact a schoolmaster in Jamaica regarding the capacities of mixed-race children. Schuster was probably only moderately interested in eugenics. In July 1905 he had been given permission not to attend the institute weekly, owing to living in Oxford. His salary was reduced and Miss Elderton's role upgraded at the same time.

In 1907 Schuster's three-year appointment came to an end and he resigned from the position. The likely reason is that the work had been statistical rather than biological. He returned to New College, Oxford, and work on zoology, although he agreed to serve on the research committee of the Eugenics Education Society when it was established in 1912 and published the book *Eugenics* in 1913. For Galton, however, Schuster's resignation was a blow. There was no obvious successor. He recognized Miss Elderton's capability and felt that with further statistical training, she would be able to do independent work. But he did not envisage her as the manager of the Eugenics Record Office. About to turn 85, Galton himself was too frail to supervise it.

Karl Pearson Takes Charge

Galton was therefore relieved when Pearson, after some hesitation, agreed to take charge of the Eugenics Record Office. The doubt was due to Pearson's believing Galton wanted immediate results from the research and fear that any disagreement between them would affect their friendship, which he valued highly.[15] Pearson was an armchair socialist and seems to have thought similarly about eugenics, wanting to advance the doctrine academically to the exclusion of practical action. Galton saw the studies as a prelude to actual reforms. After eugenics had reached the stage of influencing policy, he hoped it might be introduced into the national conscience in the same way that it provided guidance to individuals. Pearson applied statistics to the natural world in what he,

[15] Karl Pearson, *The Life, Letters and Labours of Francis Galton*. Volume 3a (Cambridge: Cambridge University Press, 1930), p. 299.

Weldon and Galton called biometry. Eugenics was aligned with biometry or mathematical genetics.

Galton therefore suggested that the Eugenics Record Office might be associated with Pearson's Biometric Laboratory. This did not reach fruition in 1907, but concurrently with planning for a new eugenics fellow, Galton let Pearson know that he intended to leave £ 30, 000 for a chair in eugenics in his will. He wished Pearson to be the first holder and was drafting a codicil to ensure this. Although there were underlying differences of opinion between the two, Galton put his trust in Pearson.[16] The Eugenics Record Office was renamed the Eugenics Laboratory and was to be supervised by Pearson, with Galton keeping a distant eye on it. Schuster joined the new advisory committee for the Laboratory alongside participants in the informal discussion meetings. These were William Palin Elderton, an actuary and brother of Ethel Elderton, Frederick Mott and the anthropologist James Tocher. A few newcomers were also appointed. Pearson's supervision permitted the fellowship to be readvertised with minor changes, including the reduction of the salary by £ 50 to £ 200 and the official conferring of the title 'Galton scholar' upon the assistant. The new Galton fellow was David Heron, who had graduated in 1905 from St Andrews University in mathematics and natural philosophy. Miss Elderton, as the Galton scholar, had similar duties except for academic leadership and her salary was set at £ 120. With Elderton receiving what was in effect a promotion, Amy Barrington was employed to assist with routine calculations, at £ 80 for four days' work a week.

The Work of the Eugenics Laboratory

Galton's original institute had been hosted by the University of London. The proximity of the Eugenics Record Office to University College made it gravitate towards the latter. Pearson's becoming involved as supervisor strengthened the connection since he was a professor of applied mathematics there. University College was mainly a teaching institution, and between February and May 1909 weekly lectures on eugenics were held. The audience, which included doctors and social workers, had risen from fifty to a hundred by the end of the course.[17] The lectures accompanied

[16] Pearson, *Life, Letters and Labours*. Volume 3a, p. 297.

[17] Pearson/4/10 Report on the work of the Francis Galton Laboratory folio 2v.

the new Eugenics Laboratory Lecture Series, which were publications intended for a wide readership. The first work appearing, *On the Scope and Importance to the State of the Science of National Eugenics* by Pearson, had been the Robert Boyle Lecture held at Oxford University Junior Science Club in 1907.[18] Originally published in *Nature* in 1909, it was reissued for the series the same year. Also in 1909, Pearson contributed *The Groundwork of Eugenics* to the series with another five publications issued concurrently. The most noteworthy were by Ethel Elderton: *The Relative Strength of Nature and Nurture* and *On the Marriage of First Cousins*. Pearson provided overarching visions of what eugenics stood for, its methods and problems. Unlike the parallel offerings of the Eugenics Education Society, these lectures were intended as sources for further research. The content in the *Eugenics Review* tended to be more polemical. It did, however, print the annual Galton Lecture after it was established in 1913, which often included scholarly content.

Separately, the Eugenics Laboratory Memoirs continued making findings relating to heredity known. David Heron's first contribution to it, sitting between Schuster's studies of Oxford men, concerned the inheritance of, and predisposition to, insanity in 1907. Three years later he published a study of what effects defective physique and a deprived home environment had on the development of children's intelligence. It was a statistical analysis of data obtained by the London County Council. The memoirs were often based on a plethora of raw facts, though written with a particular agenda in mind. The Education Committee of the L. C. C. and the Scottish Education Office provided 30, 000 cases for Heron's volume on physique and home environment and a similar one by Elderton which never saw the light of day, 'The Influence of Parental Occupation and Home Conditions on the Physique of the Offspring'. The results, unsurprisingly, were meant to show environmental conditions having little effect and a high correlation between social status and health. Heron returned to the issue of intelligence in 1911 in another series, Questions of the Day and of the Fray, because the statistician George Udny Yule had questioned the validity of using teachers' estimates of pupils for ascertaining this trait. Yule acted as a lecturer for the summer school convened by the Eugenics Education Society in 1912. Heron's rebuttal formed part of a series which would also later be used for answering critics of the major

[18] Pearson/4/10 folio 2v.

Laboratory publications. Another Eugenics Laboratory Memoir, on the relative influence of heredity and the environment on defective sight by Amy Barrington and Pearson was published in 1909.[19]

1909 was a busy year for the Eugenics Laboratory. Apart from the above activities, it issued the first instalment of its monumental undertaking, the Treasury of Human Inheritance. The work entitled *Pedigrees of Physical, Psychical and Pathological Characters in Man*, featured a preface by Pearson and an introductory note by Galton, while the actual pedigrees had been compiled by commissioned investigators. The pedigrees were intended to shed light on the inheritance of specific conditions and disabilities. As the Eugenics Laboratory was wedded to biometry rather than Mendelian genetics, there were illustrations and accompanying information on individual cases. Although qualifying as medical literature, it is possible the practice served as a smoke screen for the validity of eugenic projects. There were many unfortunate cases described and photographed. A cure for these debilitating diseases would clearly be difficult to find. That foregrounded eugenics as a solution.

The tenth volume of the Eugenics Laboratory Memoirs by Elderton became controversial upon publication in 1910 because its conclusion flew in the face of conventional wisdom. It examined how parental alcoholism affected the health and ability of the offspring. A study from Edinburgh carried out by the Charity Organisation Society and data from Manchester supplied by the pioneering educator of the 'feebleminded', Mary Dendy, formed the basis. Dendy had joined Galton and two other eugenists in warning about the proliferation of this group in 1909.[20] The material from Edinburgh showed that alcohol was a significant factor in at least 282 of the 449 cases where a family relied on charity.[21] The report stated that the death rate of children with alcoholic parents was higher than for those with sober parents, but otherwise drink did not lead to measurable disadvantages. Parental alcoholism was not the cause of mental defect in the children, and the offspring of sober parents in fact had slightly worse health than their counterparts affected by drink. The results were consistent with the hypothesis that alcoholism might be

[19] Pearson/4/4/2 Report by Professor Karl Pearson pp. 6–7.

[20] *The Problem of the Feeble-Minded. An Abstract of the Report of the Royal Commission on the Care and Control of the Feeble-Minded* (London: P. S. King, 1909).

[21] Galton Laboratory/3/3 Edinburgh Charity Organisation Report on Physical Condition of School Children, p. 8.

a hereditary condition. Provided children survived, their less favourable environment did not seem to have an effect on them statistically. The hereditary factor was four to ten times more potent, and larger samples would have been required to discover an environmental factor. Poor heredity could *manifest* itself in alcoholism, but drunkenness was not the cause of certain families struggling.

It is obvious why such counterintuitive findings should rankle with others, especially campaigners against alcoholism. The negative reactions to the volume necessitated the issuing of a second report on the same question later in 1910, in which Pearson tried to counter the criticism the Galton Laboratory had attracted. The debate initially raged in the *British Medical Journal* and temperance journals. Two volumes of Questions of the Day and of the Fray by Pearson were also devoted to the dispute. Sir Victor Horsley, the pioneering neurosurgeon, was a temperance advocate who objected strongly to the study. Dr Mary Sturge, who had authored a book on alcoholism with him, also disagreed profoundly with Elderton's findings. A symposium was held at the Society for the Study of Inebriety on 10 January 1911 to discuss the matter. Pearson declined to attend, but Horsley was there to explain his criticism. He accused the researchers of having manipulated the Edinburgh data and said they had ignored the work of Sturge. Pearson replied in *The Times* the next day. It was untrue his team had made up data and, comically, he denied having grouped workmen into 'drunken' and 'sober' since few labourers drank while at work.[22] (Fig. 3.1).

Pearson and the Eugenics Education Society

The debate had an additional and significant consequence. One of the temperance campaigners dismayed by the original report was Caleb Saleeby, a leading figure in the Eugenics Education Society. Already a critic of the Eugenics Laboratory, his opinion now deteriorated further. Montague Crackenthorpe, president of the Society, was another temperance advocate who shared Saleeby's opinion. Galton was forced to write to the *British Journal of Inebriety* in 1910 denying a rift between the Laboratory and the Society.[23] He had hoped the Society could be used

[22] Galton Laboratory/4/1 folio 2.

[23] Galton/3/3/7/20. Enclosure in Galton to Pearson 27 Oct 1910, folio 749.

Fig. 3.1 The Galton Laboratory common room (reproduced from the Galton Laboratory, UCL Special Collections)

for disseminating the findings of the Laboratory. Galton was nevertheless worried that the popularizers might claim too much for the doctrine, an attitude shared *par excellence* by Pearson. When other publications by the Eugenics Laboratory met with a less than admiring reaction in the *Eugenics Review*, Pearson's dislike of the Society intensified.

Pearson's hostility to the Society had echoes of a similar conflict on a related issue. He wanted to turn statistics into a reputable branch of mathematics. Traditional Euclidean or Cartesian practitioners might stand in the way of this goal. The problem was compounded by the poor work of some who described themselves as statisticians. Equally, Pearson had ambitions for eugenics to be recognized as an academic discipline. Enthusiastic amateurs with insufficient knowledge pronouncing on behalf of the doctrine compromised this.[24] Leonard Darwin had the distinction of the

[24] Pearson, *Life, Letters and Labours*. Volume 3a, pp. 362–363.

Darwin name behind him, but he was an army officer and former MP rather than a scientist. Although he contributed generously to the Laboratory appeal for funds to house Pearson's new department in 1911, it did not appease Pearson. He thought Darwin did not understand the genetic basis of eugenics. His incorrect assertions would make it harder to put eugenics on a sound footing, because the public would associate them with the head of what was by then known as the Galton Laboratory.

The antagonism was not limited to Darwin personally. In a Fabian pamphlet of 1907, Sidney Webb repeated Pearson's claim that a quarter of all parents produced half of the next generation.[25] However, Webb personally was responsible for the statement that the quarter consisted of foreigners on the one hand and British casual labourers on the other. Pearson felt ill-used as it was far too early to say who the prolific parents were. Webb did not even join the Eugenics Education Society when it was founded, but Pearson saw the mistake as damaging the eugenic cause. Although the Society's research committee went into hibernation during the First World War, it became active again in 1923.[26] This created overlapping roles for it and the Galton Laboratory. Some Society members were able to develop the doctrine. Saleeby, who had classified the main versions of eugenics as 'positive' and 'negative', added a third category called 'racial poisons' to the schema. It consisted of alcohol, drugs, lead and venereal diseases, which stayed in the body and, Saleeby believed, weakened the germ plasm.

Like Pearson, Galton wanted eugenics to be turned into an academic subject, though not to the exclusion of progress in practical terms. When he founded the Eugenics Record Office, he insisted that the fellow should have freedom of action to argue the case for eugenics, even against unsympathetic members of the University of London's governing body.[27] Later, when expanding the scheme to the endowment of a professorship, Pearson was in correspondence with the university's academic registrar, Philip Hartog, about the legality of having to appoint a candidate to a position which had been advertised. The former was acting on Galton's behalf or perhaps his own, since Galton had decided by then that Pearson

[25] Sidney Webb, *The Decline in the Birth Rate* (London: Fabian Society, 1907), p. 17.

[26] Pauline M. H. Mazumdar, *Eugenics, Human Genetics and Human Failings. The Eugenics Society, its Sources and its Critics in Britain* (London: Routledge, 1992), p. 91.

[27] Galton Laboratory/1/1 folder 1 of 2. Folio 11. Galton to Pearson 19 Jan 1907.

would be the first eugenics professor. This can be seen from the 1909 codicil to his will. Hartog answered that one ought not to be forced to fill a chair unless there were qualified candidates.[28]

The Galton and Biometric Laboratories

The Galton Laboratory rose to greater prominence in 1911, upon the death of its founder. His will of 1908 bequeathed the residue of his estate to the University of London for the purpose of establishing the chair in eugenics. The bequest amounted to £ 45, 000 and put the Galton Laboratory on a much sturdier foundation. It was a gift of a different magnitude to the £ 500 Galton had provided annually since 1904. Galton was not overly demanding of the professor. The position involved gathering data relating to eugenics, analysing it and publicizing the results. An associated central office would undertake 'occasional publications' and 'occasional public lectures' as well as training staff in eugenics.[29] By the time his will was executed, the 'central office' had already published about 25 books or pamphlets and given some important lectures.[30]

The codicil agreed and even specified that Pearson should continue supervising the Biometric Laboratory at University College alongside the duties incumbent upon the eugenics professor.[31] This was essential for Pearson, who had built up the Biometric Laboratory and could not let go off it. Henceforth the Biometric and Galton Laboratories were to be connected through the Department of Applied Statistics and Eugenics, all led by Pearson. This was confirmed by the Board of Advisors in Eugenics on 26 May 1911, acting for the University of London.[32] There had been a complete changeover in this committee. Leonard Darwin had joined on the strength of recently becoming president of the Eugenics Education Society. Two members who were fellows of the Royal Society, Frederick Mott and Charles Sherrington, accompanied him onto the committee.

[28] Galton Laboratory/1/1 folder 2 of 2. Folio 8. Hartog to Pearson 3 May 1909. Pearson passed on this advice to Galton on 7 May 1909. Pearson, *Life, Letters and Labours*. Volume 3a, p. 383.

[29] Galton/2/4/12/1/1 folios 8–9.

[30] Joe Cain, 'Publications Produced by the Francis Galton Laboratory for National Eugenics', *The Library* 22:4 (2021), p. 529.

[31] Galton/2/4/12/1/1 folio 13.

[32] Pearson/4/8/3 Board of Advisors in Eugenics.

Sherrington would become president of the Royal Society in 1920. Edward Galton Wheler, Galton's nephew, was another new member.

The reorganized department had internal consistency as Galton and Pearson had advanced eugenics through biometry. Pearson had begun his academic career at the Department of Applied Mathematics and Mechanics. In 1895 the statistics course which he taught became a regular feature. This was probably the first appearance of statistics in an English university. Postgraduates and former undergraduates who had learned from Pearson gathered in a room in the College later termed the Biometric Laboratory. They hoped to make statistical findings and to increase the standing of their subject within applied mathematics.[33] In 1903 a grant of £ 1000 from the Drapers' Company, a livery company of the City of London, was allocated to the Biometric Laboratory by the University's Senate. The drapers had only specified that they wished to support University College, but in 1905 they gave another £ 2, 000 signalling their acceptance of the arrangement.[34] This funding would continue for the rest of Pearson's tenure as director.

Galton's bequest was a considerable sum of money. In 1907 the whole University of London had declared an income of £ 95, 000. It freed Pearson from various chores and gave him the opportunity to concentrate on the statistical labours which he saw as essential to eugenics.[35] It involved dealing with masses of information describing conditions in society. The Biometric Laboratory had spent seven years on an investigation of schoolchildren and six years on a project measuring family resemblance.[36] Delays relating to number crunching were at the heart of the late publication of a few of the Laboratory's findings. Occasionally it could take fifteen years before a project was written up in its final form. Pearson was aware of the problem; therefore when it came to eugenics, he preferred to use actuarial death rates and examination of family pedigrees. He was hard-working and dedicated, taking a personal interest in the work of his staff.

[33] Pearson/4/8/2.

[34] Pearson/4/4/2 Report by Professor Karl Pearson, p. 2.

[35] Pearson/4/4/3 Report by Professor Karl Pearson, pp. 1–2.

[36] Pearson, *Life, Letters and Labours*. Volume 3a, p. 298.

The first organizational issue for the new department in 1911 was finding a building in which it could function. Galton had not foreseen the need for expenditure on accommodation. Not unreasonably, he had imagined that the University of London, if he gave it much of his wealth, would be amenable to providing space for the proposed research. This seemed not to follow and thus a separate appeal for funds had to be undertaken. The acceptability of the cause can be seen from *The Times*'s editorial on 7 October 1911 entitled 'A Laboratory for Eugenics'. The newspaper heartily recommended the appeal published in that day's edition. The recent strikes and the health of a rising industrial population were examples of the need for eugenics. A healthy and wise society could not be built up at present, owing to 'the unchecked multiplication of the unfit'.[37] Supposedly benign interventions were having the effect of restricting the reproduction of the 'wiser and stronger portions of the community'.

There was a familiar feel to the funding drive (even literally). It produced £ 100 from Leonard Darwin, £ 500 from another son of Charles Darwin, the banker William Erasmus Darwin, and £ 250 from Edward Galton Wheler. Former or current beneficiaries of Galton's largesse were Pearson, who donated £ 500, £ 20 came from Edgar Schuster, whose uncle Professor Arthur Schuster contributed £ 250. The project had support in other quarters too as can be seen from the £ 100 donated by Rupert Guinness, a former MP who belonged to the famous brewing family, and £ 50 from the newspaper proprietor Lord Northcliffe.[38] Lord Rothschild contributed £ 50, which ironically was later allocated to the study of Jewish immigrants concluding that they were not an asset to Britain.[39] An anonymous benefactor, known to Lord Reay, the chairman of the council of University College, then stepped forward to offer the University a whole new building for the purpose of accommodating the department. He was later revealed as the businessman Sir Herbert Bartlett. The gift enabled the construction of a separate building on a site north-west of the College quadrangle in Bloomsbury. It would provide a home for architecture, sculpture and applied statistics. Construction started in 1912, with the Department of Applied Statistics

[37] Galton Laboratory/4/9/2.

[38] Galton Laboratory/1/3 Outline of the History of the Department, pp. 6–7.

[39] Pearson/4/9/4 Henry A. Miers to Pearson 8 Dec 1911.

and Eugenics finally moving into it in October 1919. From 1907 and until that date, it was to be found in the South Cloisters of the main building in the College, sharing the space with a military hospital for the duration of the war.[40]

Pearson was undoubtedly capable but perhaps like many great minds, he was not easy to get on with. A precursor of what lay ahead occurred as early as the funding drive. Pearson had requested that his department should not publicly be associated with the anonymous offer made by Bartlett. He was unsure if the building would be suitable and even if it was, doing so might reduce donations from other sources. This was disregarded, which Pearson labelled the 'destruction of the Galton appeal'.[41] He thought the assigning of the department to the new building would only be temporary. Lord Reay replied that he did not think the report on the appropriation of college sites was 'discourteous' or contained 'misstatements'. He wrote that the College committee had done its utmost to satisfy Pearson throughout the establishment of the Galton chair. He also offered a special meeting of the College committee if Pearson thought there was anything to be gained from it.[42]

Ten months later, in March 1913, Pearson wrote to the provost of University College, Gregory Foster, resigning from the advisory committee of the Galton Laboratory.[43] It was a consequence of that day's meeting which had approved plans for the basement of the new building and a lecture theatre. Pearson disagreed with the plans and felt they had been sprung unawares on the committee by one of its members, Professor R. J. S. Simpson, a specialist in tropical medicine. Pearson also disagreed with the College on other matters. In 1914 he found the chosen building to be noisy and said it would be expensive to maintain. The one advantage of its location was that it would attract public notice, which, however, was vitiated by the lack of an exit to the street. A door would have made it possible to bring in visitors to the museum and anthropometric laboratory housed in the department. Pearson was critical of Bartlett for having cut back his funding, apparently due to the war, which would lead to more

[40] Pearson/4/4/4 Report by Professor Karl Pearson, p. 1.

[41] Pearson/4/11/8. Pearson to Lord Reay 17 May 1912.

[42] Pearson/4/11/9 Reay to Pearson 16 May 1912.

[43] Pearson/4/11/8. Pearson to Foster 5 Mar 1913.

of his own budget being spent on electrical and other fittings. Furthermore, the building was inadequate for breeding animals, which he felt to be essential in fulfilling Galton's wishes.[44]

That was not his final misgiving. Pearson said the Board of Advisors in Eugenics for the University was not working well, with many of the members habitually absent. This included Leonard Darwin, a fact which Pearson attributed to the conflict between the Eugenics Education Society and his own laboratory.[45] The board was in any case becoming less influential as additional functions were devolved from the University to the College. Pearson would have preferred to be under the University rather than the College and said Galton had shared that opinion. (Although as Galton's biographer, Pearson sometimes used the older man as an authority for what were in fact his own views.) Pearson did seem demanding and critical of the input of others, an attitude seen especially in regard to Bartlett, who after all was donating a great deal of his money to the College voluntarily. And the basic scheme was tailor-made to Pearson's requirements. There would nevertheless be further tension later.

After 1907, when Pearson became supervisor, the work on eugenics may meaningfully be seen as a strand within the Biometric Laboratory.[46] Pearson was probably at least equally interested in statistics, and it may only have been his friendship with Galton which made him turn towards eugenics. But his eugenics studies too were impeccable and aimed at influencing government policy. The department also sought to provide data to scientists and opinion formers. Thus, even before the chair had been established, in his report for 1903–1909 to the Worshipful Company of Drapers, Pearson explained that work on Scottish Pauper Lunacy Rates, varying meteorological conditions affecting time fuses for the Royal Artillery and inoculation against enteric fever in the Indian Army and South African Army had been part of the department's brief.[47] The wide span of topics would continue in the years to come.

[44] Pearson/4/4/3. Report by Professor Karl Pearson, p. 2.

[45] Pearson 4/17. Journal of the Galton Laboratory 1915–, p. 3.

[46] Pearson 4/17. Journal of the Galton Laboratory 1915–, p. 1. Shows that payment accounts were overlapping with Amy Barrington being paid from both laboratory funds.

[47] Pearson/4/4/2. Report by Professor Karl Pearson p. 5.

By the time Pearson next reported to the Drapers' Company, covering the years 1909–1914, he had taken up his new professorship. Thus, he could include eugenics among the subjects that the department studied alongside statistical methods and social, economic and medical problems. Eugenics pursued hereditarian explanations of criminality, with Pearson able to commend a long report called *The English Convict. A Statistical Study* by Dr Charles Goring, published by H. M. Stationery Office in 1913. It had benefited from the support of H. M. Prison Parkhurst, which had allowed Goring as a medical officer of this institution to work at the Biometric Laboratory, even providing him with two assistants to gather data. It comprised the physical and mental characteristics of 3, 000 criminals, making it the largest treatise on the topic ever compiled.[48] Although hardly progressive in nature, the Laboratory was superseding previous research which seems fundamentally misconceived today. In the report Goring disposed of the notion that criminals were a special category of humans, as had been put forward by Cesare Lombroso in *L'uomo delinquente* (1876), *La donna delinquente* (1893) and elsewhere. Nor could criminals be detected by stigmata on their bodies. Goring regarded the inclination to criminal behaviour as inherited in much the same way as other pathologies. The criminologist Enrico Ferri, who had worked under Lombroso, lauded the report in the *New Statesman*.[49]

Less emotive and genuinely hereditarian is the condition of albinism. Between 1911 and 1913 the Biometric Laboratory published three volumes on it. The methods used in the gathering of data on albinism shed light on the compiling of all the publications emanating from the department. The presence or absence of medical characteristics was a matter for the observation of field workers, as was the grading concerning traits which appeared in varying degrees such as health. Subjective feelings among the field workers might therefore affect the results and there was even less scope for controlling the pedigrees provided by contacts. Albinism is a case in point as the rarity of the condition meant that

[48] Pearson/4/4/3. Report by Professor Karl Pearson p. 1.

[49] *The New Statesman. A Weekly Review of Politics and Literature.* Vol II No. 39 Saturday, January 3, 1914 (London: The Statesman Publishing Company, 1914), pp. 398–399. Donald C. Bellomy, '"Social Darwinism" Revisited', *Perspectives in American History* new series 1 (1984), p. 43.

data often had to come from abroad. A variety of foreign intermediaries were put to work. Some of these operated in smaller, probably more transparent countries such as Iceland and Norway.

Pearson obtained pedigrees of albinos from across Norway in 1909 through the neurosurgeon Vilhelm Magnus. Magnus had worked under Victor Horsley, the critic of the Galton Laboratory, but was on good terms with Pearson. A letter from Magnus dated 26 May 1909 explains how he was in pursuit of an albino man with two children, who were affected by the disorder too. He had consulted the doctor, vicar, schoolmaster and bailiff on the matter though hitherto with no success. Three days earlier Magnus had managed to provide such a pedigree from Orkdalen, near the city of Trondheim.[50] This became plate 51 in the department's treatise. Plates 52 and 53 were also supplied by Magnus, whom it is clear was paid for his services. On New Year's Day 1910 he thanked Pearson for the receipt of a cheque to the value of a pound and seventeen shillings.[51] In terms of verifying the findings, photographs were often sent and there is copious correspondence relating to the cases.[52] Attempts to order the data and produce an hypothesis of the associated genetic traits were inconclusive on albinism, but the combination of blue eyes and dark hair was provisionally marked down as significant.[53]

By being linked to other studies, eugenics gained in scientific credibility. It combined aspects of both a science and an ideology. Its methods involved gathering facts, investigating a single question exhaustively, providing the data and procedure so that they could be verified by others and formulating a conclusion based on what had been discovered. This is how science proceeds. Bias was nevertheless built into the studies because they were designed to showcase the need for eugenics. This entailed finding the hereditarian rather than the environmental causes of phenomena. It made the results less plausible than if they had been based on a posteriori investigations. The Galton Laboratory did not recommend specific solutions to the medical and social problems it investigated, but it provided ammunition to convinced supporters calling for regulations or

[50] Galton Laboratory/2/1/1/4 folio 24 r-v, folio 17 r-v.

[51] Galton Laboratory/2/1/1/4 folio 52 r-v.

[52] Galton Laboratory/2/1/2/1-5 and Galton Laboratory/2/1/1/5 respectively.

[53] Galton Laboratory/2/1/3/2 folder 1 of 4, folio 1.

legislative action. Eugenics as practised by it was a development of how Galton and Pearson had worked in the late nineteenth century.

In 1911 the Laboratory hosted Alexander Carr-Saunders and Harold Laski. The former had graduated from Oxford, the latter was about to matriculate there.[54] Both would remain anchored to the eugenic creed in later life, while concurrently wishing to focus on the effects of the environment. They were therefore 'reform' eugenists and both became social scientists. Laski was invited to tea at Galton's home following the publication of a youthful article in the *Westminster Review*. Carr-Saunders was appointed to the Charles Booth chair at Liverpool University in 1923 and Laski began a long-standing affiliation with the London School of Economics in 1926. Eugenics was an overweening tendency for many natural scientists into the 1930s and had impact on fields like sociology, political science and economics too. William Beveridge was the director of the London School of Economics between 1919 and 1937. He founded its Department of Social Biology in 1931 with financial support from the Rockefeller Foundation. Carr-Saunders succeeded him as director from 1937 until 1957.

In 1914 the first part of Elderton's *Report on the English Birthrate*, covering the areas north of the river Humber, was published. She calculated whether subgroups within the general population had higher or lower fertility than the average. Of the reports released by the Galton Laboratory, this had the closest connection with how eugenics was conventionally understood. Pearson described the work of the Biometric Laboratory as studying 'variation and inheritance in man and other forms of life quantitatively and mathematically'.[55] That there was no obvious dividing line between the two laboratories can be seen from the biometric one producing a series called Studies in National Deterioration. In addition to the various series mentioned above, the department published the journal *Biometrika* and some Galton memorabilia. These included a reprint of his *Life History Album* (1884), and in 1914 the first volume of Pearson's biography *The Life, Letters and Labours of Francis Galton* appeared.

Galton had specified that the chair should have a role in propagating eugenics. In 1914–1915 evening courses on the subject were held

[54] Pearson/4/4/5. Report to the Court of the Worshipful Company of Drapers, p. 2.

[55] Galton/2/4/12/1/2 folio 1. The Standard 3 Jan 1910.

over three terms. This was a joint undertaking with the Department of Zoology, revolving around biological heredity and associated theories.[56] One course was designed to give teachers a grounding in statistical methods. This series of ten lectures with David Heron would help them interpret the data provided by school medical officers. Since the Laboratory itself relied on teachers' judgments, it clearly sought to produce findings as accurate as possible. Of course the scale of the task was greater than a single course could do justice to.

The First World War

The war years saw the South Cloisters turned into a military hospital on the orders of the government. New projects ceased after 1914. The Department and its staff instead worked for the authorities, including compiling unemployment statistics, matching imports with shipping in different seasons and in 1916 studying bombing trajectories for the Admiralty Air Department.[57] As the conflict showed no sign of abating, it produced charts, tables and ballistic data of use to anti-aircraft gunnery for the Ministry of Munitions. Statisticians found government employment and pay preferable to what they were offered in the Laboratory, with many leaving during the conflict. David Heron joined an insurance company in 1915 as chief statistician and never worked for the Laboratory again.[58] He did, however, serve on committees for University College, having been appointed a fellow in 1918.

All the tasks undertaken during the First World War offered a chance to extend the influence of the Department into Whitehall. During the summer, it had been co-opted by the Board of Trade to produce graphs on unemployment in various districts. The letter stating this explained that the government wished Pearson's team to keep in readiness for future work. A noteworthy interlude then occurred on Hampstead Heath in December 1914, when Pearson bumped into William Beveridge, at the time the head of the labour exchange department within the Board of Trade. Beveridge made no mention of the graphs which Pearson was

[56] Pearson/4/17 Journal of the Galton Laboratory 1915–, pp. 2–3 of booklet at p. 10.

[57] Pearson/4/4/4. Report by Karl Pearson, p. 2.

[58] E. S. Pearson, 'David Heron, 1881–1969', *Journal of the Royal Statistical Society. Series A (General)* 133:2 (1970), pp. 288–289.

compiling, until asked directly, whereupon he said that they were not of any use in relief efforts.[59] As the Department had been compiling these to the exclusion of its own priorities for nearly six months and with little respite, Pearson was disappointed. He must have seen these statistical assignments as a patriotic duty and may also have had the ulterior motive of increasing his own sway in government circles. In answer to Pearson's inquiry, Beveridge then clarified that since unemployment was low owing to the present conflict, the graphs were not of burning relevance. They were, however, being used and would need to be constructed by others if Pearson's team did not continue producing them.[60] This reply cannot have brought much solace to the Department of Applied Statistics and Eugenics. As a scientific unit, it was attempting to create a groundswell for eugenics through laying research findings, copiously documented though not necessarily impartial, before practitioners and opinion formers.

After the First World War and the relocation of the department to the Bartlett building, the contact with the public continued. A lengthy course on statistics began in October 1920 with Elderton as the lecturer. Intended for teachers, clerks and others wanting a better knowledge of what figures meant, it comprised twenty lectures over ten weeks. They did not promote eugenics directly, unlike the talks given by the Eugenics Education Society. Pearson was always sceptical of this organization, declining to be its president when it was founded in 1907. There was only a slight connection between the two institutions. Heron served on the Society's council and Elderton lectured to it once.[61] Pearson felt that the progress of eugenics would be hampered by a grouping of amateurs sharing the limelight with his own Laboratory. This was despite the Society containing acclaimed scientists in their own right, one of whom, Sir Charles Sherrington, would win the Nobel prize for physiology or medicine in 1932.

[59] Pearson/4/17 Journal of the Galton Laboratory 1915–, pp. 14, 20.

[60] Pearson/4/17 Beveridge to Pearson 6 Jan 1915 at p. 22.

[61] Pearson/4/9/12 The Eugenics Education Society folio 1r.

Mendelism and Biometry

Once the Eugenics Education Society got going, it allied itself to the Mendelian approach to genetics, especially through its pauper pedigrees. In 1913 Reginald Punnett, a Mendelian geneticist, held a course of twelve lectures on heredity to the Society. The impact was reinforced in 1918 by a similar course given by another Mendelian, the zoologist Ernest MacBride. The fact that the Laboratory and the Society disagreed on how to research and present eugenics to the public did not aid interaction between them. For Pearson, who was cantankerous anyway, the use of biometry to advance eugenics was a matter of honour. It permeated his entire scientific output because he had started out as a statistician, utilizing mathematical techniques to obtain results in zoology and anthropology. It followed that he must pursue eugenics in the same way. Galton had devised statistical methods for what became biometry. Pearson thought of others and modified Galton's work, for instance on correlation. Not only had the two of them used biometry for decades, but they had faced hindrances on account of it. *Biometrika*, founded in 1901, was an outlet for work along these lines, in particular evolutionary topics. It came in useful when Pearson experienced difficulties in getting papers on biometry accepted by the Royal Society. In fact, it passed a resolution urging him to separate biology and statistics so that each could be published in the appropriate journal.[62] Pearson was so incensed by its lack of appreciation for biometry that he considered resigning his fellowship, only relenting when Galton urged him to remain and fight his corner.[63]

It was from working on similar topics that Galton and Pearson formed a strong bond. They began to rely on each other. In 1908, after Pearson had begun supervising the Eugenics Laboratory, Galton wrote to Hartog, the academic registrar, to encourage him to keep Pearson on for as long as possible. He said it would be 'disastrous' if Pearson stepped down.[64] Through his interaction with the older man, Pearson also became interested in eugenics. As a socialist, he was a reformer of society. As a scientist, he saw how eugenics would allow him to help form humanity in his image through simply following his vocation. He believed Galton to be nothing

[62] Pearson to Galton 13 May 1906. Pearson, *Life, Letters and Labours*. Volume 3a, pp. 282–283.

[63] Galton/3/3/7/20 folio 405 r–406 v. Galton to Pearson 14 May 1906.

[64] Pearson/1/8/2. Galton to Hartog 31 Jan 1908.

less than a genius. Pearson thus became his acolyte, despite being able to cast his net deeper than Galton (Fig. 3.2).

Biometry held the advantage over Mendelian genetics that it was better established, since the latter only became known in Britain in 1900. Having developed biometry for years, it was unlikely that the two men would change how they studied biology. Instead of using the concept of dominant and recessive genes, Galton relied on statistics. Two-thirds of an individual's stature was determined by inheritance, he claimed. Like Mendel, he had conducted experiments with peas. In 1875 he weighed thousands of sweet peas and sorted them in envelopes, which he then sent to friends and associates, including Charles Darwin. The next year he

Fig. 3.2 Karl Pearson working (reproduced from the Pearson papers, UCL Special Collections)

received 490 offspring of these in return, which confirmed normal distribution on weight and enabled him to formulate a law of regression.[65] The weights of the offspring were 1/3 closer to the mean than were those of the parent peas. Galton was nevertheless open-minded about whether Mendelian genetics would also prove worthwhile.

Eugenics and Society

An article in the *British Medical Journal* published in 1920 shows the extent to which the apprehensions of eugenists were recognized within that profession. Entitled 'The Declining Birth Rate', it stated that the fertility of the upper and middle classes was falling. As these were the leading citizens, the question was how it would affect civilization. A paper given at the Royal Statistical Society on 20 April 1920 by Thomas Stevenson, superintendent of statistics at the General Register Office, provided an opportunity to reflect on such matters. Stevenson noted the trend and left it to specialists on eugenics to find solutions. In discourse within various sciences, eugenics had largely convinced the ranks that it was valid and apposite. Graduates with basic scientific training also tended to agree with eugenics in the 1920s, helped along by the ideology being so easily comprehensible. It seemingly allowed them to apply their knowledge. The main opponents were dyed-in-the-wool reactionaries such as G. K. Chesterton or working-class figures who did not require theory. In *Eugenics and Other Evils* (1922), Chesterton rejected science as offering solutions to society's problems. He called that Prussianism and viewed the rationally organized state as stoking war. He had hoped that state intervention would end when Britain won the First World War. On eugenics, he reckoned it was not possible to breed for quality. The 'degenerate' man was an exception and not the rule. The state should take a back seat and the scientists were only promoting themselves. Working-class politicians were mainly interested in changing outer conditions rather than bodies internally. A few outliers existed, such as Will Crooks, the Labour MP for Woolwich 1900–1921, a Fabian who believed in social Darwinism.

The conflict between the Eugenics Education Society and the Eugenics Laboratory made Saleeby describe the latter as a waste of Galton's

[65] Chris Renwick, 'From Political Economy to Sociology: Francis Galton and the Social-Scientific Origins of Eugenics', *British Journal for the History of Science* 44:3 (2011), pp. 359–360.

money.[66] This view was not shared by Galton himself. In a letter of 1 January 1910, he wrote, 'I rejoice in your work all the more as it covers and includes much that I dearly wanted to see done, but I had not strength or capacity to do...' This commendation was quoted in the department's second anniversary dinner taking place on 17 January 1921, after it had moved into its permanent home.[67] The rooms had only become ready to be put to use in 1920. The opening ceremony, in which the minister of health Christopher Addison cut the ribbon, took place on 4 June.[68] Not only was eugenics acceptable as an academic subject, but it was regarded as potentially auspicious nationally. In 1922 Lewis Haslam, like Addison a Liberal MP, died and the Galton Laboratory received a bequest of £ 1000 from his estate. It facilitated research on the heredity of dogs, including coat colour and size and shape of the head.[69] As eugenics does not apply to animals, this must be seen in the context of wishing to improve knowledge of genetics (Table 3.1).

Table 3.1 Staff of the department in 1920

Name	*Role*
Karl Pearson	Galton professor
Ethel M. Elderton	Galton research fellow
Julia Bell	Assistant
Edmund Cecil Rhodes	Assistant
Adelaide G. Davin	Assistant
Harry Seegar	Assistant
Noel Karn	Assistant
Herbert Edward Soper	Research assistant
Miriam Tildesley	Fellow (Benington student)
Margaret V. Child	Secretary of Galton Laboratory

Source Galton Laboratory/1/3. Iris Clever, 'Miriam Tildesley and the Anthropological Politics of Standardizing Racial Measurements', *Perspectives on Science* 30:1 (2022), p. 23

[66] Daniel J. Kevles, *In the Name of Eugenics. Genetics and the Uses of Human Heredity* (Berkeley CA: University of California Press, 1985), p. 103.

[67] Galton Laboratory/1/3-4 folio 3.

[68] Galton Laboratory/1/3-4. University of London Opening of the New Building, p. 2.

[69] Pearson/4/4/7. Report for the years 1922, 1923, and 1924, p. 1.

From the beginning, there had existed a desire in the department for advances in genetics. Under the supervision of the ophthalmologist Edward Nettleship, who died in 1913, and his former student Charles Usher, about 200 dogs had been bred for this purpose.[70] This was about half as many as Haslam's later bequest made possible. The absence of a dedicated animal station before the First World War had hindered efforts, according to Pearson. Animal husbandry was a philosophical basis for justifying evolution and an argument, though a double-edged one, for eugenics. Humans were part of the animal kingdom and farming practices had shown what could be achieved with controlled breeding. On the other hand, Galton feared this observation might lose eugenics support, as it caused anxiety that humans would be bred like cattle. Galton's utopia Kantsaywhere and his proposed competitions from 1865, in which the prize-winners were encouraged to marry each other, saw society take control of aspects of breeding. He said that personal choice would always be maintained and that there was nothing unnatural about it. Galton's attitude was shared by *The Times* in a leading article of 1922. Clearly favourable to eugenics, as in previous years, it agreed that one must first allow the exact laws of heredity to be discovered before the ideology could be applied properly. This was in the spirit of Galton, who would have disdained many of the 'faddists' and 'rash reformers' giving themselves the title of eugenists.[71] The article was written to commemorate the centenary of Galton's birth.

Research of the 1920s

In 1925 the Galton Laboratory began publishing an additional journal, complementing *Biometrika*. The new *Annals of Eugenics* hoped to attract wider attention for the work taking place. Inaugurating the journal, Pearson affirmed that the proper function of eugenics lay with the entire nation, not just with individuals or families. That was why Galton had used the term 'national eugenics'. Probably the other publications were appreciated for their rigour and scientific methods, but seldom actually read. As a number of libraries began taking the new journal, it had the potential to reach readers who were less committed. Other than university

[70] Pearson/4/4/3. Report by Professor Karl Pearson, p. 2.

[71] *The Times*, 18 Jan 1922.

libraries, the subscribers were professional eugenists and sundry enthusiasts in a mix which even included the Sun Life Assurance Company. David Heron, the widow of W. F. R. Weldon, Alexander Carr-Saunders and the eugenist Basil Bramwell gave a homely feeling to the list. Raymond Pearl, of the Eugenics Record Office on Long Island, the American equivalent of the Galton Laboratory, and the Portuguese anthropologist Eusebio Tamagnini showed the interconnectedness of eugenics across continents.[72] The Galton Laboratory hoped to lead by example. There should be additional eugenics professors, permitting a caste of proponents of the new faith to come into being.

The directory of subscribers to the *Annals of Eugenics* expanded as the journal gained traction. By the ninth volume, achieving imprint in 1934, the range had become broader. The reception embodied how eugenics was recognized across the world and in different settings. University libraries were still the chief subscribers alongside museums, but H.M. Stationery Office, the Australian Department of Health, the Rockefeller Foundation, the New York Public Library, another New York institution, the Metropolitan Life Insurance Company and the League of Nations showed the breadth.[73] The Rockefeller Foundation, the American philanthropic organization with a traditional penchant for hereditarian studies, would in due course also fund the Galton Laboratory. However, the Red Cross had been early subscribers and then stopped taking the journal in 1930.[74]

Among the early articles was 'The Problem of Alien Immigration into Great Britain' by Pearson and Margaret Moul, of which the first two parts were published at once. The third part of this long study was published two years later in 1927, with the fourth and fifth instalments appearing in 1928. The research had started more than a decade earlier in 1913 and finally came to fruition in this form. Based on studies of pupils at the Jewish Free School, the aim was to determine whether immigration into Britain was beneficial to the nation. The Jewish children were compared to indigenous ones on a vast number of categories. Details were taken down about school placement and the child's relations with

[72] Galton Laboratory/3/1/3. Folder 1 of 5. Annals Correspondence vol I, folio 17.

[73] Galton Laboratory/3/1/1. Brown book marked 'Annals of Eugenics', folio 2 v. and folio 3 r.

[74] Galton Laboratory/3/1/3. Folder 4 of 5, folio 21.

the school. The child was rated for intelligence (again based on teachers' views), length of time preparing homework and level of persistence.[75] Similar forms of classification gave information on heredity and illnesses. The overall conclusion was that Jewish children were not abler than indigenous ones. Especially the girls were less advanced than their native counterparts. On this basis, Pearson could advise that immigration into Britain did not confer advantages. Galton had regarded thrifty immigrants as an injection of talent into the country. Although Pearson accepted that hard-working and well-integrated Jews existed, he said it would be a mistake to import an entire 'herd' just for the sake of finding one or two star specimens within it. He did not see the research as antisemitic since he was willing to lecture on it to the Union of Jewish Literary Societies in 1920. The audience was stimulated by the lecture, positively or negatively, as it led to a two-hour discussion.[76]

Among the Laboratory's most ambitious projects was the Treasury of Human Inheritance, to which new data especially on medical conditions was added periodically. Its first volume had appeared between 1909 and 1912 and the series would eventually extend to five volumes, each containing many parts. The final part appeared in 1958. Julia Bell, a mathematics graduate, became the main contributor. She had started working under Pearson in 1908 and was given credit as an author in the first volume published the following year.[77] Her work in the department was assisted by the fellowship she was awarded by the Medical Research Council in 1921, worth £ 300 per annum for four years. During the First World War, she had studied medicine. The Treasury's content was useful to physicians. The first volume covered conditions such as brachydactyly, deaf-mutism, hare-lip and cleft palate as well as favourable endowment: legal ability and general ability. The second volume began appearing in 1922 and almost all of it dealt with diseases and anomalies of the eye.

Most of the pedigrees for the Treasury and the other series were gathered by field workers known to the department. It was of course easier to reach conclusions based on direct measurement. In 1922 the new Anthropometric Laboratory, established by Ethel Elderton, became operational.

[75] Galton Laboratory/5/3. Blue folder 5 of 5, yellow folder inside (1 of 2), folio 2 v. and folio 8 v.

[76] Pearson/4/4/7. Report for the years 1922, 1923, and 1924, pp. 1–2.

[77] Peter S. Harper, 'Julia Bell and the Treasury of Human Inheritance', *Human Genetics* 116 (2005), pp. 422–423.

Students made up a vital group as they were the backbone of the future salaried class, tasked with many responsibilities in a complex society. In its first six years, it assessed more than 1100 students.[78] The Anthropometric Laboratory used blanks to fill in their personal details which leant heavily on those Galton had designed for the International Health Exhibition in 1884. They provided a reassuring continuum for eugenists. Galton's investigations and measurements were not simple quantification for its own sake. He was hoping to obtain results which helped to justify social distinctions as rooted in the body. Eugenics may be seen as classifying bodies into 'greater' and 'lesser' according to ability and health.[79]

Driven along by their intrinsic interest, Galton carried on with such studies independently of their original goal. Most famously with fingerprints, he continued deepening his knowledge, although it was clear from 1892 that no racial or social information could be read into them.[80] The Galton Laboratory asked directly for what concerned it. The forms were among the most searching in its catalogue of tools. They included enrolment details, family background and relatives also in the College. Then came personal characteristics like height and weight, in addition to musicality, ability, health, the results of five mental tests, the senses such as touch, olfaction, respiration and colour discrimination.[81] All these categories were on a single page and then followed head measurements, colour of hair, skin and eyes, physical tests, circulation, vision and hearing. Next came fingerprints of all ten digits. Finally, even graphology was harnessed in asking the subject to copy out a short passage from a text. The inspection of students at the College potentially offered rich pickings in terms of studying future achievements based on somatic data.

In 1926 or 1927 Pearson wrote a report on the income and expenditure of the Galton and Biometric Laboratories. Several frustrating issues existed between his department and University College.[82] The Galton bequest, as administered by the College, generated an annual income

[78] Pearson/4/4/9 Report for the years 1925, 1926, 1927, 1928–1929, p. 2.

[79] Marius Turda, *Modernism and Eugenics* (Basingstoke: Palgrave Macmillan, 2010), p. 20.

[80] Francis Galton, *Memories of My Life* (London: Methuen, 1908), pp. 252–253.

[81] Galton Laboratory/5/3. Folder 1 of 5, folio 1 r-v., folio 2 r.

[82] Pearson/4/4/8. Report on the Galton and Biometric Laboratories especially with regard to their Income and Expenditure, pp. 3–5.

which was allocated to the department, but was not enough for all its priorities. The University supplemented it by £ 300 a year from its general funds.[83] The College was geared towards undergraduate teaching, whereas the department was research-active and primarily for postgraduates. Pearson thought this was the underlying reason why Galton had made his bequest to the University of London, seeing it as a more impressive institution. The University, however, had not been willing to provide the Galton Laboratory with rooms, which was one of the reasons the department had ended up as part of University College. The University was in the process of channelling up to £ 200 a year to the Galton Laboratory, though an unfortunate side effect was that the College would probably cut its subsidy, believing that the Laboratory was going to be offered enough funds. Plans for an appeal to raise money for Pearson's laboratories had not extended beyond the drawing board over five years. As the Statistics in the department was unrelated to the Galton bequest and just happened to be a continuation of Pearson's former chair, greater clarity on finances may have resulted from the department being split upon Pearson's retirement in 1933.

It had taken fifteen years to research and then publish the study of alien immigration fully. Separate from the long time lag between the initiation of a project and its conclusion, the Galton Laboratory undertook several longitudinal studies. An example was children being monitored during their growth in Croydon and Willesden from 1928.[84] It related to the effects of infectious diseases on a child's development, with fieldwork taking place at various stages in the 1930s. It continued as late as 1953.[85]

Further insight into the investigative methods of the Galton Laboratory can be found in the instructions on how to fill in the schedule relating to major illnesses. The onus was on finding out to what extent they were inherited and whether a particular disease was associated with any others. Recorders of the information were told that even an incomplete schedule was of value, and that it could be filled in for an individual or encompass a whole fraternal group. The latter inevitably compromised accuracy. Since the data was anonymous, the

[83] Pearson/4/4/5 Report to the Court of the Worshipful Company of Drapers, p. 4.

[84] Galton Laboratory/2/3/8/2. Folder 1 of 2, folio 1.

[85] Galton Laboratory/2/3/8/2. Folder 2 of 2.

recorder was asked to provide his or her name and an identifying mark in relation to the subjects, permitting further inquiry if necessary. The information requested comprised age, birth order, occupation, general health, ailments, cause of death (if filled in posthumously), ability and self-control.[86] Ability was a subjective category for the recorder. The department had long used a scale of seven levels for this. They were defined in the notes for inspectors, but it was impossible to prevent personal feelings affecting the classification. Other categories were also difficult to assess, for instance was temper 'sullen', 'quick', 'even' or 'weak' (specified as 'not "even" but weak good nature')?[87]

Actual testing was only in use for drawing ability, with the subject being asked to trace a path through three mazes. This might have been seen as an intelligence test by others. In his 1916 Stanford revision of the Binet scale, the American Lewis Terman included tracing a path to find a lost ball.[88] Earlier attempts by psychologists Charles Spearman and Cyril Burt involved aligning motor-sensory tests to intelligence. This built on Galton's work and was also the approach adopted by another American, James McKeen Cattell, who had assisted Galton at the Anthropometric Laboratory.[89] Spearman wrote an important article in 1904 on how to measure intelligence. Many of the tests which he constructed involved the senses, such as discriminating sound, light and weight.[90] These were criticized by another expert on mental tests, the psychologist William McDougall, in 1906.[91]

[86] Galton Laboratory/2/3/7 folio 89.

[87] Galton Laboratory/2/3/7 folio 8 v.

[88] Lewis M. Terman, *The Measurement of Intelligence. An Explanation of and a Complete Guide for the Use of the Stanford Revision and Extension of the Binet-Simon Intelligence Scale* (Boston MA: Houghton Mifflin, 1916), pp. 210–213.

[89] *Eugenics Review* 44:4 (1953), p. 191.

[90] Charles Spearman, '"General Intelligence," Objectively Determined and Measured', *American Journal of Psychology* 15:2 (1904), pp. 243–246.

[91] Pearson/4/5. Advisory meeting January 31st 1906.

Intelligence

A few years later Burt and Spearman began using proper intelligence tests, in which performance was evaluated by written or spoken answers to questions requiring the use of mental powers. Each had shown awareness of the Binet-Simon tests in the *Eugenics Review* in 1914 and 1915, respectively.[92] This French innovation became public in 1905 and consisted of tasks wholly divorced from motor-sensory aptitudes. Alfred Binet's methods received a better reception in the United States than in Britain initially, and the Galton Laboratory preferred informal judgements for ascertaining this trait. In 1927 Pearson and Margaret Moul published a paper in *Biometrika* which rejected Spearman's latest argument for his famous two-factor theory of intelligence. Formulated between 1904 and 1914, it stated that mental ability consists of a general factor, which Spearman called *g* and a multitude of specific aptitudes which he labelled *s*.[93] He thought that the mental tests existing at the time did measure general intelligence *g* through drawing upon a range of tasks which required the exercise of this function. However, they could be refined through eliminating the effects of specific aptitudes called for in the questions, by choosing the tasks scientifically and balancing out the lopsidedness of tests which initially relied too much on verbal ability. Among the efforts to show this, Spearman published a paper in the *British Journal of Psychology* in 1925 with an American, Karl J. Holzinger, which calculated correlations between multiple tests of intelligence taken by the same human subjects. He also published a book in 1927 called *The Abilities of Man. Their Nature and Measurement*.

Pearson and Moul argued that for Spearman's theory to be verified, it was not enough to show high correlations between the scores of the subjects on the tests. That such correlation existed was virtually guaranteed, but in order to show that an underlying factor *g* had been measured, the correlations when subtracted from each other should tend to zero. (Otherwise, one had merely uncovered correlation between specific aptitudes *s*.) Spearman also claimed that his results displayed

[92] Cyril Burt, 'The Measurement of Intelligence by the Binet Tests. Part I', *Eugenics Review* 6:1 (1914), pp. 36–50, *idem* 'The Measurement of Intelligence by the Binet Tests. Part II', *Eugenics Review* 6:2 (1914), pp. 140–152. C. Spearman, 'The Measurement of Intelligence'', *Eugenics Review* 6:4 (1915), pp. 312–313.

[93] Stephen Jay Gould, *The Mismeasure of Man* (London: Penguin, 1992), p. 257.

close approximation to the normal curve of distribution, popularized by Adolphe Quetelet after first being identified by Carl Friedrich Gauss. This would lend support to his having succeeded in measuring an actual entity. Re-calculating Spearman and Holzinger's results and exposing them to statistical analysis, the Galton Laboratory researchers dismissed this. Spearman had additionally used too small a group (37 subjects) for his findings to be valid generally.[94] The department's research did not rule out a general factor underlying mental aptitudes which might appear at a later date, but said the case had not yet been proved. Many non-specialists shared their scepticism about the nature of the tests. What exactly was being measured in finding one's way out of a maze for instance?[95]

This was a technical issue which might nevertheless be important. No later than the 1920s, British eugenics started relying on intelligence as a desideratum which it was possible to rank. In the 1930s the Eugenics Society, as it had become, toned down its exhortation for a greater number of births among professional people to offset the multiplication of the poor. Instead, it began advising that everyone of good ability and health should have children. Leading up to 1936, a number of books and articles had been written which together had created a depopulation scare. In late September 1936 *The Times* published a two-part article on 'The Dwindling Family', animated by the publication of three books on the phenomenon by Oxford University Press.[96] It exacerbated the unease which already existed and led to other newspapers taking up the issue. What would happen when the British population started falling in a few years' time? The three authors were Alexander Carr-Saunders, David Glass and Rene Kuczynski, all with a connection to eugenics. However, eugenists were not letting go of the adage that quality trumped quantity. Focusing on intelligence instead of social class had the advantage of sounding less prejudiced. It silently allowed a continuation of the differential birth rate discourse, if needed, because almost all the research

[94] Karl Pearson and Margaret Moul, 'The Mathematics of Intelligence. The Sampling Errors in the Theory of a Generalized Factor', *Biometrika* 19:3–4 (1927), p. 282.

[95] L. T. E. Jones, 'Psychology and Vocational Fitness: The Inherent Danger of Intelligence Tests', *Socialist Review* 41 (June 1929), p. 39.

[96] Richard A. Soloway, *Demography and Degeneration. Eugenics and the Declining Birthrate in Twentieth-century Britain* (Chapel Hill NC: University of North Carolina Press, 1995), pp. 241, 391.

showed that professionals exceeded manual workers in intelligence. Physical attributes mattered too. In 1937 a National Fitness Council was established to encourage people to exercise and become healthier.[97]

Heredity and Social Class

The issue of social class mattered more to the Eugenics Society than to the Galton Laboratory. The former adopted the popular view as it did on Mendelian versus biometric inheritance. The latter's work generally found hereditarian rather than environmental causes for phenomena. This was epitomized as early as the third publication in its Eugenics Laboratory Series, *The Relative Strength of Nature and Nurture* (1909) by Ethel Elderton. It claimed that correlation between parents and offspring for various characteristics was 0.5 compared to only 0.1 between the environment and those same traits. However, the Galton Laboratory had not been preoccupied with the differential birth rate to the same extent. Therefore, it had less reason to pivot towards intelligence as the next aspiration. In 1927, not for the first time, the Galton Laboratory dismissed a common argument theoretically.

Most British interpretations of eugenics until the Second World War emphasized social class. Galton had proposed using an actuary to measure the net benefits of a baby born in each of his statistical divisions. On the credit side would be employment provided for self and others, patents achieved, profits from business ventures and charitable donations, while the debit side might include the effects of crime and receipt of charity. He said the net worth of a baby in the highest statistical class was thousands of pounds.[98] Workers generally belonged to his average statistical division. The Galton Laboratory monitored social status with more categories than usual. It distinguished the upper and professional classes, the trading class, farmers, clerks and shop assistants, skilled workmen, unskilled workmen, factory workers, miners and agricultural labourers. These nine divisions were the same as used in the census, suggesting that the Laboratory was

[97] Ina Zweiniger-Bargielowska, *Managing the Body. Beauty, Health, and Fitness in Britain, 1880–1939* (Oxford: Oxford University Press, 2010), p. 280.

[98] Galton, 'The Possible Improvement of the Human Breed', pp. 11–12.

hoping for official recognition sometime in the future.[99] During the First World War, it had undertaken work for the government and state.

Pearson, controversially, was not against child labour and believed that restrictions on it lowered the economic value of children to their parents. They reacted by cutting back on the number of offspring they had, a fact which he regretted in the early volume *The Problem of Practical Eugenics* (1912). Pearson said this resulted from parliamentarians not understanding working-class life.[100] If children were allowed to make a contribution to the household economy, it increased their importance and made it likely that they would be maintained in health. This reasoning was flawed because children's health probably suffered from being at work rather than at school or at home. A popular eugenist, the architect and athlete William Marshall, wrote that the type of families who saw their children as economic assets should be discouraged from parenthood.[101] The implication was nevertheless that Pearson was in favour of increasing the fertility of all classes who were economically self-sufficient. Every nation was struggling for supremacy in a world inhabited by hostile rivals. As Pearson put it in *The Grammar of Science* (1892): 'Individual societies have the strongest interest in educating, training, and organizing the powers of all their individual members, for these are the sole conditions under which a society can survive in the battle for life'.[102] Alfred Russel Wallace had argued that selection favoured societies with rational and altruistic members.[103]

Eugenics and Race

Galton shared the general outlook of his time and regarded people of refinement as better than the masses worldwide. His scheme of justifying from accomplishments why this should be so was hardly needed

[99] Galton Laboratory/5/3. Yellow folder marked 'Blank Heredity Schedules'. Folder 2 of 2, folio 7.

[100] Karl Pearson, *The Problem of Practical Eugenics* (London: Dulau and co, 1912), pp. 9–10.

[101] Wm. C. Marshall, 'The Education Bill and Eugenics', *Eugenics Review* 10:1 (1918), p. 21.

[102] Karl Pearson, *The Grammar of Science* (London: Walter Scott, 1892), p. 433.

[103] Diane B. Paul, *Controlling Human Heredity. 1865 to the Present* (New Jersey: Humanities Press, 1995), p. 27.

to convince others. It involved ranking the median member of particular groups up against each other, allowing an 'objective' comparison. He suggested this method could be used for finding the 'worth' of different races. It was not pursued and the work of the Galton Laboratory did not extend to creating or confirming a racial hierarchy. David Heron was dismissive of Charles Davenport's Mendelian race-crossing study in 1914.[104] Later, Pearson severely criticized Davenport's *Race Crossing in Jamaica* (1929), in which slight effects on mentalities were found.[105] During Fisher's tenure as director, the Galton Laboratory dismissed scientific racism.

A division of Europeans into Teutons, Alpines and Mediterraneans had been proposed by American anthropologist William Z. Ripley in 1899.[106] It became orthodox within the field in Germany, but never caught on in Britain. The pioneering ethnologist John Beddoe had written *The Races of Britain: A Contribution to the Anthropology of Western Europe* in 1885. It argued that colouring was the most permanent marker of racial differences.[107] Typical of the British tradition was his essay in 1891, 'The Anthropological History of Europe', which attempted to show that the upper classes were superior to the masses and that clever people tended to gravitate to cities.[108] Even a racist like H. G. Wells had no use for distinctions such as 'Teutonic' and 'Keltic' in 1902.[109] Britain had the largest empire in the late nineteenth and early twentieth centuries, which might offer adequate proof of the merits of Britons. The four nationalities within that identity did not need further splitting.

Pearson also held the ordinary view of human variation common to his time. He called 'undue' concern for the welfare of Native Americans

[104] Pearson 4/17. Journal of the Galton Laboratory 1915–, p. 8.

[105] Nils Roll-Hansen, 'Eugenics and the Science of Genetics' in Alison Bashford and Philippa Levine (eds.), *The Oxford Handbook of the History of Eugenics* (Oxford: Oxford University Press, 2010), p. 87.

[106] Marius Turda and Maria Sophia Quine, *Historicizing Race* (London: Bloomsbury Academic, 2018), p. 83.

[107] Review by H. W. H., 'The Races of Britain', *Science* 7: 155 (1886), p. 84.

[108] Elazar Barkan, *The Retreat of Scientific Racism. Changing Concepts of Race in Britain and the United States between the World Wars* (Cambridge: Cambridge University Press, 1992), p. 22.

[109] H. G. Wells, *Anticipations of the Reaction of Mechanical and Scientific Progress upon Human Life and Thought* (Leipzig: Bernard Tauchnitz, 1902), pp. 198–199.

and Australian aboriginals to be a false sort of humanitarianism.[110] The white man was able to make better use of their land, and since these races had not added much to the stock of human knowledge, it justified their inferior position. In 1881, the year before his death, Darwin predicted that the 'lower races' would relatively soon be eliminated by the civilized ones.[111]

For the Eugenics Education Society, social class was paramount and other nations only featured as additional venues for eugenics. The Galton Laboratory studied the mechanisms of heredity closely. Pearson regarded race as fixed traits passing from generation to generation, unrelated to where a person was born. It was the result of selection over long periods of time. This did not rule out the environment or natural surroundings affecting race to some extent.[112] The genealogy of distinguished men pioneered by Galton lost out when the Eugenics Record Office transformed itself into the Eugenics Laboratory. Genealogical charts were henceforth drawn up primarily as aides to understanding genetic illness. The laboratory worked from the hereditarian assumption that inborn factors played a significant role in the functioning of individuals. The blanks for studying heredity asked for information about health and ability. The theory was that the complexity of tasks performed by educated people had led to their being selected for brains. Weak performers among them had sunk to a lower level over time. But there were also signs that the Galton Laboratory was interpreting data to prove perceptions favourable to the eugenic creed.

In 1928 the *Annals of Eugenics* published the final parts of Pearson and Margaret Moul's old research from before the First World War pertaining to alien immigration. It was a comprehensive investigation, long enough for a monograph at 290 pages. The eyesight of the Jewish children was found to be poor, providing a rationale on eugenic grounds for their exclusion.[113] It referred to earlier work by the Galton Laboratory stating that myopia was largely a hereditary trait, especially in the case

[110] Pearson, *The Grammar of Science*, p. 438.

[111] Gregory Claeys, 'The "Survival of the Fittest" and the Origins of Social Darwinism', *Journal of the History of Ideas* 61:2 (2000), p. 239.

[112] Turda and Quine, *Historicizing Race*, p. 14.

[113] Karl Pearson and Margaret Moul, 'The Problem of Alien Immigration into Great Britain Illustrated by an Examination of Russian and Polish Jewish Children', *Annals of Eugenics* 2:1–2 (1927), p. 111.

of Germans. Many Jewish immigrants in Britain spoke German and had assimilated into that culture. As for the children who had better eyesight, the article found it to be correlated moderately with lower placement in the school class. The boys who were near the top of their form at school showed the poorest vision. It conceded that the examination had been conducted perfunctorily by a nurse. As before, the girls were judged to be less intelligent than the boys on average, and there was a stronger association between visual acuity and cognitive functions for them.[114]

Bias in the Research

This could be an example of biased work because the initial assumption that intelligence and vision were correlated was probably meant to provide a comprehensive reason for why Jewish immigration was undesirable. Not only did it cause poorer eyesight among the general population, but the effect would continue in the future generations. That the average cognitive level would also decline was not the actual result when the numbers were crunched. The correlation turned out to be negative, but for opponents of immigration at least it showed that Jews would be dragging the stock of characteristics down one way or another. It was only necessary for the Galton Laboratory to conclude that the immigrants collectively were not an asset for the disadvantages of admitting them to outweigh any conceivable benefits. Therefore, on eugenic grounds, immigration restriction could be represented as the correct policy. Pearson accepted that there might be individual exceptions, however, eugenics had to deal with large-scale outcomes.

This argument was used by others too. In his 1929 Galton Lecture, Leonard Darwin referred to research from the United States which concluded that it would take 48, 000 children born of unskilled workers to produce an additional entry in *Who's Who*, but only 1, 600 children of skilled workers.[115] Therefore, why not encourage the latter to have children in preference to the former? Galton had averred that probability should be the foundation of eugenics. Statistics, biometry and eugenics were all new subjects in the late nineteenth century. Pearson claimed

[114] Pearson and Moul, 'The Problem of Alien Immigration', *Annals of Eugenics* 2:3–4 (1927), pp. 290,295.

[115] Leonard Darwin, 'The Society's Coming of Age. The Growth of the Eugenics Movement', *Eugenics Review* 21:1 (1929), p. 16.

that Florence Nightingale had seen a connection between them.[116] She had called statistics 'the most important science in the world'.[117] The Galton Laboratory therefore aligned well with the Biometric Laboratory, which was researching a biological version of the same subject. These developments coincided with the early concern that social protection and philanthropy had led to an imbalance in the affairs of civilization. People who were 'undesirable' were having too much of an impact on society, a complication which needed to be put right urgently. Galton was mobilizing educated opinion to pay greater heed to itself.

Many of the underlying assumptions went beyond the truisms Galton mentioned, such as health and vigour being better than illness or thriving in an environment being better than struggling in it.[118] In Edwardian Britain, with great disparities of wealth, it was hardly surprising that the rich, who benefited from proper medical care and rigorous schooling, should enjoy better health and be cleverer than the poor and downtrodden, who did not. Eugenists recognized this only to a limited extent. Because lads of humble station were sometimes upwardly mobile, they theorized that this had been an ongoing process through the generations. It had resulted in the present situation, whereby the upper classes were 'fit' and workers often 'unfit'. This was adding a value judgement to Darwin's theory of evolution, as both 'reform' eugenists and Galton himself admitted that being unscrupulous could confer an evolutionary advantage. The typical eugenist, however, believed that ability, wealth, health and rectitude were signs of 'fitness', and if their possessors were outbred by those lacking such characteristics, civilization would decline. In *Hereditary Genius* (1869) Galton argued that men with at least enough ability to 'win the ordinary prizes of life' were generally not repressed no matter how lowly their birth. For those with greater gifts, it was even less so.[119] He viewed the Brontë sisters as supremely talented, but did not at this stage reflect on how differing expectations of women

[116] Pearson/4/4/5. Report to the Court of the Worshipful Company of Drapers, pp. 3–4.

[117] Roy Porter, *The Greatest Benefit to Mankind. A Medical History of Humanity from Antiquity to the Present* (London: HarperCollins, 1997), p. 378.

[118] Francis Galton, 'Eugenics: Its Definition, Scope and Aims' in *idem*, *Essays in Eugenics* (London: Eugenics Education Society, 1909), p. 36.

[119] Francis Galton, *Hereditary Genius. An Inquiry into its Laws and Consequences* (London: Macmillan, 1892), p. 39.

could lose the nation some outstanding performers. There were nevertheless contradictions in the theory which sometimes were raised, for instance the debate at the 1912 International Eugenics Congress about whether it was preferable to be the son of a healthy burglar or a consumptive bishop.[120]

An earlier example of possible bias, though not hurting anyone's interests, was Pearson's findings relating to the handicapping of the first born. According to a 1914 title published in the Eugenics Laboratory Lecture Series, two individuals with the same ancestry were differentiated by the order of their birth. The older had a higher chance than the younger to be affected by mental defect, epilepsy, insanity and tuberculosis.[121] Pearson thought it might be due to the mother being biologically less capable the first time she gave birth, with the effects more marked the older she was. If these conclusions were acted on by parents who had ambitions for their children, it might increase the birth rate among them—precisely what eugenists wanted. The findings were known before 1914. Galton took cognizance of the work, typically promoting his agenda for meritocracy on this basis. He suggested that the ablest rather than the eldest of a peer's sons should take the family seat in the House of Lords. Pearson wished to turn the upper chamber into an appointed senate. The dearth of meritorious peers at the present time was due to the failure to marry into the right stocks.[122]

The Galton Laboratory wielded some influence from its early days. It never sought to mould public opinion directly, concentrating on institutions that matter, such as government. The medical establishment preferred the Laboratory's scientific work to what appeared in the *Eugenics Review*. The former was authoritative and contributed to medicine. A leading article in the *British Medical Journal* in early 1928 bore the heading 'The Right of the Unborn Child'. It reasoned that all children had the right to be born free of hereditary taint. Consequently, it was wrong for those afflicted by or invisibly carrying a disease to beget children.[123] Legislative action might be required, or civilization was at risk

[120] Wellcome Library, London. Eugenics Society Archive. SA/EUG/B5. *Pall Mall Gazette* 3 August 1912.

[121] Karl Pearson, *On the Handicapping of the First-born* (London: Dulau and Co, 1914), pp. 62–64.

[122] *The New Statesman* Vol VI. Saturday 30 October 1915, p. 237.

[123] *British Medical Journal* 28 January 1928, p. 145.

of disappearing in the same way as previous cultures with no knowledge of heredity. Although the Galton Laboratory had considerable support among physicians, it is possible that it over-emphasized the genetic factors at the cost of mutations. Both are causes of debilitating conditions. Their relative prevalence was an important factor in the rationality of eugenics. Yet mutations had not been encompassed within the original scope of the doctrine. Genetics was still in an incipient phase, while eugenics originally had no connection to medicine. It may have been an answer to a supposed threat to Galton's social class. Others took up this tendency and magnified it. The falling reproduction rates of the upper classes commingled with the rising trend towards democracy and the demand for the redistribution of wealth. The People's Budget of 1909, National Insurance Act 1911 and other policies of the New Liberals were beginning to target capital. As eugenics gained traction, it began incorporating medical and social problems of a wider kind.

Eugenics is popularly thought to be linked to fascism, owing to the Nazis implementing heavy-handed eugenic measures after 1933. Pearson is sometimes described as being a 'national socialist' more than a socialist, usually in a literal sense.[124] His department did produce a little research with potentially antisemitic conclusions, though this was neither sustained nor overt. In his final report to the Drapers' Company, covering the years 1925–1929, Pearson drew attention to Mussolini's liking of statistical science. He quoted a passage in a speech by *Il Duce* before summarizing 'The present leader of the Italian people has a very keen appreciation of the importance for national progress of the modern statistical movement, and sees how it is spreading throughout the world'.[125] His point was that his own Biometric Laboratory had inspired similar institutions abroad through its international postgraduates. The Drapers' Company had facilitated its important research, which was slowly being recognized. He gave no commendation of Mussolini beyond that. The fascist leader was in any case regarded as an ordinary statesman before the Second World War. Both France and Britain were seeking his help as late as the *Anschluss*

[124] Paul, *Controlling Human Heredity*, p. 35. Dan Stone, *Breeding Superman. Nietzsche, Race and Eugenics in Edwardian and Interwar Britain* (Liverpool: Liverpool University Press, 2002), p. 103. Stephen Trombley, *The Right to Reproduce. A History of Coercive Sterilization* (London: Weidenfeld & Nicolson, 1988), p. 43.

[125] Galton Laboratory/4/4/9 Report for the Years 1925, 1926, 1927, 1928–1929, p. 3.

in 1938, despite the League of Nations' sanctions over the invasion of Abyssinia in 1936. In bidding for further funds from the drapers in 1930, Pearson also wrote that the Biometric Laboratory had been contacted by the German Army on how to interpret the many physical and mental tests which its recruits underwent.[126] He did not say whether this occurred before the end of parliamentary government in July 1930 or after, nor is it clear whether the inquiry resulted in anything tangible.

Pearson Retires

Apart from the actual establishment of the Galton Laboratory, the most dramatic event which it underwent in the pre-1945 period was the change of leadership in 1933, when Pearson resigned. As he had been the leading light of eugenics at University College, the committee formed to handle the transition to a new director resolved to alter the scheme significantly. Taking cognizance of Galton's will, which acted as a constitution for the chair, it was decided that the professor should devote himself or herself entirely to eugenics.[127] It was not certain that Pearson's successor would be able to straddle eugenics and statistics in the way he had. The committee also proposed that Ethel Elderton, by now reader in eugenics, should have her field broadened to 'social statistics'.[128] However, she too chose to retire at this juncture. Galton and Pearson had been practitioners of biometry. By 1933 Mendelian genetics had a predominant position worldwide and it seemed likely that Pearson's successor would be researching eugenics on this basis.

This was in fact the case. The successor Ronald Fisher happened to be an expert in both statistics and genetics, usually working in the tradition of Mendel, though also refining biometry. In a paper from 1918 published by the Royal Society of Edinburgh, he had reconciled Mendelian genetics with biometry.[129] He invented the genotype-environment equation for biometry, a tool for apportioning the respective influences of heredity

[126] Galton Laboratory/4/4/10. To the Court of the Worshipful Company of Drapers, p. 2.

[127] Galton Laboratory/4/29/1. Report of the Professorial Board on the Resignation of the Galton Chair by Professor Karl Pearson.

[128] Pearson/4/29/1. Agenda of Galton Sub-Committee 19 June 1933, p. 2.

[129] Peter M. Wisscher and Michael E. Goddard, 'From R. A. Fisher's 1918 Paper to GWAS a Century Later', *Genetics* 211:4 (2019), p. 1125.

and variance.[130] He was reluctant to ascribe variance to the environment, preferring heterozygous inheritance. The Galton Laboratory had originally resisted the use of Mendelism as applicable to mental defect, particularly disliking the simplistic schema whereby the condition was thought to be triggered by two recessive genes. David Heron had written the first part of *Mendelism and the Problem of Mental Defect. A Criticism of Some Recent American Work* in 1913. He stated that such claims did damage to eugenics as a field and were not backed up by the facts. After taking over, Fisher changed the sub-title of the *Annals of Eugenics* from 'A journal for the scientific study of racial problems' to 'A journal devoted to the genetic study of human populations'. Pearson was displeased by the appointment. They were at odds over Mendelian genetics and Pearson was personally hostile to Fisher from their interaction in the Royal Society and the academic world generally.

The Department of Applied Statistics, now split from the Galton Laboratory, would in turn be led by Egon Pearson, who was promoted to reader to facilitate such a leadership role. Fisher was not asked to teach statistics. The younger Pearson had joined his father in the Galton Laboratory in 1921 straight from Trinity College, Cambridge. He had received £ 300 for his work in the Department in 1923, almost as much as Ethel Elderton.[131] Clearly being groomed for a career, he attended statistical courses and became an assistant editor of *Biometrika*. In 1936 W. F. R. Weldon's widow Florence Tebb Weldon died and a new chair in biometry was funded from a legacy in her will. The next year J. B. S. Haldane was appointed to the Weldon chair of biometry. He had joined University College in 1933 as a part-time professor of genetics. With two departments in place of one, the staff and equipment were split according to whether they had a bigger role in eugenics or statistics. The little museum which had been set up in the early 1920s, following the move to the Bartlett building, had contained both exhibits of human remains and calculating devices. Fisher mentioned only the skins, skulls and skeletons

[130] James Tabery, 'R. A. Fisher, Lancelot Hogben, and the Origin(s) of Genotype-Environment Interaction', *Journal of the History of Biology* 41:4 (2008), pp. 724, 728.

[131] Galton Laboratory/1/2. Blue folder marked 1 of 4. Yellow folder inside marked 'Bills, correspondence to K. Pearson about laboratory accounts', Wage bill 24 April 1923.

in May 1935 in his letter to Karl Pearson, which suggests that the exhibits relating to statistics had already been relocated.[132] (Fig. 3.3)

Fig. 3.3 J. B. S. Haldane (reproduced from the Haldane papers, UCL Special Collections)

[132] Jenny Bangham, *Blood Relations. Transfusion and the Making of Human Genetics* (Chicago: University of Chicago Press, 2020), p. 58.

It was in the context of Pearson's retirement that he painted the scientific movements in which he had been involved with a broad brush. At a dinner held in his honour on 23 April 1934, Pearson gave the following summary of his career:

> The culmination of 'biometrics' lay in Galton's preaching of eugenics and in his foundation of my Eugenics Professorship. But was this really the 'culmination'? No, that lies rather in the future, with Reichkanzler Hitler and his proposals to regenerate the German people. In Germany at present a vast experiment is in hand, and some of you may live to see its results. If it fails, and we must hope it does not, it will not be for the want of enthusiasm, but rather because the Germans are only just embarking on the systematic study of mathematical statistics in the modern sense.[133]

Some scholars, among them Clyde Chitty and Donald MacKenzie, have seen these remarks as support for Nazi eugenics.[134] The Law for the Prevention of Hereditarily Diseased Offspring took effect on 1 January 1934. It ordained the compulsory sterilization of any member of the public where there was a risk of children being born suffering from alcoholism or any of eight specified illnesses. It was by far the most wide-ranging eugenic enactment introduced anywhere.

Theodore Porter thinks Pearson's comments may have been a joke gone awry, while Bradley Hart regards what was said as no more than giving the Nazis 'the benefit of the doubt' on eugenics.[135] It is possible to discern Pearson's real attitude from what he thought about so-called voluntary sterilization and the attempt to introduce it in his home country in 1931. In a letter to Eugenics Society general secretary C. P. Blacker, he wrote that the effect of the bill would be to permit 'involuntary sterilization'. For tactical reasons, this was unnecessary as the law did not prevent sterilization by consent as it stood. He strongly favoured

[133] Clyde Chitty, *Eugenics, Race and Intelligence in Education* (London: Continuum, 2007), p. 137.

[134] Donald MacKenzie, 'The Zero-Sum Assumption. Reply to Sullivan', *Social Studies of Science* 29:2 (2009), p. 227.

[135] Theodore M. Porter, *Karl Pearson. The Scientific Life in a Statistical Age* (Princeton: Princeton University Press, 2004), p. 284. Bradley W. Hart, 'Watching the "Eugenic Experiment" Unfold: The Mixed Views of British Eugenicists Toward Nazi Germany in the Early 1930s', *Journal of the History of Biology* 45:1 (2012), pp. 50–51.

the sterilization of mental defectives and was 'not afraid' of compulsory measures.[136] Therefore, it is likely that he agreed with the initial steps of 'racial hygiene'. In Germany plenty of eugenists of other political persuasions had joined the Nazis by the late 1930s.[137] Pearson maintained an interest in that country from having studied there and having adopted German theoretical socialism. He was acknowledging a connection between the eugenics which he had promoted and the sharpest legislative action to which it had given rise. Hoping the measure would succeed, he was sceptical about its not being based on statistical science. As this was his common rejoinder to popularized eugenic schemes, he was to some extent parodying himself in an appearance to mark his retirement. Pearson died just two years later, which spared him the embarrassment of unequivocally supporting the policies of the Nazi state. Fisher, however, was perhaps surprisingly candid about his views in 1948. He wrote that he would have been in favour of the regime's efforts to diminish the numbers of mentally deficient and that the Nazis had genuinely sought to improve the German stock.[138]

Conversely to the Nazi taint, it has also been argued that Pearson was not necessarily that interested in advancing eugenics. Delays in publication and the thoroughness with which each investigation was handled, lost for eugenics some of the momentum it was achieving in the early twentieth century. These were among the reasons the hotheads of the Eugenics Education Society did not value the work of the Galton Laboratory. In support of this view, it may be added that Pearson did not initially wish to supervise the Galton Laboratory. When he nevertheless became its director, he made that conditional upon remaining in charge of the Biometric Laboratory and he continued to publish on statistics and biometry.[139] Galton and Pearson agreed this was how eugenics needed to be encapsulated scientifically, but much of the work had no obvious

[136] Eugenics Society Archive. SA/EUG/C268. Pearson to Blacker 30 June 1931.

[137] Paul Weindling, *Health, Race and German Politics between National Unification and Nazism, 1870–1945* (Cambridge: Cambridge University Press, 1993), p. 486.

[138] Sheila Faith Weiss, 'After the Fall: Political Whitewashing, Professional Posturing, and Personal Refashioning in the Postwar Career of Otmar Freiherr von Verschuer', *Isis* 101: 4 (2010), p. 745.

[139] M. Eileen Magnello, 'The Non-Correlation of Biometrics and Eugenics: Rival Forms of Laboratory Work in Karl Pearson's Career at University College London, Part 1', *History of Science* 37:1 (1999), pp. 84–85.

implications for eugenics. Pearson suspected that Galton wanted quick results, about which he was probably right. Galton, after all, envisaged eugenics becoming nothing less than a grand collective purpose in British society. This would be accomplished by convincing doubters through the scientific results promulgated by his laboratory and in making the eugenic perspective count in various areas. Although Galton advocated neither the use of contraception nor sterilization, he did wish to skew reproduction towards the healthier and abler classes at the expense of the criminal and shiftless. The fact is that the Eugenics Education Society was promoting this agenda fervently, whereas the Galton Laboratory risked becoming a sideshow. It seemed to bring its influence to bear only on the civil service, state institutions and other well-informed occupations. Relatively little was achieved for eugenics during Pearson's leadership. What might be the reasons?

Pearson was an ivory-tower intellectual. He supposedly believed in socialism and eugenics, yet hardly ever disseminated these doctrines to ordinary members of the public. He would have functioned superbly as a specialist in the type of state the Fabians were hoping to build.[140] Social democracy became a marriage of policy experts with the labour movement. Pearson believed in permeation and rejoiced in the expanded role of the state in the early twentieth century. He is likely to have seen eugenics as an applied form of statistics and biometry, which were his first passions.[141] If so, it might be more important to make progress on those two, whereas the case for eugenics could almost make itself.[142] Pearson took a close interest in the work of his employees. He assuredly wanted the output of his department to be of the highest possible quality. But did even the staff care so much for the doctrine? The first two fellows, Schuster and Heron, both resigned within a few years of taking up the position, to become active in other fields. All of Pearson's successors except Fisher would tone down the eugenic aspects of their research. Pearson deliberately chose female researchers to keep salaries down and make better use of resources. Ethel Elderton, Amy Barrington and Julia Bell were unmarried bluestockings probably as interested in their jobs as

[140] L. J. Ray, 'Eugenics, Mental Deficiency and Fabian Socialism between the Wars', *Oxford Review of Education* 9:3 (1983), p. 218.

[141] Galton/2/4/12/1/2 folio 1 *The Standard* Monday January 3, 1910.

[142] Cf. Kevles, *In the Name of Eugenics*, p. 104.

anything. Eugenics may only have provided intellectual stimulation for them. Miss Bell continued her own research and helped others into the post-war era, but in practice eugenics as an ideological current in science lasted only until 1943, when Fisher resigned after a decade in the chair.

Fisher and the Serological Laboratory

The appointment of Fisher to the Galton chair ended the rivalry between the Galton Laboratory and the Eugenics Society. Fisher was a member of the Society's council and had been honorary secretary as late as 1930. In 1932 the government had appointed a committee led by the civil servant Laurence Brock to investigate sterilization in regard to mental deficiency and mental disorder. It followed a debate on the issue in the House of Commons the year before. Fisher served on the committee, whose unanimous recommendation was to allow 'voluntary' sterilization, even for carriers not themselves affected by such conditions. It reported in 1934 after hearing evidence from about sixty physicians, psychiatrists, biologists and geneticists. The committee was packed with Eugenics Society appointees and its chairman Brock was in correspondence with Blacker. Among the members serving were Alfred Tredgold, a University of London lecturer in mental deficiency, and E. O. Lewis, a psychologist and physician, who had surveyed the numbers of 'feebleminded' for the earlier Wood Committee, reporting in 1929. Fisher also participated in the Eugenics Society's own non-official grouping, the Committee for Legalizing Eugenic Sterilization. It argued that defectives could be released in the community if they had been sterilized, as opposed to the costly practice of institutionalization. As the new Galton professor had taken part in pressure-group action and the Eugenics Society was conducting research, the distinction between Laboratory and Society was in any case breaking down.

Owing to the splitting of the department, Fisher's remit as director was smaller than his predecessor's. This freed up time for writing grant applications. Fisher succeeded with the Rockefeller Foundation, whose financial support was administered by the Medical Research Council. The project, about human genetics in relation to mental disorders, got underway in late 1935. It was the most substantial funding eugenics had received for some time. In 1921 the entire Department of Applied Statistics and Eugenics had secured support of £ 1000 annually for four years from the Universities Grant Commission, £ 300 per annum from the Medical Research Council in addition to direct government funding. The latter constituted £ 300 a year from the Department of Scientific and Industrial Research. The London County Council gave £ 1000 per annum starting in 1921 and lasting for at least two separate three-year periods.[143] These grants covered a medical officer and his assistant. The L. C. C. then funded a fellowship in 1932. Much of the regular income came from the University of London and University College. Lastly, the Rockefeller Foundation had provided funds from about 1927 to 1934.[144]

For the new project, the money was kept on account with the Medical Research Council, and it was to this body that Fisher directed his quotidian inquiries. The focus was on how blood groups could be used to detect mental disorders. For this purpose, a serological laboratory was established as part of the Galton Laboratory. Blood groups had been discovered in the first decade of the twentieth century by the Austrian biologist Karl Landsteiner. In 1909 he classified these as A, B, AB and 0. Blood has always been invested with mystical qualities. Inherent conditions are sometimes described as 'being in the blood'. Fisher's investigation was regarded as potentially groundbreaking and a total of £ 7, 000 was awarded by the Rockefeller Foundation on 1 July 1935 to cover five years of research.[145] In early 1937 another grant of £ 2, 000 was conferred to enlist the services of a dedicated serologist, the funds to be dispersed over three years. The serologist who was hired to detect blood groups was George Lees Taylor, who held both an M. D. and a Ph. D. There was even financial provision for an assistant, the physician Robert Russell Race. The serologist was paid £ 440 per annum. In March 1938 Fisher received a further £ 500 upon request, which was then deducted from the account the Rockefeller Foundation held with the Medical Research Council. The Laboratory's hopes of finding traces of various diseases in the blood had not been realized by 1940. At the conclusion of his three-year contract, Taylor returned to the Department

143 Pearson/4/4/6. Report by the Director on the Work of the Drapers' Company Biometric Laboratory 1919–1921, p. 1. Pearson /4/4/7 Report for the years 1922, 1923, and 1924, p. 1.

144 M. Eileen Magnello 'The Non-Correlation of Biometrics and Eugenics: Rival Forms of Laboratory Work in Karl Pearson's Career at University College London, Part 2', *History of Science* 37:2 (1999), p. 143.

145 Galton Laboratory/2/3/5. Blue folder 1 of 2. Folio 4, folio 7 and folio 1.

of Pathology at Cambridge.[146] Fisher also moved to Cambridge University, where he had studied, becoming a professor of genetics there in 1943.

In the early days of the department, Pearson had complained about the lack of opportunities to keep animals. It is certain that the research on blood groups included experiments on living beings. In the summer of 1941, Fisher wrote to Taylor about some cages not really belonging to the department as they had been funded by the Rockefeller Foundation. Six months later Taylor told Fisher that Geoffrey Vevers, superintendent of the Zoological Society of London, had offered two monkeys and cages for them. The reply throws some light on the cruelty involved. Fisher was glad that Vevers had acted quickly. One of the two monkeys had tuberculosis and was not expected to live long. They cost £ 1 each from the zoo and Taylor was instructed to pay that amount. The team at the Galton Laboratory also had access to the animal house of University College to keep rabbits which had been immunized. Since eugenics by definition relates only to humans, these monkeys and rabbits were incidental to the work being carried out.

The establishment of the serological laboratory represented a departure from the usual practice in the Department of Eugenics. It was the first time that actual live medical discoveries were attempted, as opposed to detecting causes filtered through statistical analysis. In furthering medical science, it utilized Mendelian genetics rather than biometry, thus linking its research more closely to contemporary scientific practice. In the department, Fisher worked with Haldane and Julia Bell. He collaborated externally with colleagues such as John Fraser Roberts, at a home for 'feebleminded' children in Bristol, Lionel Sharples Penrose, then director of psychiatric research for the Canadian province of Ontario, and Lancelot Hogben, of the London School of Economics. From investigations at two mental hospitals, the scientists found that patients' blood samples reacted differently to stimuli than control samples. Detecting mental disorder (which has always been difficult to diagnose conclusively) through bloodwork thus seemed a realistic prospect. Later, the method might conceivably be extended to finding traces of physical diseases like Huntington's chorea, acholuric jaundice (very fragile red blood cells), hereditary ataxia (lack of control of muscles) and brittle bone disease

[146] Galton Laboratory/2/3/5. Yellow folder marked 'Fisher correspondence, 1941', folio 52 v., folio 60 and folio 80.

in the blood. The serological laboratory was able to advance research on blood groups in Britain, which had begun in 1927–1928. Over time, though, it became clear that diseases and mental disorders are not necessarily visible in the blood.[147] In 1939, with war approaching, the serological laboratory was instead used for building up stocks of anti-A and anti-B for the AB0 groups, doubling as part of the Emergency Blood Transfusion Service.[148] As the war began, the serological laboratory was relocated to the Department of Pathology at Cambridge and lost its connection to eugenic research.

Anthropological Work

Apart from the work on blood groups, the 1930s witnessed anthropological research on race in the Galton Laboratory. Since its early days the Department of Applied Statistics and Eugenics had published on craniology and physical anthropology. Among the contributors were Pearson, Alexander Carr-Saunders and Julia Bell. Geoffrey Morant had been the Benington student in the department, gaining a master's degree in applied statistics in 1922.[149] After graduation, he stayed on as a temporary member of staff. He became an anthropologist by writing in *Biometrika* on craniometry and a series on Palaeolithic man in *Annals of Eugenics*. Although he had preferred conditions under Pearson, it was during Fisher's tenure that Morant began to make a name for himself. Morant lauded the Institute of Race Biology in Uppsala in an article in 1934. The Swedish Parliament believed a priori that the population which it served had harboured good quality 'racial material' from ancient times.[150] However, the Institute was a scientific eugenic endeavour and Morant was not a racist. He was a voice of moderation when a study of miscegenation started up in the mid-1930s. Some of his colleagues had a greater belief in extensive racial disparities regarding mental ability and cultural accomplishments.

[147] Mazumdar, *Eugenics, Human Genetics and Human Failings*, p. 173.

[148] R. R. Race, 'Some Notes on Fisher's Contributions to Human Blood Groups', *Biometrics* 20:2 (1964), pp. 361–362.

[149] Iris Isabelle Clever, 'The Lives and Afterlives of Skulls. The Development of Biometric Methods of Measuring Race (1880–1950)', Ph. D. thesis, University of California Los Angeles, 2020, pp. 178–179, 183.

[150] Turda and Quine, *Historicizing Race*, pp. 74–75.

Fisher would later reject the UNESCO declarations of race in 1950 and 1951, while a third person involved, Reginald Ruggles Gates, a Canadian geneticist working at King's College London, might be described as an outright racist. Ruggles Gates, a polygenist, was always trying to find a racial angle to innocuous-sounding studies in biology.[151] He was disappointed when it did not prove possible to classify races into 'higher' and 'primitive' from typical blood groups. In 1936 the anthropological project welcomed Jack Trevor as a temporary colleague. He had won the Darwin research scholarship of the Eugenics Society, which covered his salary of £ 250 per year. Trevor was investigating a topic in which Leonard Darwin had a genuine interest. Yet his approach, like Morant's, was academic rather than normative. He conducted a literature study of mixed populations, ranging from coloured people in Jamaica to Norfolk Islanders who were the descendants of the mutineers from the *Bounty* and local Tahitian women. Trevor concluded that these were intermediate in physical characteristics between the ethnic groups from which they derived.[152] Variability among them was neither particularly high nor particularly low and their traits matched normal distribution. In other words, there were no special features in such groups distinguishing them from others.

Ruggles Gates was also involved with the serological laboratory, as he published on the blood groups of Eskimos and Tibetans, blood transfusion and about recent progress in the field in 1935 and 1936.[153] It was typical that he should take an ethnic perspective, whereas Fisher and the other researchers were primarily seeking to identify 'feeblemindedness' and diseases by means of the blood. As for Morant, he took a stand against German anthropology, disagreeing with its stance on pure races and arguing that within-group variability exceeded between-group variability. In 1939 his book *The Races of Central Europe* laid down the gauntlet to scientific racists. He agreed that there were differences in the cephalic index (a measure of the shape of the skull) and stature, but showed that Germans were not always the most dolichocephalic (long headed) nor the tallest.[154] In a blow to Nazi aspirations, he argued that

[151] Bangham, *Blood Relations*, p. 246.

[152] J. C. Trevor, 'Some Anthropological Characteristics of Hybrid Populations', *Eugenics Review* 30:1 (1938), p. 31.

[153] J. A. Fraser Roberts, 'Reginald Ruggles Gates 1882–1962', *Biographical Memoirs of Fellows of the Royal Society* 10:10 (1964), p. 103.

[154] Clever, 'The Lives and Afterlives of Skulls', pp. 194–196.

central Europe did not contain any people which could be described as 'Nordic'.

In 1939 the seventh International Congress of Genetics was held in Edinburgh. This was the high summer of traditional racial science with Nazi eugenists called upon to give papers. The congress took place 23–30 August, though many European delegations, including the German and Polish, sensed the approach of war and left within days of arriving. The Galton Laboratory presented its research on blood groups and linkage to genetic conditions through a joint paper by Taylor and an assistant called Aileen Prior, and one by Ronald Fisher.[155] Haldane attended from the Department of Zoology, speaking at the same session as Fisher, and then on 'natural selection' two days later. The paper on blood groups put forward that there was great variety not just within but between races. Ruggles Gates presented a paper arguing that blood groups were probably caused by mutation. They could therefore not be used for racial classification. The prevalence of blood group A (hypothetically a sign of 'Nordic' descent) among some 'primitive' races showed that this type had appeared early in humanity's evolution.[156] In fact, more than 40% of Australian aboriginals, whom Gates categorized as 'lower' than Native American, had blood group A compared to blood group 0 predominating among the latter. This was a racist assumption, but Ruggles Gates was representing King's College London. Even before the onset of antiracist thinking, caused by a re-evaluation of attitudes in the wake of defeating National Socialism, the Galton Laboratory did not stand out as particularly racist. It was at this congress that the Geneticists' Manifesto written by Hermann Muller was adopted.[157] Signed by him, other left-wing eugenists Julian Huxley, Haldane, Hogben, Joseph Needham and eighteen others, it called for social betterment to accompany measures to raise the biological quality of mankind. Published in *Nature*, it explained why it would only become possible to evaluate the genetic worth of an individual once environmental conditions had moved towards equality.[158]

[155] R. C. Punnett (ed.), *Proceedings of the Seventh International Genetical Congress* (London: Cambridge University Press, 1941), pp. 31–32, 281.

[156] Punnett (ed.), *Proceedings of the Seventh International*, p. 125.

[157] Diane Paul, 'Eugenics and the Left', *Journal of the History of Ideas* 45:4 (1984), p. 583.

[158] F. A. E. Crew et al., 'Social Biology and Population Improvement', *Nature* 144 (1939), p. 521.

Conclusion

The second restructuring of the Galton Laboratory took place as a result of Fisher's resignation in 1943. The Galton chair became vacant until the end of the war, when the psychiatrist Lionel Sharples Penrose took it up. Penrose had led the Medical Research Council's important Colchester survey of mental defect between 1930 and 1938.[159] In the late 1920s, Fisher, like other Mendelians, had believed that 'feeblemindedness' was usually caused by a single recessive gene. Sometimes other factors were involved, not necessarily hereditary, and there could also be genes in existence instigating diseases which had impaired reasoning powers as a side effect. Penrose's studies both confirmed and widened this hypothesis. With juvenile amaurotic idiocy, the distinction between normal and afflicted members of a family was absolute, and therefore the trigger was the single recessive gene. Beyond that, mental defect was complex. He investigated six possible aetiologies, being (1) specific family incidence, (2) consanguinity of the parents, (3) hereditary mental defect, (4) maternal age or similar causes in which the order of birth and number of offspring mattered, (5) an association with twins and (6) specific gender inclination based on diseases which relate to the sex of the offspring.[160] These were all biological reasons for the phenomenon and almost all were hereditary too. However, after eight years of laborious work, including interviews with the patients' relations, Penrose concluded that mental deficiency arose from a variety of genetic, pathological and environmental causes. These usually interacted to produce the effect. In the case of Down's syndrome only one factor mattered, namely the age of the mother. Because such children were often born last, there had been speculation that the condition resulted from the parents having used up their reproductive potential. Penrose was able to show that mothers above the age of 35 had a much higher risk of giving birth to a Down's syndrome child.

Upon Fisher's resignation, J. B. S. Haldane became head of the Department of Eugenics, which henceforth incorporated 'Biometry and Genetics' in its title. He continued in the Weldon chair of biometry,

[159] Mazumdar, *Eugenics, Human Genetics and Human Failings*, p. 147.

[160] David C. Watt, 'Lionel Penrose, F. R. S. (1898–1972) and Eugenics. Part Two', *Notes and Records of the Royal Society of London* 52:2 (1998), p. 343.

which he had occupied since 1937. Though Haldane believed in upper-class genetic superiority, he combined this with membership of the Communist party and a stance against the 'class legislation' of sterilization. The greater number of 'feebleminded' among the poor was due to their inability to maintain a non-working member of the family, not genetic predisposition. Penrose was a liberal-minded pacifist, who had been a conscientious objector in the First World War. With both Haldane and Penrose disagreeing with many aspects of the 'mainline' eugenic creed, there was bound to be less focus on ideological goals. Pearson and Fisher by contrast, though they held some forward-looking views, represented the 'mainliners' within eugenic thought. The leadership devolving onto Haldane and Penrose would transpire to be a sea change.

CHAPTER 4

The Eugenics Society Until 1945

In the late nineteenth century eugenics was an inchoate ideology pioneered by Francis Galton with various thinkers and scientists lending support. It derived from the theory of evolution and resembled social Darwinism. Because it built on simple notions, it could interact with already existing literary and political conceptions. In the early twentieth century eugenics started taking shape as a social movement. The aging Galton was joined by a herd of new believers. These adherents were usually organized in the Eugenics Education Society, while Karl Pearson, who wished to extend Galton's scientific studies, was to be found in the academic environment of University College with his staff.

The typical member of the Eugenics Education Society was a professional who agreed with laissez-faire economics and social Darwinism. Although collectively they represented the 'popular' branch of eugenics, many were distinguished enough to be listed in biographical works of reference. Eugenics itself was authoritative between 1901, the year of Galton's Huxley Lecture, and 1945. Across the world different versions of the ideology emerged, each based on concerns within a national context. In Britain a debate was ongoing about whether biometry or Mendelian genetics was the best method of studying heredity. Political voices on the left wished to incorporate environmental and social conditions into

D. Redvaldsen, *A History of British Eugenics since 1865*, Medicine and Biomedical Sciences in Modern History,
https://doi.org/10.1007/978-3-031-72290-5_4

eugenics.[1] In 1930 Ronald Fisher published *The Genetical Theory of Natural Selection*, which reconciled Mendelian genetics and Darwinist natural selection. Although Fisher was a thorough-going hereditarian, he created formulae in population genetics which potentially could take the effects of the environment into account. Left-wing geneticist Lancelot Hogben helped make environmentalism a credible scientific alternative in the 1930s.[2] A coherent critique of eugenics did not exist before that decade, only opposition to individual aspects of it.[3] The doctrine could be grasped easily and was gathering momentum, receiving support first from science graduates and intellectuals, later from a middle-brow audience. Eugenics was widely covered in the press as if it were self-evidently correct.[4]

The question was whether eugenists would be able to take advantage of this situation. They managed to influence measures to tackle inebriety and venereal diseases. Of greater moment was implementing the institutionalization of mental defectives. In the late 1920s, they succeeded in getting rebates for tax-paying parents. At about the same time, official eugenic opinion began recommending sterilization as a cure for degeneration. This was the matter on which eugenists overreached themselves. Contraception, which the leading tendency in the Eugenics Education Society had accepted from 1922 or so, was controversial enough. Sterilization was even more radical because it was a surgical method of contraception and not always consensual. The campaign to legalize sterilization markedly strengthened the opposition to eugenics. Sterilization over time diminished its kudos as a rising movement and it did not help that the Nazis had embraced the doctrine. Another effect of the proposals was to anger the Catholic Church, which henceforth would resist the doctrine intellectually and practically.

[1] Pauline M. H. Mazumdar, *Eugenics, Human Genetics and Human Failings. The Eugenics Society, its Sources and its Critics in Britain* (London: Routledge, 1992), pp. 46, 145–146.

[2] James Tabery, 'R. A. Fisher, Lancelot Hogben, and the Origin(s) of Genotype-Environment Interaction', *Journal of the History of Biology* 41:4 (2008), pp. 734–737.

[3] Diane B. Paul, *Controlling Human Heredity. 1865 to the Present* (New Jersey: Humanities Press, 1995), p. 11.

[4] Ian Brown, 'Who were the Eugenicists? A Study of the Formation of an Early Twentieth-century Pressure Group', *History of Education* 17:4 (1988), p. 305.

While it had been realized that eugenics was a pastime of the affluent, it was sterilization which brought home to many observers that it was dangerous to the poor and downtrodden. Working-class people would be at risk from such measures, the drive for which occurred just as the Depression got going. Did a plan exist to sterilize the long-term unemployed to make it easier for governments to balance the budget? In 1931 the biologist Julian Huxley suggested that unemployed men receiving the dole should be barred from reproducing. Infringement would result in being sent to a labour camp for a few months.[5] Henceforth the Trades Union Congress and most of the Parliamentary Labour Party would abjure eugenics. But conversely for eugenists, the labour movement gained in importance just as it started distancing itself from them. The Society's 1931 sterilization bill had been rejected largely on the back of Labour MPs voting against and it genuinely, in any case, wished to convince as many strands of opinion as possible. Consequently, it extended its outreach to working-class people, in the hope of gaining their trust and support.

The Eugenics Education Society Is Founded

The connection between eugenics and social reform was of long standing. The event most conducive to the founding of the Eugenics Education Society was Galton's presentation to the Sociological Society on 16 May 1904 entitled 'Eugenics: Its Definition, Scope and Aims'. Eighteen academics, doctors, scientists and writers spoke in the ensuing debate, some of them household names. By 1907 the Sociological Society had a new secretary in the American psychologist James Slaughter.[6] That year he became chairman of the provisional council of a new organization, to be called the 'Eugenics Education Society'. The core of it consisted of members of the Moral Education League, a pressure group agitating for secular moral instruction in schools. A young widow, Sybil Gotto, was the initiator of the new venture, liaising with Galton and recruiting several prominent early enthusiasts.

[5] Paul, *Controlling Human Heredity*, p. 73.

[6] University College London (UCL). Galton Archive. Galton/2/4/19/9/2 folio 25.

The society formally came into being on 15 November 1907 at Caxton Hall, Westminster. The members included scientists and physicians. Unlike the Galton Laboratory staff, however, only a minority would make contributions advancing eugenic knowledge. Most saw it in basic terms as encouraging the wealthy or talented to have many children, while hindering the poor or anyone with congenital afflictions from reproducing. Galton had remarked on how civilization 'artificially' kept poor stocks alive. The objective therefore was to rebalance the birth rate in favour of the 'fit'. The Society immediately involved itself in opposing the closure of homes for inebriate women in the London area, saying this would have a detrimental effect on the race.[7] Correspondence relating to the fertility of drunkards was published in *The Times* on 18 April 1908. The Home Office established a Departmental Committee to consider how to administer the 1898 Inebriates Act, to which the Society contributed a memorandum (Fig. 4.1).

In 1908 Galton wrote a set of policies for the Society, which it printed as a leaflet. In May that year he joined and was made honorary president.The leaflet described checks on the population as 'salutary', but said civilization affected these so much that they sometimes were even reversed. An insane, imbecilic, criminal or seriously ill person left on average one child equally 'unfit' as himself or herself plus many others who were carriers of the misfortune. There needed to be restrictions on marriage to prevent such 'degenerates' from starting families. On the other hand, people of good stock who were poor should be provided with adequate housing, subsistence and employment. Like before, Galton recognized that eugenic worth did not always equal social status, although he was now paying greater attention to 'negative' eugenics. Typical of Galton's designs was that the leaflet listed one of its aims as being the propagation of the discoveries of the Eugenics Laboratory and similar institutions.

The Society's own lectures aimed to explain how heredity worked and its impact on the race.[8] The pressure of public opinion would be the eugenists' handmaiden in their efforts to change laws and habits. Its

[7] Wellcome Library, London. Eugenics Society Archive. SA/EUG/A1. Annual Report 1908, pp. 16–18.

[8] UCL: Pearson Papers, Pearson/4/19/2. Folio 3 v., folio 4 r.

Fig. 4.1 Sybil Gotto (reproduced from the Eugenics Society Archive, Wellcome Library, London)

council scrutinized the by-laws of local authorities as well as parliamentary proposals to consider their eugenic effects. What constituted desirable legislation was always whatever encouraged professionals to increase their fertility or the inadequate to reduce theirs. The latter aim might be attained by keeping those concerned in prisons, workhouses or otherwise under the supervision of the authorities. Alcoholism, recidivism, major ailments and poverty seemed to be crosscurrents which, if laid at the door of heredity, were all encompassed by a simple solution. For a privileged and publicly minded person, eugenics embodied a panacea equivalent to how a Labour supporter might think of 'socialism'. All manner of social problems would disappear once the polity had been put on the correct hereditarian basis. Although tensions were on the verge of arising between the Society and the Eugenics Laboratory, the various protagonists for eugenics were initially of one mind. Galton spoke to the Society in 1908 as did the researchers Ethel Elderton and David Heron from the Laboratory.[9]

Early Concerns

Anxiety about the unhealthy urban proletariat led to the setting up of a Royal Commission on the Poor Law in 1905. As it failed to reach a unanimous conclusion in 1909, two separate reports were issued. The Majority Report, written by C. S. Loch and others, typified the views of the Charity Organisation Society, seeing poverty as caused by either bad morality or indiscriminate charity.[10] It recommended changes to the 1834 Poor Law and a system whereby charities granting relief would be registered, which would facilitate cooperation with public assistance. Labour exchanges across the country would help in the task of finding employment. The Minority Report brought a structural and scientific angle to the same issues. It was authored by the vicar Russell Wakefield, trade unionist Francis Chandler, Labour MP George Lansbury and, not least, the social reformer Beatrice Webb. She was the only eugenist on the commission and the Society had doubts about both reports. The Poor Law as it stood could not control the 'feebleminded' adequately. They were undetectable unless they applied for relief, while mixed workhouses

[9] SA/EUG/A1. Annual Report 1908, p. 17.

[10] Mazumdar, *Eugenics, Human Genetics and Human Failings*, p. 72.

led to 'unhealthy' mingling and there was too much reliance on outdoor relief.[11] At this stage, the remedy for the perceived problems was segregation, with Galton arguing that 'undesirables' should be cared for in workhouses or asylums but not be allowed to breed. In 1911 Sidney and Beatrice Webb devoted a whole chapter to eugenics in *The Prevention of Destitution*, written as an alternative to the Minority Report.[12]

The interest generated by the publication of the Poor Law reports sparked a debate about whether eugenics could act as an explanation for pauperism. A committee of the Society explored the issue and read the two reports. It decided that only a thorough examination of the evidence would yield an answer to the question.[13] This aligned with Galton's attitude to eugenics and shows that the Society too had scientific pretensions. A research committee was formed in November 1912 with Edgar Schuster, the former Galton fellow, as chairman.[14] By conducting research, the Society was challenging the informal arrangement whereby it was meant to proselytize for eugenics, while the Galton Laboratory made the scientific discoveries.[15] The committee began collecting pauper pedigrees. It was granted permission to use the Poor Law records and had the support of relieving officers in its surveys. Among them was Ernest Lidbetter, soon to become a well-known figure in the eugenics movement. He uncovered that most recipients of the dole had relations in the same circumstances. Lidbetter believed that there was a class of chronic paupers in existence, expanding due to the well-meaning activities of philanthropists.[16] The environment was not to blame, as it had steadily improved over the past fifty years, while the number of destitute had not fallen.

Apart from investigating pauperism, the research committee took an interest in the Mental Deficiency Bill, which the Society had instigated, and the Criminal Law Amendment Act. The latter it admitted had no

[11] Sidney Webb, 'The Majority Report', *Eugenics Review* 2:3 (1910), p. 233.

[12] Christopher Shaw, 'Eliminating the Yahoo. Eugenics, Social Darwinism and Five Fabians', *History of Political Thought* 8:3 (1987), p. 527.

[13] SA/EUG/A3. Annual Report 1910–11, p. 18.

[14] SA/EUG/A5. Annual Report 1912–13, p. 18.

[15] Francis Galton, *Memories of My Life* (London: Methuen, 1908), p. 321.

[16] G. R. Searle, *Eugenics and Politics in Britain 1900–1914* (Leyden: Noordhoff International Publishing, 1976), p. 62.

obvious connection to eugenics, but its hope was that the Act would raise the standard of public morality.[17] The Society claimed to stand for 'moral, mental and physical improvement'.[18] Eugenics was a rising ideology with a political agenda. As a pressure group, the Society saw its role as inferring from biology how social legislation should be fashioned. Facts would guide it, especially the birth rate among social classes. Before the census of 1911, it contacted the chief medical statistician at the General Register Office, Thomas Stevenson, and persuaded him to classify the census data according to occupational categories. The 1901 census had listed the section of the economy in which an individual worked (agriculture, industry, retail, etc.).[19] The new descriptions provided a clearer indication of socio-economic status and enabled Stevenson to publish a table of fertility in the various social grades in 1920.[20] The Society wanted such information to be included in the official census results, allowing it to monitor class stratification with trusted statistics.

Class and the Political Basis of the Society

There is no doubt that in class terms the Society represented the bourgeoisie. In 1914 the largest single group among the membership consisted of gentlemen (23.6%), one in ten was a doctor (9.6%) or academic (9.6%), one in twenty (4.7%) had a title and more than four in ten were women (41.4%), hailing from similar family backgrounds as the men.[21] A Marxist reading was the first explanation for the *raison d'être* of eugenics.[22] Eugenics helped to justify bourgeois society and its institutions in a similar manner to political economy.

In a variant of this theory, eugenics stood for the world view and interests of the professional middle class. On the one hand, it had the traditional ruling class above it. It needed evidence for why landowners

[17] SA/EUG/A5. Annual Report 1912–13, p. 22.

[18] Pearson/4/19/2. The Eugenics Education Society, folio 1r.

[19] SA/EUG/B11. 'Notes on the early days of the Eugenics Society', p. 11.

[20] Simon Szreter, *Fertility, Class and Gender in Britain, 1860–1940* (Cambridge: Cambridge University Press, 2002), p. 13.

[21] Brown, 'Who were the Eugenists', p. 297.

[22] Anton Pannekoek, *Marxism and Darwinism* (Chicago: Charles H. Kerr & Co, 1912), pp. 28–29.

and old money were not suited to be in charge anymore. Science and modern thinking purported to show that the middle-class intelligentsia was of a finer mettle than those who had merely been born into wealth for generations. On the other hand, the rising working class constituted a challenge that had to be overcome.[23] Eugenics gave a cogent reason for why the new social order of which socialists dreamt could never be attained. It would only offer the 'unfit' even better chances of surviving, leading to yet more degeneracy.[24] Although a self-proclaimed socialist, Pearson had no connection to the labour movement or workers after the 1880s. He exemplified Bernard Semmel's and Geoffrey Searle's theory about social imperialism.[25]They asserted that in order to compete with rival imperialist powers, Britain had to introduce limited social reforms to ensure that its troops remained healthy and that its citizens did not get too disgruntled. According to this interpretation, the Fabians who concurred needed to temporarily put their socialism on hold to support an urgent national revival.[26]

A non-Marxist, but on the face of it similar, approach which doubts that eugenists truly could be socialists is Greta Jones's social hygiene explanation. It sees eugenics as a rival to socialism because eugenists claimed that it would be possible to improve society and eradicate venereal diseases, inebriety, mental deficiency and congenital illnesses without the parallel redistribution of wealth.[27] In 1925 the National Council for Combating Venereal Diseases changed its name to the British Social Hygiene Council. It cooperated with the Society and there was an overlap of members, in particular the leading light Sybil Gotto. This was also true for other organizations of the same type. Pauline Mazumdar duly explained the Society's founding as middle-class social reformers coming together to find an answer to the question of the urban poor. Its distinctive facet in this network was that it laid the blame for social deprivation

[23] D. MacKenzie, 'Karl Pearson and the Professional Middle Class', *Annals of Science* 36:2 (1979), p. 137.

[24] As argued by the scientist William C. D. Whetham as early as 1910. *Eugenics Review* 2:3 (1910), p. 245.

[25] Gary Werskey, *The Visible College* (London: Penguin, 1978), p. 28.

[26] G. R. Searle, *The Quest for National Efficiency. A Study in British Politics and Political Thought, 1899–1914* (Oxford: Blackwell, 1971), p. 63.

[27] Greta Jones, *Social Hygiene in Twentieth Century Britain* (London: Croom Helm, 1986).

on a single cause, namely heredity. To many adherents the implication was that no social reforms beyond eugenic ones were required. And with the differential birth rate in force, whereby the poorest had more children than professionals, the future looked bleak without eugenics.

When Geoffrey Searle published the first monograph on British eugenics in 1976, he regarded the movement as addressing anxieties held by the professional classes in the first two decades of the twentieth century. He did not belabour this point as much as the Marxists and did not see eugenics as complementary to Fabianism.[28] His work revealed the interconnectedness of eugenics and Edwardian politics. Eugenists reacted to colonial setbacks such as occurred during the Boer War 1899–1902, what was called 'deterioration' and the proposed reform of the Poor Law in 1909. As the Liberals turned leftwards and began implementing the type of policies contained in the People's Budget of 1909, it raised hackles among the well-to-do, who found a solace in eugenics. The budget introduced several taxes on wealth, as did the National Insurance scheme in 1911.

All the interpretations agree that the Society encapsulated a potent and modern ideology in which the bourgeoisie could believe. Although Galton did not participate directly at first, he was sympathetic to the idea of creating such an organization.[29] His friend and neighbour Montague Crackenthorpe, a barrister, became its president in 1909. Crackenthorpe was a Liberal Unionist and held some progressive views, such as being against the arms industry and supporting birth control as well as women's emancipation.[30] In time, the Society would adopt concomitant policies. From the beginning, it held regular meetings for members and supplied lecturers to outside groups for talks on eugenics. An ambitious early venture was convening a major international conference, planned for 24–30 July 1912. Likeminded associations had been established in other European countries and the United States, allowing the proposed conference to go ahead with the support of American, French and German eugenists.[31] Major Leonard Darwin, a former Liberal Unionist MP,

[28] G. R. Searle, 'Eugenics and Class' in Charles Webster (ed.), *Biology, Medicine and Society 1840–1940* (Cambridge: Cambridge University Press, 1981), p. 233.

[29] SA/EUG/B11. 'Notes on the early days of the Eugenics Society', folio 4.

[30] Geoffrey R. Searle, 'Eugenics: The Early Years' in Robert A. Peel (ed.), *Essays in the History of Eugenics* (London: Galton Institute, 1998), p. 27.

[31] SA/EUG/A3. Annual Report 1910–1911, p. 19.

became the president of the Society in 1911, which also made him the president of the proposed conference to be held at the University of London.

The First International Eugenics Congress

This event was the First International Eugenics Congress. In addition to the countries listed above, delegations came from Belgium and Italy, with eugenists from other European countries and Australia also attending. The congress was meticulously organized, including special discounts negotiated with all British and many foreign railway companies.[32] The acceptability of eugenics is shown by the patronage afforded to the congress. The Duchess of Marlborough, the American ambassador and the lord mayor of London all held receptions for delegates, while former Prime Minister Arthur Balfour spoke at the opening banquet. Balfour urged eugenists to stand together, as the public needed to be persuaded about the utility of science in solving the problems of the day. Establishment support for the congress was apparent from the lord chief justice, two bishops and serving members of the Liberal government (the home secretary Reginald McKenna and the first lord of the Admiralty Winston Churchill) along with international luminaries from the worlds of science and medicine acting as vice-presidents. Numerous local councils, scientific bodies, universities and the Ministry of Health were also present. The deliberations of the congress saw its executive committee turned into the Permanent International Eugenics Committee, preserving the cooperation which had been established by the delegates. It would sit between congresses and lay the groundwork for future joint efforts.

All who participated in the congress studied eugenics or took an interest in it, but delegates were not in complete agreement on individual issues. With hindsight, the most noteworthy paper was probably one on genetics and eugenics by Reginald Punnett, professor of biology at Cambridge University. Punnett stated that ‘feeblemindedness’ was overwhelmingly due to Mendelian inheritance. Therefore, strict segregation would rapidly put an end to a condition which eugenists considered among the worst social problems. This slant contrasted with the biometry practised by the Galton Laboratory. Three sequential pamphlets of

[32] SA/EUG/A83. First International Eugenics Congress, pp. 5–7.

the series Questions of the Day and of the Fray were devoted to countering the claims of the Mendelian geneticists. All bore the title *Mendelism and the Problem of Mental Defect* and were written by David Heron (part I from 1913), Pearson and the Swedish physiologist Gustav Jaederholm (part II published in 1914) and Pearson on his own (part III also published in 1914).[33] Especially part I was polemical, with the Galton Laboratory being totally absent from the congress. Punnett's paper was remarkable because he later furnished the strongest scientific argument against sterilization. In 1917 he showed that, provided mating occurred randomly, it would take 22 generations of sterilization to bring the incidence of feeblemindedness down from one to one-tenth of a per cent of the population.[34] J. B. S. Haldane and Sewall Wright, in the United States, gave support to this view.[35] Ronald Fisher, however, argued in 1924 that Punnett's assumptions were faulty. Punnett had based his calculation on random mating in the population, whereas Fisher more realistically estimated the number of feebleminded as depending on what percentage of them had two parents themselves classed as such. In addition, mental deficiency was polygenic, i.e. caused by a combination of genes. Under those assumptions, Fisher computed that segregation or sterilization of the feebleminded would lead to a 36% reduction in a single generation (Fig. 4.2).

Social Hygiene

The Society wielded influence as a pressure group in its early days. It teamed up with the National Association for the Care and Control of the Feeble-Minded in 1911 to produce a bill on that subject. The bill sought to lock up the mentally impaired who were a hazard to themselves or others, permitting segregation in institutions rather than indoor relief in workhouses. During the first general election of 1910, the Society had contacted scores of sitting MPs to ask if they supported checks on the

[33] Hamish G. Spencer and Diane Paul, 'The Failure of a Scientific Critique: David Heron, Karl Pearson and Mendelian Eugenics', *British Journal for the History of Science* 31:4 (1998), pp. 442–443.

[34] Daniel J. Kevles, *In the Name of Eugenics. Genetics and the Uses of Human Heredity* (Berkeley CA: University of California Press, 1985), p. 165.

[35] Cf. Elof Axel Carlson, *Genes, Radiation, and Society: The Life of H. J. Muller* (Ithaca NY: Cornell University Press, 1981), p. 125.

Fig. 4.2 Dining at the First International Eugenics Congress (reproduced from the Eugenics Society Archive, Wellcome Library, London)

fertility of this group.[36] Because the bill was drawn at random in the private members' ballot in 1912, it was introduced and voted on. This led to the Liberal government bringing in its own Mental Deficiency Bill later that year.[37] When it took effect in 1913, it provided for the mentally deficient and 'moral imbeciles', a term for those with criminal impulses who were judged also to be mentally deficient, to be institutionalized. Segregation in mental institutions was a useful form of control to prevent reproduction and moreover provided care for these vulnerable people. The Society had mobilized support in Parliament in advance and the bill passed easily on its second reading by 242 votes to 19. Labour MPs saw

[36] SA/EUG/B3. Feeblemindedness 1910–1911.

[37] Mathew Thomson, *The Problem of Mental Deficiency: Eugenics, Democracy, and Social Policy in Britain c. 1870–1959* (Oxford: Clarendon Press, 1998), pp. 38,43.

it as a sensible precaution and not as class legislation. *The New Statesman* also supported the bill.[38]

With the First International Eugenics Congress taking place in London and the passage of the Mental Deficiency Bill through Parliament in 1912, the doctrine had erupted onto the national stage. Chapters of the Society had been formed in Liverpool, Haslemere, Glasgow and Belfast. Eugenics was discussed briefly at the Labour Party conference in 1913.[39] The same year a Jewish committee was organized for members of that ethnicity.[40] The most famous committee member, the anthropologist Charles Gabriel Seligman, became its chairman. It also included Solomon Herbert, a Manchester doctor, who gave lectures to the Workers' Educational Association on biology.[41] He had promoted eugenics as complementary to socialism in the Independent Labour Party's journal the *Socialist Review* in 1909 and wrote the article 'Eugenics and Socialism' for the *Eugenics Review* in 1910.[42] Herbert did not believe that talented persons from the masses were necessarily upwardly mobile because, as he wrote, under capitalism rising out of one's class occurred through a combination of favourable circumstances over which the individual had little control.[43] All the members in the Jewish committee were doctors or academics. But some of their poorer brethren may have been barred from migrating to Britain through the influence of social hygiene. The Aliens Act of 1905 enabled the government to ban passengers on immigrant ships from entering if it was thought they were diseased, criminal or potentially a

[38] *The New Statesman. A Weekly Review of Politics and Literature*. Volume I. April 12, 1913 to October 4th, 1913, inclusive (London: The Statesman Publishing Company, 1913), pp. 134–135.

[39] *Report of the Thirteenth Annual Conference of The Labour Party* (London: Labour Party, n.d.), p. 97.

[40] SA/EUG/A5. Annual Report 1912–13, p. 30.

[41] See S. Herbert, *The First Principles of Heredity* (London: Adam and Charles Black, 1910). S. Herbert, *The First Principles of Evolution* (London: A&C Black, 1915).

[42] S. Herbert, 'Socialism and the New Science', *The Socialist Review. A Monthly Review of Modern Thought* I March–August 1908 (London: Independent Labour Party, n. d.), p. 308.

[43] S. Herbert, 'Eugenics and Socialism', *Eugenics Review* 2:2 (1910), pp. 118–119.

drain on the public purse.[44] There was an exemption though for those fleeing religious or political persecution.[45]

The social hygiene aspect of eugenics, whereby it was thought that urban poverty and ill health could be tackled without the redistribution of wealth, meant that many Labour stalwarts had prejudices against the doctrine. Intellectual left-wingers, however, often thought that eugenics could aid socialism, as long as social improvements also took place. This point of view, common among Fabians, was shared by other middle-class socialists, such as Harold Laski, Havelock Ellis and the physician Eden Paul. In 1909 Caleb Saleeby published a book called *Parenthood or Race Culture*. It was dedicated to Galton, who liked it, but regretted that it contained 'too much denunciation'.[46] Saleeby criticized right-wing eugenics for supporting laissez-faire economics, identifying wealth with evolutionary fitness and for its lack of interest in social amelioration. Such opinions were endemic within the Society, of which Saleeby was a leading member until he withdrew a few years later.[47] One of his opponents there had been the nationalist journalist Arnold White.[48]

In spite of the attitudes of most of the membership, it would be wrong to conclude that the Society was entirely reactionary. An editorial in the *Eugenics Review* in 1910 claimed that 'the large proportion of the children of the very poor are by no means sprung from a degenerate stock'.[49] The Society put forward some proposals for solving social problems. After it had achieved its initial goal on mental deficiency, its next focus was countering venereal diseases. Members of its council served on the Royal Commission on Venereal Diseases from 1913 and, on 2 April 1913, it assembled a delegation to meet with the education minister Charles Trevelyan to ask for eugenics or sex hygiene to be taught in schools.[50] It wanted teachers to be trained to give pupils information about the facts

[44] Kevles, *In the Name of Eugenics*, p. 98.

[45] Aliens Act 1905 Sect. 1 (3).

[46] Galton/3/3/7/20. Galton to Pearson 24 Sept 1908 folio 633 r. and 634 v.

[47] Richard A. Soloway, *Demography and Degeneration. Eugenics and the Declining Birthrate in Twentieth-century Britain* (Chapel Hill NC: University of North Carolina Press, 1995), p. 174.

[48] Daniel Becquemont, 'Eugénisme et socialisme en Grande Bretagne 1890–1900', *Mil neuf cent* 18:1 (2000), p. 56.

[49] *Eugenics Review* 2:3 (1910), p. 166.

[50] SA/EUG/A5. Annual Report 1912–13, p. 18.

of life on a one-to-one basis. Trevelyan said that experiments along these lines could go ahead. The Society was furthermore beginning to realize that war was calamitous from a eugenic perspective.[51] David Starr Jordan, an American eugenist known for his pacifist views and a vice-president of the British society, gave a talk to members on this subject in July 1913. When war erupted the following year, his misgivings were found to be correct since the middle classes were disproportionally killed on active service.[52] On inebriety, the Society's policy was severe: the use of indeterminate sentences for habitual drunkards. This was meant to act as a strong deterrent and to keep such people off the streets.[53]

The First World War

The outbreak of the First World War on 31 July 1914 would have a profound effect on the eugenics movement. The annual general meetings of 1916 and 1917 were cancelled.[54] There is no mention of the Jewish committee after 1914 and a ten-week course designed for social workers could not go ahead owing to pressure of work, unlike in 1913. The international cooperation recently formalized suffered a blow when the Permanent International Eugenics Committee was unable to meet in war-torn Brussels on 8 August 1914. The French society went into abeyance and there was no contact with the German society either until the 1916–1917 session, when news was received about its activities.[55] It appeared that the latter's energies had been thrown into the same causes as the British society: campaigning against venereal diseases and urging a high birth rate.[56] Both had also been working vigorously against alcoholism during the war.

In 1918 the Society was presented with a request for lectures to the Army on the part of Lord Gorell, the deputy director of staff education

[51] Paul Crook, *Darwinism, War and History. The Debate over the Biology of War from the 'Origin of Species' to the First World War* (Cambridge: Cambridge University Press, 1994), pp. 119–120.

[52] J. M. Winter, *The Great War and the British People* (Basingstoke: Palgrave Macmillan, 2003), p. 2.

[53] SA/EUG/A6. Annual Report 1913–14, p. 6.

[54] SA/EUG/A10. Annual Report 1917–18, p. 1.

[55] SA/EUG/B11. 'Notes on the early days of the Eugenics Society', p. 18.

[56] SA/EUG/A9. Annual Report 1916–17, p. 12.

at the War Office.[57] It was soldiers on the front line who had supposedly asked for the topic. This can be understood in the context of eugenics appearing in popular discourses which fascinated many. Health, popular science, politics and maybe visions of the future were all connected to it. The Society saw the desire for lectures as betokening increased racial responsibility among the public. Its policies intended to recreate Britain and its empire through adjusting the differential birth rate. Life should be made better for those it felt mattered to the wellbeing of the country. For instance, it tried to persuade the government to change the regulation that staff of the regular Army, reserves and territorial force who married were not entitled to a leaving allowance.[58] Eugenists believed in marriage and valued soldiers, especially in the reserves and territorial force, which were primarily made up of middle-class officers.

The war again confirmed the class alignment of organized eugenics in Britain. The Society accepted that the conflict would mean tax rises, but hoped that parenthood would be taken into account in their distribution.[59] In the second decade of the twentieth century, direct taxation was only levied on people with high incomes.[60] The Society circulated a letter to all MPs warning that income tax hikes could lead to a diminution of the fertility of this group. It tried to secure rebates for those on high incomes who had several children. It also played a leading role in the Professional Classes War Relief Council. They were likely to have offspring of greater than average value to the nation, it was argued. The Council's management consisted of well-known national figures and others representing allied associations. The lord mayor of London served as president and the archbishop of Canterbury as vice-president.[61] Leonard Darwin was chairman and Mrs Gotto took charge of perhaps its most notable activity, the provision of maternity assistance. The famous American banker and supporter of eugenics John Pierpont Morgan had died in 1913. His estate was undergoing probate and his son of the same name agreed to lend the

[57] SA/EUG/A11. Annual Report 1918–19, p. 7.

[58] SA/EUG/B11. 'Notes on the early days of the Eugenics Society', p. 20.

[59] SA/EUG/A8. Annual Report 1915–16, pp. 4,6.

[60] Greta Jones, 'Eugenics and Social Policy between the Wars', *Historical Journal* 25:3 (1982), p. 718.

[61] *Professional Classes War Relief Council. 13 & 14 Prince's Gate, S. W.* (no place of publication or date), p. 4.

Professional Classes War Relief Council two houses in Kensington, which were used as a nursing home during the war. By 1917 more than 300 babies had been born there.[62] Meanwhile, school fees had been paid for 617 children and many educational establishments persuaded to waive or reduce theirs.[63]

Sybil Gotto also took part in National Baby Week, in raising awareness of venereal diseases and in producing new posters for the Society. National Baby Week helped uncover bodily differences between the offspring of the affluent and the poor from very early in life.[64] Eugenists hoped to quantify these in order to justify social inequality as stemming from heredity. Among the tasks of the Professional Classes War Relief Council was facilitating emigration to the Dominions. This was a topic in which the Society took an interest. On 6 November 1917 it held a meeting on emigration. It requested information on three topics especially: the intermarriage between races, the composition of ethnic groups in each Dominion and the impact climate might have on fertility. Later, its policy became that it wished to encourage the emigration of family units and not of British men on their own. This was due to scepticism about intermarriage with other ethnic groups and because an emigrant would have fewer children if it took time to find a suitable spouse. If it preferred British men to marry women of the same race, it was largely because it sought to preserve the Empire and the mother country's bonds with it. Geographical knowledge of the kind sought could assist in determining suitable stocks for overseas colonization. The Society had branches in New Zealand (located in Christchurch, Dunedin, Nelson, Timaru and Wellington) and Australia (Melbourne and Sydney).

The report of the Royal Commission on Venereal Diseases was published in 1916. In conjunction with the National Council for the Control of Venereal Diseases, the Society was engaged in preventing ailments of this type. Major Darwin put a resolution before the Society proposing that their treatment should be made illegal for anyone who

[62] SA/EUG/A9. Annual Report 1916–17, p. 11.

[63] Bernard Waites, *A Class Society at War. England 1914–1918* (Leamington Spa: Berg Publishers, 1987), p. 51.

[64] C. D. Whetham, 'Eugenics and National Baby Week', *Eugenics Review* 9:3 (1917), p. 234.

was not a qualified medical practitioner.[65] The intention was to prioritize effective cures, as venereal diseases were often treated informally with quack remedies. The embarrassment involved made it less likely that sufferers would seek medical attention. Eugenists were concerned about the spread of contagion in the population generally and these illnesses might even affect heredity. Although Caleb Saleeby did not give a paper at the 1912 First International Eugenics Congress, his suggestion that a third category of eugenics, 'racial poisons', be introduced was discussed in the morning session on 30 July. Racial poisons were thought to stay in the body and affect the reproductive functions. Two presentations on alcoholism and one on hereditary syphilis were delivered. The speakers were respectively Dr Valentin Magnan, Jon Alfred Mjøen and Dr Henri Hallopeau. Saleeby criticized Magnan and praised Mjøen, who 'excited due and unprecedented attention'.[66] Saleeby and Mjøen were both temperance campaigners.

The Society was sensitive to the criticism that it opposed social reform. A prepared statement was therefore formulated on its relations to organizations dealing with environmental amelioration. It argued that better housing, maternity assistance and infant welfare were indeed valuable, but while decent material circumstances could foster good qualities, they were nevertheless incapable of producing them.[67] Less formally, Darwin would say that there were many organizations which pursued better living conditions, but only one which was engaged in improving heredity. This was a politic formulation which improved on what Galton had said in his autobiography, namely that charity for the least fortunate should instead be directed towards those who had the potential to benefit from it.[68]

The Society wanted redistribution towards middle-class parents. A first attempt occurred on 4 June 1918, when the Conservative MP Basil Peto attempted to gain income tax relief for men earning between £ 800 and £ 1000 a year for their third and subsequent children. Colonel Sir Hamar Greenwood made a stronger eugenic case for income tax to be calculated according to the number of family members. The leader of their party,

[65] SA/EUG/B11 'Notes on the early days of the Eugenics Society', p. 21.

[66] National Archives of Norway, Oslo. Diverse sunnhet og helse. Sosialdepartementet, 3. medisinalkontor saksarkiv. RA/S-1278/D/L0279. The Discussion of Alcoholism at the Eugenics Congress by Caleb Williams Saleeby, pp. 2–3.

[67] SA/EUG/A10. Annual Report 1917–18, p. 4.

[68] Galton, *Memories of My Life*, p. 322.

the chancellor Andrew Bonar Law, would not contemplate the latter as it meant a considerable loss to the Exchequer. Progress on the tax rebate was slow. Major Darwin had the opportunity to give evidence before the Income Tax Commission in October 1919.[69] A significant tax cut for parents was then implemented in 1927.

After hostilities ended in November 1918, the Society could regain its normal footing. Most recent activities had gone ahead, including a summer school in Oxford in August 1918 with its ally, now known as the Moral and Civil Education League. The combination of civics and eugenics was judged to be useful to the two hundred teachers, lecturers and social workers who attended. The Society hoped that eugenics would permeate many areas of life, making joint efforts worthwhile. As the First World War drew to a close, it had proved itself as a forceful pressure group. Branches had sprung up in Birmingham, Brighton, Cambridge and Oxford in addition to the places mentioned above.[70] Due to the vocational networks of many members, it was a presence in professional associations, universities and civil organizations. Its lobbying had been crucial in the passage of the Mental Deficiency Act of 1913. The First International Eugenics Congress had been a success, gaining momentum for the ideology. As the Society took stock, it had reason to be optimistic about the future.

Developments of the Early 1920s

Its new statement of policies and aims was called 'The Eugenics Education Society and Social Reconstruction'. Within a few years it would go on to become the primary manifestation of eugenics in Britain, overtaking the Galton Laboratory. This was symbolized by its change of name to simply the Eugenics Society in 1926. The document began by saying that exertions should be set in motion to spread the message. It had been encouraged by the high number of MPs agreeing with its stance before the passage of the Mental Deficiency Bill. Capable of mustering support inside Parliament, it included Conservatives Neville Chamberlain and Sir William Joynson-Hicks as members.[71] The document went on to outline

[69] SA/EUG/A12. Annual Report 1919–20, p. 6.

[70] SA/EUG/A11. Annual Report 1918–19, p. 8.

[71] Soloway, *Demography and Degeneration*, p. 35.

the current policies. It wished to see income tax and death duties divided by the size of the family, expenses in connection with maternity made tax deductible, compulsory health checks before marriage, prohibition of the immigration of 'mental or physical defectives', emigration to the Dominions in family units, the discouragement or prevention of parenthood on the part of 'degenerates' at home, better conditions for parents in state employment and housing, greater financial support for students in post-compulsory education and instruction for teachers in eugenics and sex education, which they could then pass on to pupils.[72]

The knowledge of heredity was often extended in tandem with eugenics. The Society told the public that soldiers crippled in the war would make excellent husbands and countered fears that any missing limbs would affect the children born to them. This was a theme in the propaganda film 'The Twins. A Drama' produced in 1919, in which the heroine is in love with a disabled veteran and marries him, while her less sensible twin sister marries a wealthy stockbroker whose grandfather died in an asylum.[73] The moral was that one should consider the pedigree and not just outward appearances. The Galton Lecture, instigated in 1913 and held annually on or close to his birthday 16 February, was resumed in 1919. The noted pessimist William Ralph Inge, Dean of St. Paul's Cathedral spoke on 'The Future of the English Race'. Though he was known to be an archconservative, his essay 'Patriotism' published in 1915 had criticized imperialism, militarism and pseudo-scientific theories of race.[74] During the war, the production of another film had commenced, largely at the behest of Mrs Gotto, who resumed her previous role as honorary secretary after the annual general meeting of 1919. This permitted another two-year stint, during which she promoted eugenics through making preparations for the Second International Eugenics Congress in New York.

Taking place in 1921, nine years after the London event, this signal five-day conference showcased the American contribution to the movement. The delegates spent a whole day at Cold Spring Harbor on Long Island. Here the Eugenics Record Office was located, a rival to the

[72] SA/EUG/A10. Annual Report 1917–1918, pp. 8–10.

[73] SA/EUG/D68. Films: proposed propaganda films on eugenics 1919. The Twins. A Drama.

[74] Paul Crook, *Darwin's Coat-tails. Essays on Social Darwinism* (New York: Peter Lang, 2007), p. 86.

Galton Laboratory in terms of influence across the world. Directed by Harry Laughlin, it stored an impressive number of genealogies of American families. The United States had a claim to be a pioneering eugenic country, through the successes of the doctrine in affecting policy. The push for sterilization which was to engulf British eugenics in the 1920s had begun in America. In 1909 the largest state California implemented this medical procedure and by 1944 had performed 17, 012 operations.[75] Sterilization enabled legislators and doctors to make decisions on which family lines they wanted to curtail. Leonard Darwin, who gave the opening address, was re-elected as the president of what was now called the Permanent International Commission of Eugenics.[76] Its core was Anglo-American to a greater extent than in 1912, after delegates from the former Central Powers had been banned from attending the congress. They would be readmitted to the Permanent International Commission in 1927.[77] Many American geneticists took part in the Congress, enhancing the credibility of eugenics which was after all a political ideology. Until the Second World War most geneticists were supporters of eugenics.[78] Human genetics and eugenics were almost indistinguishable.

In 1922 the beginnings of what would become a eugenic campaign for contraception arose. On 14 March that year a speaker called Harold Cox addressed a members' meeting on 'The reduction of the birth-rate as a necessary instrument for the improvement of the race'.[79] The Society was engaged in small-scale endeavours to introduce sex education in schools, but had not yet made the leap to condoning sexual relations which were separate from reproduction. This would occur later in the 1920s. It was obvious that if eugenists really desired to improve the British population, piecemeal replacement of a few members of the residuum by an equivalent number of professionals would not avail much. Cox called for the use of contraception by everyone (especially poor mothers) and for the

[75] Paul Weindling, 'International Eugenics. Swedish Sterilization in Context', *Scandinavian Journal of History* 24:2 (1999), p. 187.

[76] SA/EUG/A14. Annual Report 1921–22, [p. 3].

[77] Stefan Kühl, *For the Betterment of the Race. The Rise and Fall of the International Movement for Eugenics and Racial Hygiene* (Basingstoke: Palgrave Macmillan, 2013), p. 45.

[78] Mazumdar, *Eugenics, Human Genetics and Human Failings*, p. 58.

[79] SA/EUG/A14. Annual Report 1921–22, [p. 4].

sterilization of the 'feebleminded'.[80] In the audience were Marie Stopes and Sybil Gotto (now Mrs Neville-Rolfe) who both spoke at length in the ensuing debate. Stopes was famous for her advocacy of contraception and Neville-Rolfe was of like mind, owing to her work on preventing venereal diseases.

The council of the Society sent a memorandum on birth control to the minister of health in 1922. It asked for information about these techniques to be given to all married people in state-aided homes, but also warned about other organizations becoming involved as this might lower fertility generally.[81] Here lay the essential conundrum for eugenists. Just as with other Malthusian checks, as Galton observed, if the prudent applied them and the feckless did not, the 'problem' would be exacerbated. But the eugenic implications of contraception were so decisive that the topic could not be avoided. When birth control was accepted in the Society's next policy document, safeguards such as only married women being informed about and given access to contraceptives were built in.

By the middle of the 1920s, the pressure-group politics of the Society had acquired an academic leavening. In 1923 this was exemplified by the Eugenics Education Society being permitted to hold members' meetings at the headquarters of the Royal Society. Sir Charles Sherrington, the latter's president, was a member of the Eugenics Education Society and had been on the first advisory committee of the Eugenics Laboratory. He was a guest at the Galton Dinner that year alongside the rector of Imperial College, Sir Thomas Holland, and Sir William Beveridge, director of the London School of Economics.[82] Both institutions offered courses with eugenic themes, the latter convening a course called 'Heredity and society' in the first half of 1924. Its lecturers Julian Huxley, then a fellow of New College in Oxford, Ronald Fisher and Donald Ward Cutler, Fisher's colleague at the agricultural institute Rothamsted Experimental Station in Hertfordshire, were biologists. Cyril Burt was a psychologist and Major Darwin was still in charge of the Eugenics Education Society. Perhaps Darwin's lecturing was the reason the Society received a grant from the London School of Economics that year. Research conducted under the

[80] Harold Cox, 'The Reduction of the Birth Rate as a Necessary Instrument for the Improvement of the Race', *Eugenics Review* 14:2 (1922), pp. 91–92.

[81] SA/EUG/A14. Annual Report 1921–22, [p. 5].

[82] SA/EUG/A16. Annual Report 1923–24, p. 7. Mazumdar, *Eugenics, Human Genetics and Human Failings*, p. 65.

Society's auspices was regarded with enough promise to be awarded £ 100 from the Medical Research Council. The following year the Royal Society gave £ 100, while the London School of Economics maintained its previous grant.[83] In 1923 the Society added lustre to its membership roll when it began calling its members 'fellows', the term 'member' henceforth denoting those previously known as 'associates'.

The Society did not care greatly about racial differences. On the prompting of Cora Hodson, its secretary during the 1920s, and Leonard Darwin though, it did tentatively consider miscegenation. In 1924 the qualities immersed in the building up of 'the great British Commonwealth' were discussed by Darwin at the British Empire Exhibition. A week later, on 21 May, the geologist John Walter Gregory of Glasgow University gave a paper on 'White Australia', which was described as 'very interesting'.[84] Gregory said Britons were doing well even in tropical climates like Australia due to the people's great adaptability. He warned against importing another race for manual labour, as it would inevitably cause miscegenation. The following month the anthropologist E. N. Fallaize spoke at another members' meeting on the suitability of different 'native' races for specialized tasks. In 1926 Darwin argued against miscegenation in his book *The Need for Eugenic Reform*, saying that it might give rise to individuals worse than both parents stocks.[85] However, there were no ethnic barriers to the expansion of eugenics worldwide. As part of an international movement, the Society was appreciative of efforts made outside Europe and North America.

The Society now had a dual purpose as a pressure group and attempting to be a learned association. It was making progress through its efforts and innovations, among which was the completion of the film under the supervision of founder Sybil Neville-Rolfe.[86] The use of this film at propaganda meetings was effective in bringing the message to wider attention, especially due to the novelty of the format. Its lectures were a mixture of pressure-group politics and academic content pertinent to eugenics. An example of the former was Charles Vickery Drysdale's

[83] SA/EUG/A17. Annual Report 1924–25, p. 7.

[84] SA/EUG/A16. Annual Report 1923–24, p. 5.

[85] Dan Stone, *Breeding Superman. Nietzsche, Race and Eugenics in Edwardian and Interwar Britain* (Liverpool: Liverpool University Press, 2002), p. 107.

[86] SA/EUG/A18. Annual Report 1925–26, pp. 1–2.

presentation on 'Possible means of ascertaining the relative racial values of different social strata' on 10 December 1925. An example of academic content was when Bronislaw Malinowski, the anthropologist, gave a lecture on 'Mate selection in primitive society' in early 1926. The scientific strands may have provided a rationale for the London School of Economics renewing its grant and the Royal Society giving another £ 100.

A Change of Name and New Policies

The year 1926 was a crucial one for eugenics. On 15 November that year the Society became incorporated, making it a legal entity and allowing it to own property. It had moved to new premises in Belgravia in September, having previously shared accommodation with the Royal Anthropological Institute. It was advised by Wing Commander Archibald James, a future Conservative MP, on incorporation.In connection with this legal device, it changed its name from the Eugenics Education Society to simply the Eugenics Society. The new name acknowledged that its remit had expanded from merely proselytizing to conducting research and functioning as a learned association. Its output was sometimes better known than what was published by the Galton Laboratory. This was especially true for its studies of pauper pedigrees. Ernest Lidbetter's results seemed to show inherited defects for five or six generations among the destitute and intermarriage between various types of defectives, supposedly producing a wider range of undesirable hereditary conditions.[87]

It was also in 1926 that a policy document called 'An Outline of a Practical Eugenic Policy' was agreed. This set the scene for the rest of the interwar period. The credo witnessed the acceptance of contraception as a eugenic method, though it remained an awkward issue. The initial description stated both that it was 'highly beneficial' and 'constitutes a potential danger to the Race'.[88] However, with as many as five cases mentioned as justifying its use, it was clear that birth control would be a prominent feature of eugenics in the future.A dual campaign was to be undertaken, commending contraception if utilized by married women

[87] SA/EUG/D80. Genetics Congress, Berlin 1927. Secretary to Professor Baur 1 Mar 1927.

[88] SA/EUG/A18. An Outline of a Practical Eugenic Policy, p. 3.

of the poorest classes, but condemning it if used to limit families who were 'healthy both in mind and body'.[89] The phrase in reality meant the sort of people (especially professionals) whom the Society believed to be of eugenic worth. The Society would make life better for mothers of illegitimate children through making their court-ordered payments easier to obtain. This might have helped to remove some of the stigma of illegitimacy—another modernization of sexual mores.

It was a different policy which would prove dramatic to the Society. For the first time, it endorsed sterilization as a tool of 'negative' eugenics. It still recommended segregation as a means of preventing procreation by habitual criminals, confirmed drunkards and the unemployable (a danger sign to working-class people). But for sufferers from heritable diseases and the mentally defective, it henceforth recommended sterilization. This was to be 'voluntary', though the description went on to say that the consent could be given by his or her family or guardian if the person affected was not able to make the decision. Even more chillingly, a limited amount of pressure could be applied to make sure that the individual did not withhold consent 'unreasonably'. Sterilization would soon become the Society's signature policy and change its future direction. From being merely a right-wing pressure group with a modicum of progressive ideas to boot, it took on the mantle of a potentially perilous organization. It was not certain that only individuals with faulty heredity would be the targets of these measures. The policy referred to 'all afflicted with some definite and serious defect which may be inherited'. This formulation was too broad to ensure that only persons with medical conditions of a genetic nature were included. Cyril Burt claimed in 1919 that 21% of working-class parents whom he had tested in Liverpool had the intelligence of an average eleven or ten-year-old.[90] This might be construed as 'hereditary mental defect'. Sterilization was regarded as worthwhile because it would allow hereditarily defective men and women to live otherwise normal lives in the community. It was further tagged as 'voluntary' sterilization, the volition involved being in many cases that the inmate would consent to his or her sterilization in return for being allowed liberty.

[89] SA/EUG/A18. An Outline of a Practical Eugenic Policy, pp. 2, 4.

[90] Cyril Burt, 'Problems of Subnormality', *Eugenics Review* 10:4 (1919), p. 226.

In the Society's thinking, there was a hierarchy of undesirability or disease with sterilization being intended for the most serious and identifiable cases. Stocks which were no worse than 'inferior' would be encouraged to restrict their breeding by whatever means, but would not be compelled to do so. In a few instances, the state would again use pressure, this time by cutting off charitable aid, alternatively ordering segregation for those living in 'uncivilized' conditions.[91] It was noted in the policy document that not all members of the council supported everything which it comprised. It is likely that this caveat related to birth control and sterilization. Members existed who were interested in the ideology and liked discussing it, but who did not necessarily wish to implement anything contentious. Such stragglers were a minority. The majority were social Darwinists, believing either that the struggle for life was a boon to society or at least that the redistribution of wealth harmed the race. Money almost automatically accrued to whoever had the right heredity. This was despite most eugenists having an academic or professional background rather than being in business.[92] In the early years its journal the *Eugenics Review* was a mouthpiece for such attitudes. Almost all writers took for granted that middle-class people were 'fit' in the Darwinian sense, while workers were often 'unfit'. They conflated social class with 'fitness'. Herbert Spencer had stated that by 'fit' he meant only those suited to a particular environment and confirmed that what might be seen as 'inferior' often caused survival.[93] This could be lack of scruples.

The Society constituted the most visible expression of social Darwinism in Britain in the early twentieth century. A fraction of the membership in its first two decades were 'reform' eugenists who could be relatively progressive, seeing the doctrine as part of a package of social improvements. The two distinct wings were, however, equally likely to support sterilization as an instrument of 'negative' eugenics from the mid-1920s. It was especially this facet which robbed eugenics of any constructive benefits it might otherwise have had. The other policies were hardly progressive either. On immigration the Society proposed enforcing restrictions of a far-reaching kind, as anyone judged likely to

[91] SA/EUG/A18. An Outline of a Practical Eugenic Policy, pp. 1–3.

[92] Searle, *Eugenics and Politics in Britain 1900–1914*, p. 113.

[93] Robert C. Bannister, *Social Darwinism. Science and Myth in Anglo-American Thought* (Philadelphia: Temple University Press, 1979), p. 48.

lower the quality of the national stock was to be excluded.[94] Such a broad formulation could be used to effectively end immigration, as it is seldom the well-adjusted who begin new lives in other countries. The effects of miscegenation were to be studied in order to formulate a policy thereon. Meanwhile, great caution was to be exercised since if intermarriage became common, the process would be irreversible.[95] Lastly, the Society repeated its insistence that it was concerned with environmental amelioration too. It did not wish to replicate the activities of the many organizations which were doing good work in this manner, but said that it appreciated them.

The Society had made waves with its private member's bill to institutionalize the 'feebleminded' in 1912. In 1927 it urged an update to its terms. The story began when many members joined by Hensley Henson, the bishop of Durham, wrote a letter to *The Times*, complaining that there was nothing in the law to prevent a mentally defective person from marrying. This attracted the attention of Conservative MP Brooks Crompton Wood, who in his maiden speech asked for the Mental Deficiency Act to prevent this. During the parliamentary debate, it came to light that there were provisions in the Lunacy Act 1890 which would prevent people of unsound mind from marrying. But the Mental Deficiency Act was amended to encompass people who had acquired their affliction in later life.[96] On a less central issue, the Society was busy trying to bring in the abovementioned tax breaks for costs in connection with maternity and for raising children generally. Both would be of much greater value to the higher income brackets. The Society received support from Labour MPs Commander Joseph Kenworthy, Hugh Dalton and Willie Graham in 1927 for income tax to take the number of children in a family into account.[97] All three of them had middle-class backgrounds, but concurrently the Society was showing increased interest in the working class as an audience for its propaganda. Eugenics was a popular topic for meetings at all sorts of associations, and the Society

[94] SA/EUG/A18. An Outline of a Practical Eugenic Policy, p. 5.

[95] This formulation was a better worded version of the Norwegian eugenist Jon Alfred Mjøen's booklet from 1908, which Leonard Darwin endorsed. RA/S-1278/D/L0279. 'Program for Racehygiene (sic)', p. 6.

[96] SA/EUG/A20. Annual Report 1927–28, p. 2.

[97] SA/EUG/D106. Income tax amendments. Letters from secretary dated 3 May (Graham), 28 May (Dalton) and 14 Jun 1927 (Kenworthy).

was pleased to announce that more inquiries were coming from cooperators and left-wing political groupings. It was also appealing for funds to simplify the literature to make it clearer to the uneducated.

That Alexander Carr-Saunders, a 'reform' eugenist, had replaced the 'mainliners' Darwin and Burt on the 'Heredity and society' course might be seen as an omen of the change to come.[98] The leadership was recruited from both the 'mainline' and 'reform' wings, with officers usually wishing to appear fair-minded and objective, whichever tendency they favoured. Although the Society was markedly political, it did not ally with a political party and instead aimed to bring its influence to bear on any organization and the general public. When Charles Vickery Drysdale argued along neo-Malthusian lines that contraception would be beneficial in restricting the number of births among 'the less fit', he received the support of the editor of the *Eugenics Review*.[99] But when he stated that neo-Malthusianism was strictly scientific, whereas socialism was based on class hatred and devoid of any such principles, the editor felt the need to remind readers that the Society was neutral on such matters. This had not happened in the early days, when a high proportion of articles in the journal had shared Drysdale's attitude (Fig. 4.3).

Leonard Darwin Steps Down

In 1928 Leonard Darwin stood down as the president after 18 years at the helm. He had been a commanding presence and had also served as president of the Permanent International Commission of Eugenics 1912–1927, by then renamed the International Federation of Eugenics Organizations.[100] His surname Darwin had added lustre to the enterprise. His book *What is Eugenics?* published in October 1928, the same month that he resigned, showed that he intended being a continuing presence in the movement. The book sold well and was distributed to libraries across the country with financial support from the Society, generating renewed interest.[101]

[98] SA/EUG/A19. Annual Report 1926–27, p. 4.

[99] C. V. Drysdale, 'Scientific Bases of Birth Control', *Eugenics Review* 20:3 (1928), pp. 173,175.

[100] 'International Eugenics Federation: Report of the Annual Meeting', *Eugenics Review* 20:3 (1928), p. 185.

[101] SA/EUG/A21. Annual Report 1928–29, p. 6.

Fig. 4.3 Leonard Darwin, president of the Eugenics Society 1911–1928 (reproduced from the Eugenics Society Archive, Wellcome Library, London)

In summing up Major Darwin's contribution, Professor Ernest Macbride, a marine biologist, called him the greatest single influence upon the movement in the last 18 years. He had been responsible for policy successes too, such as the recent budget containing a doubling of the tax rebate available on each child. It was a result of Darwin's submission to the Income Tax Commission in 1919, his letters to the press and his correspondence with parliamentarians. John Macnicol explains the measure as the government wishing to give tax cuts to professionals after instituting social insurance for the working class in 1925.[102] By 1928 the Eugenics Society enjoyed an enhanced role in popular debate and was able to get its points across directly through being invited to give evidence to departments and public committees. Darwin addressed the Ministry of Health and the Board of Education on making the next census, due in 1931, a relevant tool for charting fertility across different social groups. As the Royal Society endorsed his request, the Registrar-General decided to implement them. It became natural to include the Eugenics Society in health and social policy commissions and parts of the membership had expertise which allowed them to represent other organizations. Governments never completely embraced eugenics though, keeping it at an arm's length.[103] The clearest example would be sterilization.

The 1930s has been described as the heyday of the eugenics movement, and it was the decade when the Society began formulating a new vision of what its doctrine entailed. Leonard Darwin's old-fashioned views were still prevalent, but received less prominence than before. Instead, it attempted to show that eugenics was compatible with a range of political stances including left-wing opinions, underrepresented at an earlier stage. There had always been a minority of eugenists who were socialists though and now that combination was put forward in seeking to disabuse outsiders of the notion that eugenics was only for Establishment types. As time went by, the emphasis on convincing left-wing opinion was redoubled.

The Society was placed on a better financial footing through the death of its anonymous benefactor Henry Twitchin in 1929. Twitchin, who died on the Riviera, was an Australian sheep farmer who had given £ 1,

[102] John MacNicol, 'The Voluntary Sterilization Campaign in Britain, 1918–1939', *Journal of the History of Sexuality* 2:3 (1992), p. 426.

[103] Chris Renwick, 'Eugenics, Population Research, and Social Mobility Studies in Early and Mid-twentieth Century Britain', *Historical Journal* 59:3 (2016), p. 849.

000 to the Society yearly since 1925.[104] It appeared that he had first made a will in favour of the Society as early as 1912. The legacy, though it took time to materialize and was reduced somewhat by the poor economic conditions worldwide, gave it much freer play in its activities. The Society was looking into founding a hospital where sterilization could be carried out. The uncertainty over the legality of the procedure pushed it towards focusing on sterilization as its key demand. It wanted the law to state unequivocally that sterilization was allowed when the patient himself or herself consented. As it applied first and foremost to mental defectives, the decision could be left to a relation or guardian, which meant that it was not 'voluntary' in reality. A propaganda committee was established to try to make the public ready for such a step.

Mental Deficiency and 'Voluntary' Sterilization

1929 was the year the Mental Deficiency Committee, led by the civil servant Arthur Wood, reported. It had been tasked with estimating the number of mentally deficient people in the population and recommending remedies. Originally set up by the Board of Education in 1924 to identify children of subnormal ability, it had been extended the following year to encompassing adults too. The psychologist and physician E. O. Lewis, later a member of the Society, worked on estimating the number of cases. Among the eugenists on the committee were Cyril Burt, at that point the London County Council's psychologist, and Alfred Tredgold, a psychiatrist who lectured at the University of London. Lewis surveyed six different areas, a mixture of rural and urban, and found that the incidence of 'feeblemindedness' was 8.57 in a 1000.[105] The Committee thereby arrived at a figure of 249, 000 mentally deficient people in England and Wales in 1927.

Having discharged this responsibility, Lewis was appointed a commissioner of the Board of Control in 1930. He described the bottom ten per cent of the population as the 'social problem group', characterized by pauperism, recidivism, alcoholism, mental illness and mental deficiency. Thus, from a eugenic point of view, mental deficiency might be a symptom of an underlying wider pathology, rather than being an

[104] SA/EUG/A22. Annual Report 1929–30, p. 3.

[105] Mazumdar, *Eugenics, Human Genetics and Human Failings*, p. 198.

independent variable. The report recommended further segregation in handling 'feeblemindedness'. This was in line with eugenic opinion, but as institutionalization could not possibly apply to everyone, the Society additionally wished to sterilize as many of those remaining in the community as possible. It judged that the high rate of mental deficiency reported and the coverage already taking place in the press, made the time ripe.

The 1929 Local Government Act provided further opportunities for the Society. An interpretation of it by the Ministry of Health authorized information about contraception to be given for medical reasons. This memorandum 153/MW was initially only available upon request.[106] After it had resolved its inner tension on the issue, the Society became an enthusiastic proponent of birth control. It would fund Walworth Women's Welfare Centre, situated in a deprived area of south London.[107] Likewise, it supported initiatives in other localities for the same target group of poor, married women. It would not brook much wider dissemination than this, an attitude replicated by the 1930 Lambeth Conference of the Anglican Church which permitted the use of birth control within marriage.[108] However, in the same year the Catholic Church prohibited contraception and sterilization, through its encyclical 'Casti Connubii'. It did not condemn eugenics outright, but it warned against eugenic grounds for preventing men and women who were physically capable of marriage from contracting it. The Local Government Act empowered local authorities to make decisions in the matter of the segregation of the mentally deficient. The Society hoped that its membership would be proactive in bringing known cases to the notice of their local councils.

As the Society's main activity in 1930, fast becoming a defining attribute, was the propagation of 'voluntary' sterilization, it attracted spirited opposition from then onwards. Eugenics and Catholicism almost became polar opposites. In 1929 the research department of the Labour Party had sent a memorandum to the Ministry of Health opposing sterilization. Those who agreed with sterilization believed naively that only people with genuine hereditary afflictions would be affected, that it would

[106] Lesley A. Hall, *Sex, Gender and Social Change in Britain since 1880* (Basingstoke: Macmillan, 2000), p. 117.

[107] London Metropolitan University (LMU). TUC Collection. HQ 766. Society for the Provision of Birth Control Clinics Annual Report, 1931–1932, p. 15.

[108] Patrick T. Merricks, *Religion and Racial Progress in Twentieth-century Britain. Bishop Barnes of Birmingham* (Cham: Palgrave Macmillan, 2017), pp. 52–53, 58.

be voluntary, or they were indifferent to people from society's lowest tiers being mutilated. Many middle-class socialists believed in sterilization as a last resort, wishing to help the unfortunate but also hoping to limit the numbers requiring assistance in the first place.

Because the matter was the Society's chief policy initiative, its subcommittee for propaganda reconstituted itself the Committee for Legalizing Eugenic Sterilization after the 1930 annual general meeting.[109] It decided that when the Society's sterilization bill was laid before Parliament, it should apply only to mental defectives. In October 1930 the psychiatrist C. P. Blacker was appointed general secretary. This would turn out to be a pivotal moment. Blacker, who was left leaning politically, would put birth control, research on sexuality and population studies at the heart of the Society's mission, though also sterilization to a less benign effect. The next few decades would bear his imprint on the Society, similarly to how Leonard Darwin had been the main figure in the 1910s and 1920s.[110]

1931 witnessed intense political drama with the end of the Labour government in the context of bank collapses in Europe and great economic uncertainty at home. The National government was created, still with Ramsay MacDonald as prime minister, soon dismissed as Labour leader. In the general election of that year, the supporters of the government decimated Labour. Before all this had taken place, on 21 July 1931 the Society was able to introduce its scheme for sterilization as a private member's bill.[111] The proposer was Archibald Church, a major in the army and a trade unionist for the Association of Scientific Workers.[112] He sat in the Commons for Labour, but could not get many of his colleagues to support the bill. He had reluctantly joined the party and subsequently attacked it in editorials in *Nature*.[113] Labour MPs instead rallied against sterilization, almost certainly frightened by Church's assertion that it was a prelude to compulsory sterilization.'Voluntary' sterilization was refused leave to proceed by 167 votes to 89.

[109] SA/EUG/A23. Annual Report 1930–31, p. 6.

[110] Richard A. Soloway, 'From Mainline to Reform Eugenics—Leonard Darwin and C. P. Blacker' in Peel (ed.), *Essays in the History of Eugenics*, p. 52.

[111] SA/EUG/C67. Letters to Archibald Church 29 Apr 1931 and 10 Jun 1931.

[112] Jeremy Hugh Baron, *The Anglo-American Biomedical Antecedents of Nazi Crimes. An Historical Analysis of Racism, Nationalism, Eugenics and Genocide* (Lewiston NY: Edward Mellon Press, 2007), p. 132.

[113] Werskey, *The Visible College*, p. 41.

The efforts to secure sterilization had become vital to the Society. It clarified that Major Church had only been speaking for himself in saying that compulsory sterilization would follow 'voluntary' sterilization.[114] The bill's main opponent was Dr Hyacinth Morgan, another Labour MP, with a practice in the poor London district of Kilburn. His argument for refusing permission to read the bill was that it constituted anti-working-class legislation. This was a correct assessment, but the Society resisted the categorization. It thought that some MPs had been reluctant to back the bill owing to Catholic opposition in their constituencies and claimed that it did not represent class legislation.[115] Its Committee for Legalizing Eugenic Sterilization would accordingly work to convince everyone that the poor had nothing to fear. A case in point was perhaps that the Women's Cooperative Guild had passed a resolution in May 1931 calling for compulsory sterilization.[116] Although some sections of the labour movement saw sterilization as an attack on them, others apparently did not. The variable which affected this the most was gender. Women's organizations were concerned about the sexual abuse of women and children. Sterilization would have been ineffective against this scourge, but many mistakenly saw it as a tool against rape.[117] Members of the Women's Cooperative Guild also often used abortifacients and birth control, which gave them a personal interest in sterilization.[118]

The Outreach to the Labour Movement

The Women's Cooperative Guild became one of the main organizations to which propagandists for the Society spoke. Its branches and other audiences of working-class women were often convinced about the merits of

[114] SA/EUG/D204. The president and general secretary to the editor of the *Spectator* 31 Jul 1931.

[115] SA/EUG/A24. Annual Report 1931–32, p. 6.

[116] LMU. HD 3423. Annual Report of the Women's Cooperative Guild 1931–1932, p. 23. Annual Report of the Women's Cooperative Guild 1930–1931, p. 24. A resolution asks for the segregation of persons found guilty of criminal assaults on children. Annual Report of the Women's Cooperative Guild 1937–1938, p. 22 asks for the segregation of 'mental defectives' guilty of assaults on young children.

[117] MacNicol, 'The Voluntary Sterilization Campaign', p. 423.

[118] Lucy Bland, *Banishing the Beast. Sexuality and the Early Feminists* (New York: The New Press, 1995), p. 208.

sterilization. If framed in the right terms, diverse associations could be brought round to support this measure. In 1932 the Cambridge Union overwhelmingly passed a resolution in favour of sterilization for mental defectives. But the new policy on sterilization alienated some groups with which the Society had earlier been on a better footing.

From the beginning, propaganda work had been undertaken among intellectual socialists, working men's colleges and a few labour movement organizations. Much of it had taken place in conjunction with the Workers' Educational Association, but in 1931, once sterilization had been put onto the agenda, it reacted wearily to the Society's overtures for closer relations.[119] The Workers' Educational Association did hope that the Society would affiliate to itself and pay a fee for this subscription. However, it was unwilling to guarantee the Society opportunities to lecture to its audiences. Speakers must be in broad sympathy with working-class ideals, it stated, regardless of the undoubted qualifications which Society members possessed in science.[120] As this had not been a problem before, it is likely that sterilization came between the two organizations, which had previously cooperated well.

The issue advanced the following year when the Ministry of Health granted a departmental committee to examine sterilization. It was appointed on 9 June 1932 and led by the civil servant Laurence Brock. Ronald Fisher was selected to serve on to the committee, which proved the Society had clout as he had not yet been made head of the Galton Laboratory.[121] Five physicians who all belonged to the Society, including C. P. Blacker, gave evidence to the committee on a single day in 1933. This hinted that the inquiry was in the Society's hands. Its political posturing had persuaded the government to establish the committee in the first place, its remit was confined to sterilization and several Society members were allowed to influence it. Brock himself had conferred with

[119] The first took place on 12 March 1913 to the Workers' Educational Association, Birmingham. Another 10 lectures followed before the end of 1920. See SA/EUG/A5. Annual Report 1912–13, p. 27. SA/EUG/A6. Annual Report 1913–14, pp. 20–21. SA/EUG/A9. Annual Report 1916–17, p. 11. SA/EUG/A10. Annual Report 1917–18, p. 7. SA/EUG/A13. Annual Report 1920–21, p. 5. SA/EUG/E16. Belfast Branch c. 1912–1913. Draft of Annual Report, p. 4.

[120] SA/EUG/D249. Workers' Educational Association 1920–1931. Ernest Green to Cora Hodson 13 May 1931.

[121] *British Medical Journal* 27 Jan 1934, p. 165.

the proposer of the Voluntary Sterilization Bill, Archibald Church, to make it as effective as possible.[122] Perhaps symbolic of a new phase starting was that the Society revisited its founding place Caxton Hall. It hosted a conference on sterilization there on 23 May 1932 with many representatives of local authorities in attendance.[123] 53 local councils had officially supported Church's bill in 1931. They tended to be southern and rural, with Norwich the only one under Labour rule.[124]

The confidence of the Society was, not surprisingly, well-placed. Laurence Brock corresponded with C. P. Blacker during the committee's work and when the report was published on 18 January 1934, it did indeed recommend 'voluntary' sterilization. It went further than the Society in also suggesting that carriers, who did not themselves display signs of deficiency, should be sterilized.[125] This made the measure even more nebulous and dangerous to individuals from poverty-stricken families. The Society was pushing for sterilization at all costs. Soon after the report came out, it formed the Joint Committee for Voluntary Sterilization with three other organizations which were involved in taking care of the 'feebleminded'.[126] The Joint Committee shared premises with the Eugenics Society and was in fact a front organization for it. Its funds came from the Society, which thought it would stand a better chance of success if a groundswell for sterilization took place.

C. P. Blacker tried to reassure the public that sterilization would not be used as an instrument of class tyranny. In a seven-minute address the day the report was published, he underlined this important point.[127] Equally, sterilization should not become a means for healthy individuals to avoid parenthood. An attempt was made to debate sterilization in the House of Commons in February 1934, proposed by Arthur Molson MP and seconded by Wing Commander James .[128] On 23 May 1935 the Joint

[122] Bradley W. Hart and Richard Carr, 'Sterilization and the British Conservative Party: Rethinking the Failure of the Eugenics Society's Strategy in the Nineteen Thirties', *Historical Research* 88: 242 (2015), p. 726.

[123] SA/EUG/A25, Annual Report 1932–33, p. 3.

[124] Jones, *Social Hygiene in Twentieth Century Britain*, p. 101.

[125] SA/EUG/A25, Annual Report 1932–33, p. 3.

[126] SA/EUG/A27. Annual Report 1934–35, p. 3.

[127] SA/EUG/D50. Summary of Brock Report.

[128] Stephen Trombley, *The Right to Reproduce. A History of Coercive Sterilization* (London: Weidenfeld & Nicolson, 1988), p. 129.

Committee and two bodies representing local authorities called on the minister of health, Sir Hilton Young, and asked him to implement a sterilization bill which the deputation had worked out in advance.[129] Young welcomed his visitors, but politely declined to fall in with their wishes, citing the lack of parliamentary time and that public opinion was not ready for sterilization. Hence the government acted as a brake in spite of having allowed the Brock Committee to go ahead. The support of the British Medical Association would have been encouraging, but it was never persuaded about the benefits of sterilization.[130] A rally against sterilization supported by 200, 000 Catholics had taken place in Glasgow in 1934.[131]

In 1935 the Society founded a separate Workers' Committee for the Legalizing of Voluntary Sterilization. It consisted entirely of members of the Labour Party and was chaired by the American physician Caroline Maule. The original idea of a separate committee for socialists, though not restricted to sterilization, may have come from Ellen Wilkinson, known as Red Ellen.[132] The members were largely unknown, except Frida Laski, the wife of Labour academic, Harold. All the committee members were female, though Ernest Thurtle MP was willing to serve on it.[133] Its task was to argue the case that sterilization would not be inimical to working-class people. Between March and May that year, Maule and the leading propagandist of the Society, the nurse Hilda Pocock, held ten meetings with Labour women's sections. In total, the Society spoke at 25 meetings of Women's Co-operative Guilds and two Labour Party League of Youth branches in 1935.[134]

Such efforts were by now typical. In 1935 the Women's Cooperate Guild abandoned its support for compulsory sterilization, in favour of the Society's version, the seemingly respectable 'voluntary' sterilization

[129] SA/EUG/A28. Annual Report 1935–36, p. 4.

[130] Clyde Chitty, *Eugenics, Race and Intelligence in Education* (London: Continuum, 2007), pp. 62–63.

[131] Thomson, *The Problem of Mental Deficiency*, p. 68.

[132] SA/EUG/C68. Blacker to Dr Stella Churchill 17 Dec 1931.

[133] SA/EUG/D150. Maule to Blacker 3 Oct 1935.

[134] SA/EUG/A29. Annual Report 1936–37, p. 10.

only.[135] The following year there was further progress. Blacker was pleased to relate to Brock that the Labour Women's Conference, held in Swansea in May, had approved 'voluntary' sterilization.[136] He probably did not know that women's organizations were marginal within the Labour Party. Even as the Women's Cooperative Guild increased its membership (77, 807 in 1935 against 66, 566 in 1930), it lost influence.[137] The Trades Union Congress was vastly more powerful and had passed a resolution against sterilization in 1934, on the prompting of George Gibson of the Mental Hospital and Institutional Workers' Union. Gibson argued, subtly, that it was impossible to tell whether a person's progeny would be defective without regard to the other parent. Moreover, even in families where one or both parents were identified as abnormal, healthy or indeed gifted children might be born.[138]

Blacker was not completely opportunistic in what he was stating. Like his teacher in biology at Oxford, Julian Huxley, he was a 'reform' eugenist. Huxley argued in his 1936 Galton Lecture that redistribution of wealth was conducive to eugenics.[139] Great differences in assets and opportunities made it impossible to evaluate who was eugenically valuable and who less so. This argument had appeared in Eden Paul's work as early as 1911.[140] Blacker and Huxley did not agree with 'mainline' eugenists that the sorting process had already taken place, and that social background was indicative, almost wholly, of heredity. Blacker wrote that the eugenic worth of artisans and probably the working classes was equal to professionals and the leisured rich.[141] The Society began to exhibit

[135] LMU. HD 3423. Annual Report of the Women's Cooperative Guild 1935–1936, p. 26.

[136] SA/EUG/D50. Blacker to Brock 25 May 1936.

[137] Andrew Flinn, '"Mothers for Peace", Co-operation, Feminism and Peace: the Women's Co-operative Guild and the Antiwar Movement between the Wars' in Lawrence Black and Nicole Robertson (eds.), *Consumerism and the Co-operative Movement in Modern British History. Taking Stock* (Manchester: Manchester University Press, 2009), p. 139. Gillian Scott, *Feminism and the Politics of Working Women. The Women's Co-operative Guild, 1880s to the Second World War* (London: Routledge, 2001), p. xii.

[138] *Eugenics Review* 26:4 (1935), p. 253.

[139] Soloway, 'From Mainline to Reform Eugenics', p. 65. Julian Huxley, 'Eugenics and Society', *Eugenics Review* 28:1 (1936), p. 27.

[140] Eden Paul, *Socialism and Eugenics* (Manchester: National Labour Press, 1911), pp. 12–13.

[141] Soloway, *Demography and Degeneration*, p. 198.

pedigrees showing desirable and undesirable biological qualities existing in rich and poor families alike.[142] These were a far cry from the old pauper pedigrees purporting to show how 'feeblemindedness' was an inherited condition within the residuum (Fig. 4.4).

Even before Blacker became general secretary, there had been appreciation of contrasting political views. Ronald Fisher was honorary secretary in 1928 and tried to avoid excessive class bias. That was the reason he asked the socialist geneticist Michael Pease to reconsider when Pease wished to resign from the council.[143] Blacker actively looked for socialists who could join the council and succeeded in getting the general practitioner and socialist Stella Churchill elected in 1931. In 1933 the editorship of the *Eugenics Review* passed from the geneticist Eldon Moore

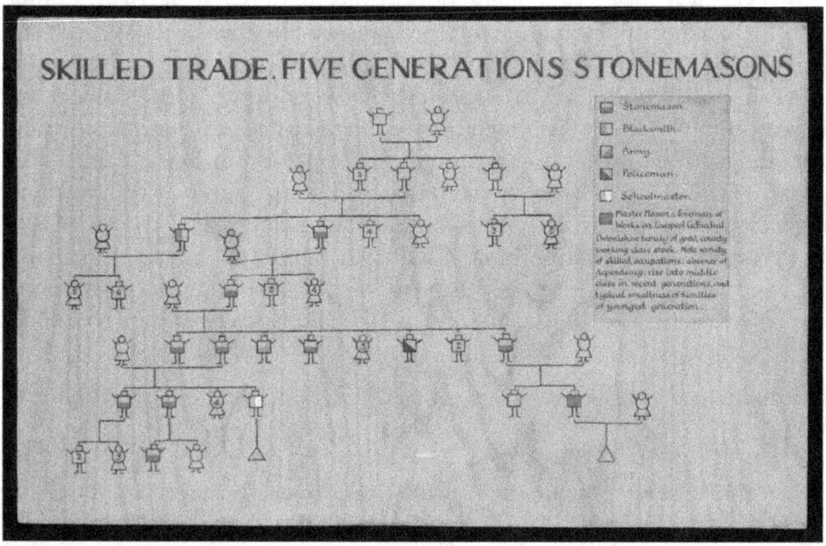

Fig. 4.4 A chart showing a respectable working-class family (reproduced from the Eugenics Society Archive, Wellcome Library, London)

[142] SA/EUG/G34/6 'Two Well-To-Do Families Manic-Depressive Insanity. Insanity and Mental Deficiency'. SA/EUG/G35/5 'Skilled Trade: Five Generations of Stonemasons'. SA/EUG/G35/4 'Two Families in an East London Area'.

[143] SA/EUG/C269. Fisher to Pease 18 Apr 1928.

to Maurice Newfield, another socialist and officer of the Workers' Birth Control Group.[144] Newfield set out to transform it from 'a parish magazine' to a serious intellectual journal covering demography (including birth control) and human genetics.[145]

The Society began emphasizing that it was neutral on political issues not connected to eugenics. This came in the form of announcements and coverage in the *Eugenics Review*. It led to a membership meeting on 19 January 1932 featuring Ellen Wilkinson, former Labour MP for Middlesborough. The meeting was described as the best attended for some time. While the socialists were naturally the most sceptical about class differences, they were 'as keen as any on sterilization and the keenest of all on eugenical birth control'.[146] Traditional supporters of 'mainline' eugenics also spoke. They were not necessarily at odds with the new focus on winning wider support. Charles Vickery Drysdale, an opponent of socialism, argued that a socialist society would need eugenics. Leonard Darwin had liked and agreed with the article 'Eugenics and Socialism' by Herbert Brewer, published in the *Eugenics Review* in 1931. Some old stalwarts were, however, implacably opposed to the new direction and resigned their membership. They included the writer Lady Askwith and the Oxford philosopher F. C. S. Schiller.[147] Cora Hodson, until recently the secretary, also left but she had earlier facilitated contact with Labourites.

The Society was genuine in trying to make amends for printing the prejudiced views of many members in the past. An example had been William Ralph Inge, who in the very first issue of the *Eugenics Review* had worried that the urban proletariat would 'cripple our civilization'.[148] Due to the National government landslide, between 1931 and 1935 the Labour Party had less than fifty Members of Parliament. The Society could have tried to ram sterilization through with the votes of the Conservative and Liberal parties, both of which had contained more supporters

[144] University of Warwick. Modern Records Centre. Trades Union Congress. MSS 292/840/1/11. Workers' Birth Control Group, p. 1.

[145] Soloway, *Demography and Degeneration*, p. 201.

[146] *Eugenics Review* 24:1 (1932), p. 5.

[147] Soloway, *Demography and Degeneration*, p. 273.

[148] W. R. Inge, 'Some Moral Aspects of Eugenics', *Eugenics Review* 1:1 (1909), p. 30.

of sterilization than opponents in 1931.[149] Instead, it chose to concentrate its endeavours on the Labour Party. In 1935 the Eugenics Alliance was founded to cater for those who could not afford the Society's dues of ten shillings a year. Members received information in the post, and it was possible for an entire branch or organization to join.[150] This mirrored the Labour Party's and trade unions' affiliation policy, possibly creating a homely feeling for working-class people.

In 1933 the Society was able to make use of the Twitchin estate. It bought the new premises at 69 Eccleston Square after moving into them the year before.[151] This would be its headquarters for the remainder of its halcyon days. In 1934, the year after Pearson stood down from the Galton chair, the Society gave a subsidy to the Galton Laboratory for the purpose of publishing its journal the *Annals of Eugenics*.[152] The fact that Ronald Fisher had taken over as editor obviously facilitated this development.

The Policies of the Late 1930s

A new policy document was agreed in 1936, replacing the one from a decade earlier. The insane, mental defectives and some criminals would be detained under the provisions of the Mental Deficiency Act (updated in 1927).[153] An important concept was officially introduced, to be expatiated the next year in a book. This was the 'social problem group'. It harked back to Galton's calculations that about 9% of the population were 'undesirables'. Later defined as making up the bottom tenth of the population, the group was characterized by 'insanity, mental defect, epilepsy, occupational instability, recidivism, inebriety and social dependency'.[154] The term had first been used by E. O. Lewis for the Wood Committee. It was a medicalized version of the old 'residuum'.[155] The Society wanted to investigate if persons with these traits formed a logically coherent unit and if there were hereditary influences at work. Lewis found

[149] Hart and Carr, 'Sterilization and the British Conservative Party', p. 724.

[150] SA/EUG/J18. The Eugenics Alliance.

[151] SA/EUG/A26. Annual Report 1933–34, p. 2.

[152] SA/EUG/A27. Annual Report 1934–35, p. 2.

[153] SA/EUG/A29. Annual Report 1936–37, pp. 2–3.

[154] SA/EUG/A29. Annual Report 1936–37, p. 3.

[155] Jones, 'Eugenics and Social Policy between the Wars', p. 723.

that the police, educational authorities, hospitals and welfare workers kept tabs on the same families.[156] The Society's line was hardening somewhat on miscegenation. Apart from asking for further study on the matter, it now alleged that 'in certain circumstances, race mixture is known to be bad'.[157] This was probably a reference to research undertaken for the Society by Rachel Fleming on children of mixed descent in Liverpool, Cardiff and East London. Her interim report from 1927 created an hierarchy of miscegenation, whereby she claimed that half-Chinese children were sometimes 'exceptionally talented'.[158] However, her conclusion was less favourable for the children who had black fathers and British mothers, and this may have prompted the Society's assertion. Immigration was seen as important with regard to settlement policy in the Empire and Britain itself, where a mistaken practice could lead to the deterioration of national standards of heredity.

The Eugenics Society adopted modern policies on birth control and abortion. Both were accepted on eugenic grounds. Abortion was linked to the sterilization policy. Those who qualified for sterilization would also receive the 'right' to have their pregnancies terminated.[159] The same was true where there had been incest or rape, which from a genetic point of view could lead to the foetus receiving unbalanced heredity. There should be legalized facilities for such procedures. The Society's cooperation with the National Birth Control Association began in late 1936.[160] Its President Lord Horder had become president of the Eugenics Society too the year before. There was still prudence about its policies on sensitive topics. For mentally defective and insane persons, sterilization was now regarded as additional to segregation, whereas previously one or the other was deemed to be enough (Fig. 4.5).

Long-standing supporter Wing Commander Sir Archibald James raised the question of sterilization in Parliament again in 1937. His motion sought to implement the recommendations of the Brock Committee,

[156] SA/EUG/D50. Booklet 'The Committee for Legalising Eugenic Sterilization'.

[157] SA/EUG/A29. Annual Report 1936–37, p. 4.

[158] Lucy Bland and Lesley A. Hall, 'Eugenics in Britain: The View from the Metropole' in Alison Bashford and Philippa Levine (eds.), *The Oxford Handbook of the History of Eugenics* (Oxford: Oxford University Press, 2010), pp. 219–220.

[159] SA/EUG/A29. Annual Report 1936–37, p. 5.

[160] SA/EUG/D23. Cooperation between the NBCA and the Eugenics Society. Minutes of Joint Meeting held 3 Dec 1936.

Fig. 4.5 Lord Horder (reproduced from the Eugenics Society Archive, Wellcome Library, London)

claiming that health and nutrition were deteriorating across the nation. The minister of health, Sir Kingsley Wood, was less pessimistic. The number of mental defectives rose as the general population rose, he said, and local authorities were getting better at identifying them. He was not certain there had been an actual increase in the numbers. He preferred to wait until the consequences of the recent sterilization acts in Scandinavia and Germany had become known (passed between 1929 and 1935) before contemplating legislation at home.[161] This in effect bought the government time. The debate moreover showed that the Parliamentary Labour Party was sticking to its original views. Three out of the four Labour MPs who spoke (David Logan, Denis Pritt and Jack Jones) argued against sterilization and only one (Ernest Thurtle) in favour.[162] Logan and Jones were Catholics. Although the Labour Women's Conference

[161] *Hansard*, 13 April 1937, column 848.

[162] Modern Records Centre, Warwick University. MSS 229/NA/4/1/10. Mental Hospital and Institutional Workers Journal 26:5 (1937), pp. 10.- 11.

supported sterilization, the only female MP the party had in 1937 was Ellen Wilkinson, now representing Jarrow.

The Menace of Eugenics

From its early days, the Society had discussed whether the working class matched their 'betters' genetically.[163] Recently, the emphasis had been that parts of it did. And, whatever one's personal views, the class made up such a large proportion of the nation that it could not be ignored in any scheme to improve heredity. The Society no longer argued that the urban proletariat presented a threat, instead the new label the 'social problem group' was being developed as the acme of unacceptability. It began carefully by putting a question mark after the title *A Social Problem Group?*, when its book on the subject was published in 1937. Edited by Blacker, it contained a chapter by Ernest Lidbetter on social dependency and another by Sybil Neville-Rolfe on prostitution. A later generation of eugenists was represented by the psychiatrist Eliot Slater on mental disorder.[164] The social problem group's fertility should be restricted.[165] The designation was a sub-category of 'rough' workers, whose separation from 'respectable' workers was operational within the class.[166] This could be a reason why such audiences occasionally went further than the Society in asking for sterilization to be compulsory.

The Society was in favour of family allowances (child benefit) but proportional to income so that the payments would have a eugenic effect. It was completely against flat-rate payments, which might leave untouched 'the fertility of the middle classes' and might instead 'constitute an enormous bonus to members of the submerged tenth'.[167] Nevertheless, the Society managed to give the impression that it supported benefits and its stance on war also helped when dealing with progressives. Armed conflict was judged to be dysgenic for two reasons. Firstly, it was probable that men who were above average in health and ability

[163] Searle, *Eugenics and Politics in Britain 1900–1914*, p. 51.

[164] SA/EUG/A29. Annual Report 1936–37, p. 9.

[165] SA/EUG/A24. Annual Report 1931–32, p. 1.

[166] Ross McKibbin, *Classes and Cultures. England 1918–1951* (Oxford: Oxford University Press, 2000), p. 198.

[167] SA/EUG/D 65. Eugenics and Family Allowances, p. 7.

would be killed or maimed disproportionately. Rejects for military service were those with less desirable bodily or mental characteristics. Secondly, war led to a reversal of parenthood.

John Maynard Keynes had been a founding member of the Cambridge University Eugenics Society in 1911 and acted as its first treasurer.[168] Nevertheless, it could be seen as indicative of the shift to the left that he was chosen to give the Galton Lecture in 1937. His topic, 'Some economic consequences of a declining population', may help in explaining why the Society was reinventing itself. As demographers, such as Enid Charles, Alexander Carr-Saunders and David Glass, now thought the British population would decline, class differentials mattered less. Charles wrote in *The Twilight of Parenthood* (1934) that differential fertility had been a temporary phenomenon.[169] The Society established a Population Investigation Committee in 1936 to examine what could be done to reverse the fall in the birth rate.[170] The need to rethink issues may have nudged more of the membership to attend than ever before—224 in total.[171] 'Reform' eugenics was also encapsulated in a joint venture with Political and Economic Planning, starting early in 1938. This organization was in favour of a managed economy along the lines recommended by Keynes. Together with the Society, it formed the Population Policies Committee, which was established to promote fertility according to eugenic thought. Influential eugenists believed that a Royal Commission on Population would be established in the near future. Cooperation with Political and Economic Planning was intended to prepare the ground before any commission was formed. This was another Blacker idea and under his tutelage the Society gave modest grants to PEP.[172]

Jack Trevor had been awarded the Society's Darwin research scholarship in 1936. He was engaged in work at the Galton Laboratory with Ronald Fisher and the anthropologist Geoffrey Morant on a literature study of race crossing. Joint efforts between the Society and the Laboratory had become easier when Fisher took over as director. Trevor's studies had progressed far enough for a talk to be given in the run-up to

[168] SA/EUG/E23. Folio 5 v.

[169] Kevles, *In the Name of Eugenics*, p. 174.

[170] SA/EUG/D161. Draft of a letter to *The Times* sent 30 Mar 1937.

[171] SA/EUG/A29. Annual Report 1936–37, p. 15.

[172] Soloway, 'From Mainline to Reform Eugenics', p. 68.

Christmas in 1937 as 'Some anthropological considerations of race crossing'.[173] The scholarship was a boon to Trevor. After completing a book on 'hybrid populations', he was appointed as lecturer in anthropology at Cambridge University in 1938. The Society continued to insist that the question of miscegenation required 'further knowledge' and the wording in 1939 was that miscegenation 'may, in some cases, be biologically harmful'.[174] Meanwhile, it took little notice of this and did not find it worthy of comment that someone involved in eugenics and birth control in India, Mrs Dutta, should be Scottish and married to Indian educator.

The Society did not condemn outright the hereditarian policies of the Nazis. It stated euphemistically that questions of race had become an issue in 'one European country'.[175] In Germany, science had long been notable for a focus on anthropological differences. In 1935 societies for racial hygiene (the German name for eugenics) had divested themselves of their Jewish colleagues and thrown in their lot with the totalitarian state.[176] The British society disagreed with this policy. Its council had earlier passed a resolution stating that it deplored 'the use of the term Eugenics to justify racial animosities'.[177] For Blacker, it was vital to put clear blue water between the Nazi Sterilization Law and his own preferred measures. Opponents of sterilization would assuredly try to tarnish the Society with references to the Nazis. He suggested that the response should be to inform the public tirelessly about what it did in fact stand for. At the same time, members were genuinely interested in finding out how eugenics was faring in a country where it had reached the distinction of being part of the regime's ideological basis.

On 17 January 1939 members gathered to hear Dr Felix Tietze on 'Eugenic measures in the Third Reich'.[178] Tietze, a paediatrician from Vienna, was a visiting research fellow at the Galton Laboratory. Germany had gone full throttle on compulsory sterilization, choosing to apply it to perhaps as many as 360, 000 people between 1934 and 1945. John

[173] SA/EUG/A30. Annual Report 1937–38, p. 17.

[174] SA/EUG/A31. Annual Report 1938–39, pp. 4, 11.

[175] SA/EUG/A29. Annual Report 1936–37, p. 7.

[176] Paul Weindling, *Health, Race and German Politics between National Unification and Nazism, 1870–1945* (Cambridge: Cambridge University Press, 1993), p. 499.

[177] C. P. Blacker, 'Eugenics in Germany', *Eugenics Review* 25:3 (1933), pp. 158–159.

[178] SA/EUG/A31. Annual Report 1938–39, p. 20.

Macnicol believes that the adoption of 'voluntary' sterilization in Britain would in turn have led to the compulsory sterilization of members of the 'social problem group'.[179] Indeed, there is no saying how the law would have been amended in future years. The mere fact of passing such a law and practising it inevitably causes abuses. But although the Society's preferred version was 'voluntary' in name only, it did not go as far as the Nazi legislation, prescribing sterilization in fewer cases and involving at least a notion of consent.

The Society argued that 'voluntary' sterilization would give a right to the poor, currently only on offer to the rich, namely the opportunity to forego further reproduction. It was a policy where opinion tended to be divided by social class rather than political persuasion. Left-wing eugenists Julian Huxley, Lancelot Hogben, Frida Laski and Stella Churchill supported sterilization, as did the *Manchester Guardian* and the *New Statesman*.[180] Except for a few MPs, including Ernest Thurtle, William Henderson, Dr Somerville Hastings, Rennie Smith and Ellen Wilkinson, the Parliamentary Labour Party did not, and the Trades Union Congress was actively opposed. In a comparative article, Leo Lucassen explains the failure of British eugenists to secure a sterilization law as arising from class antagonism. Eugenics had acquired such a reputation for snobbery that sterilization was regarded as a step too far. Hogben implicitly argued along the same lines, claiming that eugenists' hostility to working-class leaders had prevented a 'national minimum of parenthood' being agreed.[181]

The campaign for sterilization in any case went into a lull at the end of 1938. In October that year the Royal College of Physicians appointed a committee to examine the social and clinical effects of 'voluntary' sterilization. The Joint Committee on Voluntary Sterilization therefore decided to await the outcome of the Royal College's report, hoping it would point a way. The Society described 'extreme difficulties' in getting the issue debated in Parliament due to the tense international situation.[182] In reality, there was no appetite for raising such a controversial issue. It

[179] John Macnicol, 'Eugenics and the Campaign for Voluntary Sterilization in Britain between the Wars', *Social History of Medicine* 2:2 (1989), pp. 154–155.

[180] Michael Freeden, 'Eugenics and Progressive Thought: A Study in Ideological Affinity', *Historical Journal* 22:3 (1979), p. 667.

[181] Werskey, *The Visible College*, p. 109.

[182] SA/EUG/A31. Annual Report 1938–39, p. 17.

had been rebuffed in 1931, failed to become a matter for Parliament in 1934 and 1937, and the government would not have wanted to proceed until at least partial consensus had been reached relating to the measure. The Workers' Committee on Voluntary Sterilization was put on hold. The last funding it received was £ 50 for 1938, of which the final instalment was paid late in the year.[183] The campaign would never properly get going again.

At a time of mass unemployment, political upheaval at home, the rise of dictatorships abroad and a fascist presence in the streets, the sterilization proposals seemed doubly threatening. Working-class men, if not their wives, feared that they would be sterilized if they lost their jobs and became dependent on the dole. It was not reassuring that eugenists had consistently supported redistribution to the advantage of wealthy parents and had compiled pauper pedigrees, serving to damn the poor for their own problems. The pausing of the campaign for sterilization did not mean that the Society stopped its lecturing activities. Between the annual general meetings in 1938 and 1939, a total of 105 lectures was given to a variety of organizations (Fig. 4.6).

The Second World War and the Beveridge Report

The outstanding event of 1939 was of course the outbreak of the war. Blacker was called up for active duty with the Royal Army Medical Corps on 1 September, the day German forces entered Poland, and went abroad a fortnight later. It robbed the Society of its guiding light. Power devolved instead to an emergency committee of nine members, all of whom were able to attend meetings regularly. Having the power to co-opt, it would administer activities on behalf of the council for the duration of the war. The National Birth Control Association had merged with other organizations of the same type to form the Family Planning Association. Already ensconced on the second floor of the Society's building, it was told that it would not be charged rent for the duration of the war.On a similar issue, the Society's most recent policy statement had approved of abortion in cases where the sterilization of the person concerned was also mandated or if the pregnancy resulted from incest or rape. Continuing along these lines, Dorothy Thurtle, the wife of the Labour MP Ernest

[183] SA/EUG/D150. Blacker to Maule 18 Nov 1937, Business Secretary to Mrs R. Keeling 19 Aug 1938.

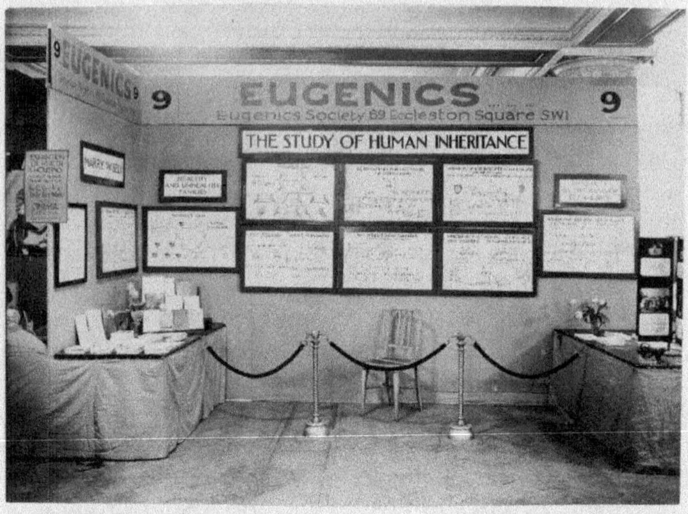

Fig. 4.6 Eugenics Society stand at a health exhibition (reproduced from the Eugenics Society Archive, Wellcome Library, London)

Thurtle, spoke on 'Abortion— Right or wrong' to members in March 1940.[184]

As in the First World War, the outbreak of hostilities led to changes in the quotidian affairs. Propaganda efforts continued at least until the middle of 1940, but on 15 October the Society was put onto a care and maintenance basis. Hilda Pocock, the chief propagandist, and the stenographer Miss Dunn were made redundant.[185] Its office stayed open under Gwendolyne Collyer. Another European war was again turning the thoughts of eugenists towards the Dominions. They had received word from the Eugenics Society of Canada that it could help with the evacuation of London children, who would be able to live on the other side of the Atlantic for the duration of the war. In August and September 1940 three such groups, totalling five mothers and 23 children, took advantage

[184] SA/EUG/A32. Annual Report 1939–40, pp. 32–34.

[185] SA/EUG/A33. Annual Report 1940–41, pp. 1–2.

of the offer. Miss Pocock played a major role on the London side.[186] Uncertainty was created in February 1941, however, when the Canadian society wrote to say that it could no longer run the scheme owing to government restrictions and regulations. These must have landed a knockout blow, as it later became clear that the Canadian society had ceased to exist.

Possibly manifesting the Society's leftward trend, the German lecturer Rene Kuczynski was engaged in 1940 and 1941 to work on the demographic problems of the overseas possessions, as part of its Population Investigation Committee. He had researched in the Department of Social Biology at the London School of Economics and was an adviser to the Colonial Office.[187] It is still noteworthy that the Society should employ him, as he had been involved in various socialist undertakings previously and had fled his native land when the Nazis came to power.[188] His statistical analysis had produced results since his manuscript of no less than a million words was soon to be published. The Society was hoping that eugenics would be a strand of post-war reconstruction, as seen in one of the few members' meetings held in 1941, in which Brian Goddard spoke on 'Eugenics and the new world order'.

But what sort of eugenics would that be? Sir William Beveridge had supported the doctrine for decades. As director of the London School of Economics, he had introduced courses on heredity and eugenics. He had been responsible for the founding of its Department of Social Biology in 1931, staffed by 'reform' eugenists. And yet when he gave the Galton Lecture in 1943, he advocated flat-rate family allowances (child benefit). His views truly mattered because his report 'Social Insurance and Allied Services', published in the autumn of 1942 would be crucial to the post-war welfare state. In it he said concern for the British race was the guiding principle for him.[189] Eugenists had generally backed family allowances since the 1930s, but not a flat-rate scheme. The 1943 annual report claimed that Beveridge wanted a scheme of graded family allowances to

[186] *Eugenics Review* 56:3 (1964), p. 131.

[187] Renwick, 'Eugenics, Population Research, and Social Mobility Studies', pp. 852, 862.

[188] Jeremy Shearmur, 'Beveridge and the Brief Life of "Social Biology" at the LSE', *Agenda: A Journal of Policy Analysis and Reform* 20:1 (2013), p. 88. Mazumdar, *Eugenics, Human Genetics and Human Failings*, p. 125.

[189] Nira Yuval-Davis, *Gender and Nation* (London: Sage Publications, 1997), p. 31.

run alongside it.[190] He did in fact suggest cuts in income tax and further selective grants to supplement the flat-rate scheme of child benefit. Such grants could be competitive scholarships. Taken together, income tax relief and scholarships would satisfy the eugenic aspects of an emerging welfare state. It was not a new idea that a welfare state could promote eugenic goals. Fabians Saleeby, H. G. Wells and the Webbs had argued thirty years earlier that improved maternity services, grants to mothers, pensions, free milk and nurseries would encourage the best members of the working class to increase their birth rate.[191] The recipients of the credit Beveridge had in mind would be chosen according to desirable eugenic criteria (high income or high ability). Even so, the volte-face on a significant policy naturally caused a spirited discussion at the lecture.

Beveridge argued on eugenic territory. Citing the work of Ronald Fisher, he claimed that the differential birth rate, whereby the economically successful members of society had fewer children, occurred *because* a small family increased the opportunities for each child.[192] If there were not so many children, they could continue for longer at school and the parents were able to invest more resources in them. That was true for any segment of the socio-economic scale. Therefore, people of greater ability, whatever their social status, chose to have fewer children. A flat-rate scheme of child benefit would help neutralize the differences not between classes but between families. It would encourage prudent parents to have additional children, in the knowledge that some of the costs of rearing them would be borne by the state. The feckless, meanwhile, would have children irrespective of whether they could afford to or not. Thus, Fisher had argued, a flat-rate scheme would operate eugenically, not dysgenically. Beveridge noted these findings which reinforced his own views.

In the 1930s the Society had funded research by the psychologist Raymond Cattell which brought a novel perspective to the old differential birth rate. In 1937 his book *The Fight for our National Intelligence* showed that it was the least capable individuals in society who were having the most children. Although he did not abandon the traditional argument

[190] SA/EUG/A35. Annual Report 1942–43, p. 2. William Beveridge, 'Eugenic Aspects of Children's Allowances', *Eugenics Review* 34:4 (1943), pp. 121–122.

[191] Soloway, *Demography and Degeneration*, p. 293.

[192] Beveridge, 'Eugenic Aspects', p. 119.

on the differential birth rate, the focus on intelligence permitted it to be formulated along more socially acceptable lines. The emphasis became that the intelligent were not having enough children, leading to declining national standards.[193] Cattell's contention could then be applied in a practicable way to promote eugenics without showing overt preference for the affluent. He described himself as a socialist and regarded it as a problem that the intelligent among the masses were also deciding upon merely having one child.[194] In a postscript to the book, he stated that *within* most of the occupations studied, it was the least intelligent who were having the most children.[195]

Fisher's, Cattell's and Beveridge's views were developments of eugenics. There had always been shades of opinions within the movement. Beveridge's presentation was at the cutting edge of what 'reform' eugenics had become. It proved it was possible to support an incipient welfare state and yet retain an anchoring in eugenics, which at least in its early years had been a dogmatic class ideology. Not everyone was willing to depart from that line. The issue came to a head on 18 January 1944 when left-wing members Dick Mitchison, who was married to J. B. S. Haldane's sister Naomi and later entered Parliament for Labour, and Eva Hubback, the principal of adult education provider Morley College, urged the acceptance of the Beveridge report at a members' meeting on the issue.[196] Ranged against them were 'mainline' eugenists Cecil Binney and Ernest Lidbetter. As the Society had previously stated that flat-rate schemes would be wholly dysgenic, they had reason to be sceptical about this change of heart. The Family Endowment Society seemed to share the Eugenics Society's traditional view in this respect and had received a grant of £ 250 from it in 1943. Later the former's position came under scrutiny. At the last annual general meeting during the war, held at the Royal Society in May 1944, retired civil servant J. P. Brander queried the grant to the Family Endowment Society. It appeared to be advocating a flat-rate child benefit and Brander questioned whether this would be eugenic. He could not find a seconder for his motion to ban further

[193] Raymond B. Cattell, *The Fight for our National Intelligence* (London: P. S. King & Son, 1937), pp. 17, 123.

[194] Cattell, *The Fight for our National Intelligence*, pp. 63, 147.

[195] Raymond B. Cattell, 'Some Further Considerations between Intelligence, Fertility and Socio-Economic Factors', *Eugenics Review* 29:3 (1937), p. 179.

[196] SA/EUG/A36. Annual Report 1943–44, p. 2.

grants to this organization, but Lord Horder as president undertook to examine the issue which had been very correctly justified with reference to Society policy. The lack of a seconder may indicate that the membership was leaning towards the new line, namely that it was ability and not social class in itself which was the desideratum.[197]

In March 1944, as predicted by eugenists in the know, the Royal Commission on Population came into being. It encompassed many members of the Eugenics Society either on the commission itself or as experts in its subcommittees. These included Alexander Carr-Saunders, David Glass, Alan Parkes and C. P. Blacker. It put the Society in good stead to influence debate in what would be the post-war reconstruction period. All were 'reform' eugenists, perhaps helping to entrench this version of the doctrine after the war. The demographer Glass had worked under Lancelot Hogben in the Department of Social Biology at the London School of Economics. Its most famous output is probably the paper by J. L. Gray and Pearl Moshinsky, published in *Sociological Review* in April 1935, which showed that the intelligence possessed by children from poorer families was often wasted. It was in tune with Hogben's emphasis. Originally, the Society had been sceptical of this research, as its own studies by Cattell had shown a marked disparity in favour of the children from higher social groups.[198] However, 'reform' eugenist Julian Huxley believed in its conclusion that although a third of children of exceptional intelligence came from the upper and professional classes, the offspring of wage earners accounted for fully half of them.[199] The finding was even independent of the wealthier children having had better scope to develop their talents. In his Galton Lecture of February 1945, Blacker tried to steer the Society towards what would be the new consensus after the war was over. He thought Hogben's attacks had damaged eugenics, as it was becoming the commonplace view that the doctrine was simply class prejudice masquerading as science.[200]

[197] *Eugenics Review* 35:3–4 (1943), p. 54. J. A. Fraser Roberts, 'Population Problems in Light of Differential Fertility', *Eugenics Review* 36:1 (1944), pp. 9–10.

[198] SA/EUG/C171. C. P. Blacker to Lord Horder 18 Aug 1936. C. P. Blacker to Mrs Scarfe 10 Nov 1936.

[199] Julian Huxley, 'Eugenics and Society', *Eugenics Review* 28:1 (1936), p. 22.

[200] Mazumdar, *Eugenics, Human Genetics and Human Failings*, p. 247.

Conclusion

By 1945 'reform' eugenics had been brewing in the Society for decades. Blacker had argued that equalizing opportunity meant attaining more precise information about eugenic worth. The hackneyed differential birth rate notion had been substituted by a less abrasive version: it was people of low ability who were having too many children, whatever their economic position. It meant that a social problem still existed which required eugenics as a solution, but it was not about snobbery. The Society's goal of persuading working-class people meant that it had to reconfigure eugenics in such a way that it would be acceptable to this target group. It said that the poor were divided between those who would benefit from better material conditions and those who would sink back into squalor.[201] The Society also gradually adopted opinions which, for other reasons, were compatible with left-wing views. Its openness to discussing sexuality may have been one reason why it attracted progressives.[202] From the mid-1920s it had been in favour of contraception. A decade earlier it had started working on introducing biology and a form of sex education in schools. From the middle of the 1930s it was willing to legalize abortion, opposed war as dysgenic, believed in social mobility and supported a form of family allowances (child benefit). Since it had substantially toned down its class rhetoric too, it might be considered to have drifted to the left in the 1930s.[203]

This trend was to continue during the war years, which saw public opinion as a whole becoming more collectivist and caring. Sir William Beveridge gave a talk to the Society in which old-style eugenics was so diluted that the ideology seemed almost unrecognizable. Some of the membership adapted easily to this way of thinking, others were slower to follow. It began cooperating with the Family Planning Association and Political and Economic Planning, in such a way that the Society too was seen to be taking on progressive and modern opinions. By 1945 it was in fact ready to support the emerging consensus stemming from the Second World War. In 1938 it had temporarily abandoned most of its agitation for

[201] SA/EUG/G29. 'One of our Responsibilities as Citizens' by Hilda Pocock.

[202] Greta Jones, *Social Darwinism and English Thought. The Interaction between Biological and Social Theory* (Sussex: The Harvester Press, 1980), p. 110.

[203] G. R. Searle, 'Eugenics and Politics in Britain in the 1930s', *Annals of Science* 36:2 (1979), p. 169.

sterilization and the moratorium became permanent as the circumstances never really became right again.

When the war ended, the Society was pleased to confirm that its premises had not been damaged by enemy action. The Family Planning Association was still in situ having had the lease of its rooms within 69 Eccleston Square extended rent free until March 1946.[204] The Eugenics Society looked forward to the appearance of the report of the Royal Commission on Population, in which it wielded influence owing to the commission's makeup. It hoped the report would set the tone for social policy in the rebuilding of Britain. The resumption of a full set of activities and the exuberance of the war being over made it optimistic about the future. But how would its image be affected by far-right eugenics being so central to the functioning of the racial state which the Allies had defeated? Would whatever respectability the doctrine had in professional circles be eroded as the full horrors of Nazi Germany became clear?

The initial reaction to the changing context came in the October 1945 issue of the *Eugenics Review*, the first to be published since the Second World War had ended. The argument put forward was that what had occurred in Germany under the Nazis was a perversion of eugenics and not typical of the doctrine itself. At the same time, in discussing the anthropologist Ashley Montagu's work *Man's Most Dangerous Myth: The Fallacy of Race*, it admitted that eugenics had been brought into disrepute by the Nazis.[205] It said, however, that the same was true for anthropology. Montague had studied under Karl Pearson at University College. The Society referred to the German version of eugenics as 'race hygiene' in an effort to distance itself from it. And it marshalled the doubts and disparities of opinion among its supporters to show that British eugenics was not necessarily based on snobbery in the way that German eugenics was based on racism. Hogben had argued that the two versions diverged precisely in this way.[206] But as before, the existence of eugenists with left-wing opinions was a convenient shield to hold up. Now there were more of these than ever.

[204] SA/EUG/A37. Annual Report 1944–45, p. 3.

[205] *Eugenics Review* 37:3 (1945), p. 88.

[206] Gavin Schaffer, *Racial Science and British Society, 1930–62* (Basingstoke: Palgrave Macmillan, 2008), p. 26.

A problem was that even 'reform' eugenists like Hogben and Huxley had supported sterilization.[207] It had been the signature policy of the Eugenics Society in the 1930s. This facet of the doctrine was the hardest to explain away. Its new statement of aims claimed that it had stood for gradual, slow-moving 'negative' eugenics based on consent and responsible parenthood in consultation with the individuals concerned. The Society also revisited the debates about eugenic worth, giving examples of people with exemplary talents who were frail in body and mind (John Keats and Fyodor Dostoyevsky).[208] Eugenics had been aware of this paradox for some time, while continuing to believe that the totality was more important than special attributes appearing on their own. This suggests the doctrine was there for political purposes, rather than fostering genius. A draft of its submission to the Royal Commission on Population, written up in early 1945, shows that it considered eugenic desirability to be a function of five traits in combination, not isolated characteristics. These were health, intelligence, social usefulness, freedom from genetic taints and belonging to a good family.[209] Because intelligence was easier to measure than the others, the Society wanted a greater number of births occurring among the most intelligent of any social class. The new version of the differential birth rate argument was used: the least intelligent families were having three times as many children as the most intelligent. And, in general terms, it envisaged that a better population would include more members of the Registrar-General's first four classes, which placed the dividing line inside the group of unskilled workers. The bar was thus set lower than before. In January 1945 the *Eugenics Review* could declare that 'Eugenically, all class distinctions are "false": the true distinctions ... cut right across all divisions of society based on wealth, status, or rank'.[210]

[207] Kevles, *In the Name of Eugenics*, p. 167.

[208] *Eugenics Review* 37:3 (1945), p. 89.

[209] SA/EUG/D160. Memorandum for the Royal Commission on Population. January 1945, pp. 3, 6.

[210] *Eugenics Review* 36:4 (1945), p. 114.

CHAPTER 5

Eugenics Between 1945 and 1990

The war was over. The Eugenics Society had not shown any predilection towards engaging with the German Society for Racial Hygiene during the conflict, unlike in the First World War when it eventually succeeded in making contact. It felt that its version of eugenics was fundamentally at variance with the racial state erected in Germany, and that the sterilization it recommended bore no resemblance to the compulsory sterilization practised by the Nazis. Eugenists in Denmark also tried to distance themselves from their southern neighbour. In November 1946 Tage Kemp, director of the Institute of Heredity at the University of Copenhagen, explained in a speech to the Eugenics Society that they used the term 'hereditary hygiene' to distinguish themselves from the racial hygienists of the Third Reich.[1] Yet Kemp and his colleagues had prioritized 'negative' eugenics and been able to implement wide-ranging sterilization as well as abortion on eugenic grounds. Eliot Slater, a psychiatrist who joined the council in 1944, was ambivalent about the Danish example. It was 'humanitarian' in comparison with what the Nazis had done, but also less

[1] Tage Kemp, 'Danish Experiences in Negative Eugenics, 1929–45', *Eugenics Review* 38:4 (1947), p. 181.

D. Redvaldsen, *A History of British Eugenics since 1865*, Medicine and Biomedical Sciences in Modern History,
https://doi.org/10.1007/978-3-031-72290-5_5

efficient.[2] C. P. Blacker wrote to Kemp to welcome him to Britain and offer him dinner.[3] The latter would go on to give the Galton Lecture in 1957.[4]

British eugenics had taken steps towards becoming meritocratic rather than being based on social class. This matched developments in the county at large. The Hadow Committee's report 'Education and the Adolescent' of 1926 had suggested selecting children for grammar, technical and secondary modern schools by ability at 11.[5] The idea was echoed in the Spens Report on Secondary Education in 1938 and supported by the 1943 Norwood Report. The 1944 Education Act implemented it. Previously the grammar schools had often been the preserve of families who could afford to pay the fees and the hidden extras. The Act did not equalize disparities between local education authorities and what sort of secondary education they could provide, but it was conducive towards making academic ability rather than financial means the selection criterion. As early as 1919, Lancelot Hogben had warned that people possessing an education were not necessarily better endowed that those who did not.[6]

The Eugenics Society had long argued for further assistance to those likely to benefit from secondary and tertiary education, and for technical education to youth whose talents were practical rather than theoretical. The 11+ examination, in principle at least, enabled a fairer selection of pupils to the three types of secondary schools. It was influenced by the theories of educational psychologists who were committed eugenists.[7] Especially important was Charles Spearman, a member of the Society until

[2] Volker Roelcke, 'Eugenic Concerns, Scientific Practices: International Relations in the Establishment of Psychiatric Genetics in Germany, Britain, the USA and Scandinavia, c. 1910–60', *History of Psychiatry* 30:1 (2019), pp. 28, 32.

[3] Danish National Archives, Copenhagen. 2752 Københavns Universitet. Tage Kemp, professor. 1941–1963 Korrespondance, faglig og administrativ. Box 3. 1945–1948. Folder marked 1945–1946. Blacker to Kemp 13 Nov 1946.

[4] Tage Kemp, 'Genetic-Hygienic Experiences in Denmark in Recent Years', *Eugenics Review* 49:1 (1957), p. 11.

[5] Clare Hanson, *Eugenics, Literature and Culture in Post-War Britain* (New York: Routledge, 2013), p. 14.

[6] Lancelot T. Hogben 'Modern Heredity and Social Science', *The Socialist Review: A Quarterly Review of Modern Thought* Vol.16 No. 89 (April–June 1919), p. 154.

[7] Stephen Jay Gould, *The Mismeasure of Man* (London: Penguin, 1992), pp. 293–295.

1930 who thereafter served on its Consultative Committee.[8] He believed that relative intellectual standing among individuals remained static after the age of eleven. Sir Cyril Burt agreed with the utility of the 11+ but thought intelligence was so overwhelmingly determined by heredity that school placement could take place at an even earlier age. Other members serving on the Royal Commission on Population welcomed the 1944 Act.[9]

By 1926 the Eugenics Society had taken over as the mainspring of the British version of the doctrine. After the war the personnel involved in academic eugenics began the journey towards abandoning the ideological framework which it had previously espoused. J. B. S. Haldane had led the new Department of Eugenics, Biometry and Genetics at University College since 1943, when Ronald Fisher took up a post at Cambridge. Haldane was a communist so could hardly be accused of being pro-Nazi, despite eugenics forming a component of their ideology. His faith in dialectic materialism suffered a blow in 1948 with the Lysenko affair, a controversy relating to whether acquired characteristics can be transmitted genetically, though he stayed in the Communist party until 1956.[10] The Galton professor Lionel Sharples Penrose, engaged in studying the effects of heredity on human characteristics, was not a zealot in the advocacy of genetic factors to the exclusion of everything else. In 1954 the *Annals of Eugenics*, the journal edited in the department, became the *Annals of Human Genetics* with the former name retained as a subtitle until 1965.[11] Human genetics, still seeking to improve health and longevity, was a more acceptable name for eugenics, bereft of the ideological connotations which the doctrine engendered. As for the Galton Laboratory, Penrose became the head of its parent body, the new department, in 1957. This occurred upon Haldane's departure for India, where he would continue to work on genetics and biometry in Orissa after first taking up a position as a research professor at the Indian Statistical Institute until 1961. In 1965 the name of Haldane's former fief was changed to the Department

[8] Wellcome Library, London. Eugenics Society Archive. SA/EU/C318. Hon. Secretary to Spearman 20 Aug 1930.

[9] *Eugenics Review* 37:3 (1945), p. 101.

[10] Jonathan Rose, *Proletarian Philosophers. Problems in Socialist Culture in Britain, 1900–1940* (Oxford: Clarendon Press, 1984), p. 104.

[11] Ruth Clifford Engs, *The Eugenics Movement. An Encyclopedia* (Westport CT: Greenwood Press, 2005), pp. 10, 84.

of Human Genetics and Biometry, becoming known as the Department of Biology the following year.

The Treasury of Human Inheritance continued to be published until 1958, when it completed its fifth and final volume. Much of it was written by Galton Laboratory worker Julia Bell. She was first or sole author for the reports making up volumes 4 and 5.[12] Volume 4 was partly and volume 5 wholly published after the Second World War. She had also written virtually all of volume 2, published from 1933 onwards. It related to diseases of the eye, volume 4 was about nervous diseases and muscular dystrophy while volume 5 concerned hereditary anomalies of the digits. The Eugenics Laboratory Memoirs reached its conclusion in 1966. Both these series were occasional after the Second World War, each with a total of five publications (not amounting to full volumes). Penrose contributed to volume 5 part II of The Treasury of Human Inheritance, on syndactyly (webbed fingers) and polydactyly (more than five fingers) and wrote some of the prefaces for the Eugenics Laboratory Memoirs. Haldane, who did not contribute to any of the Galton Laboratory series, worked primarily on mathematical statistics and Penrose on medical genetics. Thus, eugenics as such did not feature prominently. The other journal previously connected with the Galton Laboratory, *Biometrika*, was still appearing for those interested in mathematical genetics. This had been the preferred method of Galton and Pearson to investigate heredity and biological questions. It was not, however, synonymous with eugenics. Egon Pearson, the editor, stuck to statistics and did not promote eugenics in the way his father had done.

In 1949 Penrose spoke to the Eugenics Society on the work taking place in the Galton Laboratory. He said that advancing medical genetics was the prime concern of his team. Listing the research carried out by staff, he explained how it would be helpful in the treatment of disease.[13] Penrose then went on to give his personal and scientific views on eugenics, having already warned in the preamble that they were likely to diverge considerably from the opinions held by the typical member of the Society. On the 'social problem group', he had constructed a model whereby two genes determined intelligence in a Mendelian fashion. If both were

[12] Joe Cain, 'Publications Produced by the Francis Galton Laboratory for National Eugenics', *The Library* 22:4 (2021), pp. 544–546.

[13] L. S. Penrose, 'The Galton Laboratory: Its Work and Aims', *Eugenics Review* 41:1 (1949), pp. 17–19, 23, 25.

ordinary it led to average mental ability, but such people were not fully replicating themselves. (His data may have been a few years out of date, as it had been thought in the late 1930s that the British population would decline over time.) To counteract this, Penrose argued, the normal group needed to be supplemented by offspring from the 'social problem group', where fertility was higher. If the 'social problem group' had one normal and one 'deficient' gene for intelligence, a quarter of children born to them would be normal. If random mating occurred within the whole population, about half the children born to one member of the 'social problem group' would be normal. He also said that, on first principles, if the 'social problem group' was fecund and intellectuals relatively barren, it was the former who were 'fit' in the Darwinian sense (Fig. 5.1).

These observations proved that Penrose no longer believed in eugenics in the conventional sense of the term. Eleven years earlier, he had written a memorandum on population policy based on classic eugenic notions.[14] He had suggested raising the standard of living and providing better entertainment for unskilled and unemployed married couples so that they might reduce the number of offspring they had. One idea was to hand out free wireless sets. His Colchester survey of mental defect was also partly funded by the Society. Although he had abandoned the doctrine, the discussion afterwards showed that the members, including C. P. Blacker and Eliot Slater, appreciated his lecture and were willing to take these arguments into account.

In reality, the Department of Eugenics ceased to operate as Galton had intended when Fisher abandoned it in 1943. Penrose would later say that during his tenure, nobody taught eugenics and the Galton professor was not a eugenist.[15] Although Haldane maintained a belief in a left-wing interpretation of the doctrine, with class prejudice taken out, he tended not to actively advocate it. He did publish two articles, both called 'The Prospect of Eugenics', in 1955 and 1957 and another on biometry concurrently. Eugenics was sometimes a topic in his books.[16] Haldane had

[14] SA/EUG/D65. 'Note on Dr Penrose's memorandum entitled "A National Population Policy based on eugenic principles."'.

[15] Pauline M. H. Mazumdar, *Eugenics, Human Genetics and Human Failings. The Eugenics Society, its Sources and its Critics in Britain* (London: Routledge, 1992), pp. 156, 182.

[16] C. P. Blacker, 'J. B. S. Haldane on Eugenics', *Eugenics Review* 44:3 (1952), pp. 147–148.

Fig. 5.1 Lionel Penrose (reproduced from the L S Penrose papers, UCL Special Collections)

always been a great popularizer, which would have made him an efficient propagandist had he chosen this path. Penrose continued to study blood groups as he was interested in the linkage of genes as causes of disease. Human genetics also had applications for improving life, such as detecting foetal malformation. Progress was made when triploidy, a rare condition whereby the individual has three sets of chromosomes instead of two, was discovered in a spontaneously aborted foetus in about 1960.[17]

[17] Ursula Mittwoch, 'From Eugenics to Human Genetics' in *Penrose: Pioneer in Human Genetics. Report of a Symposium Held to Celebrate the Centenary of the Birth of Lionel*

The Eugenics Society in the Late 1940s

The only directly unpalatable policy the Eugenics Society stood for, 'voluntary' sterilization, did not re-emerge prominently after 1938. The Society had been hoping that the Royal College of Physicians would lend support after its committee reported in 1946. This was not the case. 'Voluntary' sterilization remained an official policy of the Society and was discussed occasionally as part of reproductive practices which included birth control, artificial insemination and abortion. The Society would never start campaigning for it again. When vasectomy became available on the National Health Service in 1972, it was genuinely voluntary and to be used for family planning only. The contraceptive pill had started being prescribed to unmarried women five years earlier. Paying heed to the reputational damage done to its doctrine by the Nazis, the type of eugenics the Society promoted in the post-war period was less abrasive. The only new issue with the potential to create serious friction was Commonwealth immigration from the late 1950s. The Society in fact trod a path between gently warning about the genetic influence and social effects of immigration and coming to terms with the new arrivals. It was changing its nature from being an outright pressure group to becoming a learned association with a mission. Eugenics offered the platform on which a range of social and scientific issues could be discussed.

The Society would still seek to advance its creed after 1945, without the messianic fervour of the early twentieth century. The metamorphosis into an emphasis on slow change, volition and reproductive rights was completed in this period. It had been argued in the relatively recent past that civilization was doomed if the differential birth rate were not reversed. The development whereby intelligence came to be seen as vital persisted after 1945. The knowledge of psychologists allowed only intelligence to be tied directly to the differential birth rate.[18] The new formulation suggested that intelligence might be declining as the higher social classes and talented individuals were not having enough children. It amounted to 'degeneration' in a much milder form.

In late 1946 the Society agreed funding for its Population Investigation Committee of up to £ 2, 000 for a joint project with the

Penrose Held 12th and 13th March 1998 (no place of publication: Centre for Human Genetics at UCL, not dated), p. 40.

[18] SA/EUG/C108. Copy of Draft Stencil to Council Members, 22.7. 37.

Scottish Council for Research in Education. The latter would survey the mental ability of 80, 000 schoolchildren, which could help determine whether intelligence was declining in Britain.[19] The educational psychologist Godfrey Thomson's Galton Lecture in 1946 related to this question. He had already conducted a study of Scottish 11-year-olds in 1932 and another was due in 1947. Other eugenists showed an interest in the forthcoming study. The Scottish Council for Research in Education had received suggestions from the geneticist John Fraser Roberts and Alexander Carr-Saunders about the survey.[20] It then met with the Population Investigation Committee. Another event of 1946 was the publication of Sir Cyril Burt's pamphlet *Intelligence and Fertility* in an occasional series maintained by the Society. He encouraged the Royal Commission on Population to implement periodic mental surveys like Thomson's. From the present facts, Burt believed that intelligence was declining by about one third of a year's mental development per generation. He saw the principal cause as being the differential birth rate. Fraser Roberts and E. O. Lewis, the latter who had conducted the Wood Committee's survey, argued along similar lines.[21] Making birth control more widely available had become the usual solution to the problem.

In 1946, when Sir William Beveridge joined the Consultative Council, it somehow represented the past history of the Society. It included the bishops of Birmingham and Durham, Sir Laurence Brock and Felix Tietze, an Austrian Jewish geneticist who had fled the Third Reich. Sterilization was their common denominator. The two bishops, Ernest Barnes and Hensley Henson, had been known for supporting sterilization in the 1920s and 1930s.[22] Brock had chaired the Departmental Committee on sterilization, while Tietze had supported the Nazi sterilization laws before becoming a refugee. A possible repercussion of a perceived association between eugenics and Nazi Germany was that Macmillan terminated its

[19] SA/EUG/A39. Annual Report 1946–47, p. 2.

[20] Ian J. Deary, Lawrence J. Whalley and John M. Starr, *A Lifetime of Intelligence. Follow-up Studies of the Scottish Mental Surveys of 1932 and 1947* (Washington DC: American Psychological Association, 2009), pp. 16–17.

[21] Richard A. Soloway, *Demography and Degeneration. Eugenics and the Declining Birthrate in Twentieth-century Britain* (Chapel Hill NC: University of North Carolina Press, 1995), p. 342.

[22] Patrick T. Merricks, *Religion and Racial Progress in Twentieth-century Britain. Bishop Barnes of Birmingham* (Cham: Palgrave Macmillan, 2017), pp. 2, 56.

contract to publish the *Eugenics Review* after 27 years.[23] In 1947 the two psychologists, American professors Lewis Terman and Louis Leon Thurstone, joined the Consultative Council.[24] Both were experts on intelligence testing, Terman having translated Alfred Binet's scale into English in 1916 and Thurstone propounding a theory about primary mental abilities. The appointments highlighted a development which helped revive eugenics as a doctrine.

It did not mean that the prejudiced type of eugenics would have no future role. In June 1947 the psychologist Hans Eysenck was co-opted to the council. He would later put forward assertions about black and working-class genetic inferiority, especially in *Race, Intelligence and Education* (1971). Eysenck saw himself as working in the tradition of Galton and wrote the introduction to a 1979 reprint of *Hereditary Genius*. A new committee on 'problem families' was formed in 1947. The dropping of the 'social problem group' terminology was an acknowledgement that they might not constitute a class.[25] A field worker would visit suspected problem families and sift them. Of those who were left in the investigation, a sample would be given mental tests. And, proving that the Society had not completely given up, a new legal opinion was sought on sterilization. 'Voluntary' measures worked well in the United States and Scandinavia, it was claimed.[26]

The Galton Lecture in 1949 was given by Ernest Barnes, the bishop of Birmingham, on 'The Mixing of Races and Social Decay'.[27] Marked by a certain amount of prejudice, the lecture nevertheless did not see miscegenation as worrying. Mixing the races had the advantage of benefitting people of 'lesser' civilization. The bishop's racism then seems to have increased in his public utterances of 1950 and 1951.[28] Racism, however, was not the reason there was an outpouring of denunciation in the press after two speeches he gave in November 1949 and September 1950.[29]

[23] SA/EUG/A39. Annual Report 1946–47, pp. 3, 9.

[24] SA/EUG/A40. Annual Report 1947–48, facing p. 1, pp. 1–2.

[25] Mazumdar, *Eugenics, Human Genetics and Human Failings,* pp. 247–248.

[26] *Eugenics Review* 42:2 (1950), p. 70.

[27] SA/EUG/A41. Annual Report 1948–49, p. 3.

[28] Merricks, *Religion and Racial Progress in Twentieth-century Britain,* pp. 102–104.

[29] *Eugenics Review* 42:4 (1951), p. 188.

These took place in Birmingham, to a Rotary club and the British Association respectively. Barnes's Christianity was heterodox and he supported both euthanasia and compulsory sterilization.[30] It may be significant that he was appointed by another professed Darwinist, Ramsay MacDonald, in 1924.[31]

Following on from Barnes, Jack Trevor spoke on 'Some problems of race'. Brought up in Africa, he claimed not to have been aware of any racial antipathy until he moved to Europe, whereupon he had tried and succeeded in making himself an anti-racist.[32] Trevor was an anthropologist whose lecture was about academic problems of classification. The Society was not intrinsically racist and proved attractive to ethnic minority members in this period. In his book *Eugenics. Galton and After* (1952) Blacker used a comparison with Nazi *Herrenvolk* ideology to argue that Galton had not been particularly racist either. Humans varied because particular climates and physical environments gave an advantage to certain characteristics.[33] But neither Galton nor Blacker believed all races had the same potential to extend Western civilization.[34] This was not a hindrance to well-qualified professionals joining (Table 5.1).

The Population Policies Committee, a joint project with Political and Economic Planning, achieved its purpose in April 1948 when it published a volume called *Population Policy in Great Britain*.[35] It resembled the Population Investigation Committee greatly, including being established in the same year 1936. The latter had taken on an independent existence through being moved to the London School of Economics in 1946. Initiated by Alexander Carr-Saunders and chaired by David Glass, the Society to which they belonged would support it intermittently until 1961.[36] In the post-war period the Society would sometimes cloak its involvement to maximize the impact of its work. It also liked to give the impression

[30] *Eugenics Review* 46:1 (1954), pp. 12–13.

[31] Merricks, *Religion and Racial Progress in Twentieth-century Britain*, p. 18.

[32] *Eugenics Review* 41:1 (1949), pp. 3–4.

[33] C. P. Blacker, *Eugenics. Galton and After* (London: Gerald Duckworth, 1952), pp. 323–324.

[34] *Eugenics Review* 43:1 (1951), p. 22.

[35] SA/EUG/A41. Annual Report 1948–49, p. 3.

[36] Faith Schenk and A. S. Parkes, 'The Activities of the Eugenics Society', *Eugenics Review* 60:3 (1968), pp. 151–152.

Table 5.1 List of new ethnic minority members 1945–1970

Name	*Qualification*	*Reference*
S. C. Chatterjee		ER 39:4 (1948), p. 155
M. M. Desai		ER 40:2 (1948), p. 88
Rupert Jupal		ER 40:3 (1948), p. 156
E. A. Achonwa		ER 41:1 (1949), p. 39
G. M. Khan		ER 41:1 (1949), p. 39
R. N. Sinha	Professor	ER 41:1 (1949), p. 39
Mrs Gob Kok Kee	M B E	ER 43:4 (1952), p. 192
V. S. Sharma	M. A. LL D. F. R. E. S	ER 43:4 (1952), p. 192
S. H. Tan		ER 43:4 (1952), p. 192
Mrs Hem Sanwal		ER 44:3 (1952), p. 158
H. M. Abdul Razack		ER 45:1 (1953), p. 46
R. G. Rajan	Doctor D. C. F. B. S. S	ER 45:2 (1953), p. 107
S. Chandrasekhar (Hon. Fellow)	Professor	ER 46:1 (1954), p. 50
Pierra Vejjabul	Doctor	ER 52:1 (1960), p. 37
P. K. Sharma	Doctor	ER 52:4 (1961), p. 235
G. N. Dev		ER 53:1 (1961), p. 40
T. Rajhathy	Doctor	ER 54:4 (1963), p. 216
S. C. Odibi	F. R. Econ. S	ER 56:2 (1964), p. 101
Saw Swee-Hock	Ph. D	ER 56:4 (1965), p. 208
Abass Ali	F. R. Econ. S	ER 58:2 (1966), p. 96
Suresh K. Arya	B. Pharm. M. S. (Wash)	ER 58:3 (1966), p. 159
M. Khalid Baig	M. Sc	ER 58:3 (1966), p. 159
L. S. Sidhu	M. Sc	ER 58:3 (1966), p. 159
D. A. Olatunbosun	M. B. B. S	ER 59:1 (1967), p. 52
Miss S. Ummat		ER 59:4 (1967), p. 265

that it was a charity rather than a pressure group. In 1950 an extraordinary general meeting was held to amend its articles of association making this clearer, with charitable status avoiding the need to pay income tax.[37] (Fig. 5.2).

The Royal Commission on Population reported in 1949. Along with its detailed fertility surveys which continued to be published until 1954, the report provided the fullest picture of demography since the Fertility of Marriage Census (1911).[38] It noted that general fertility had declined for

[37] SA/EUG/A42. Annual Report 1949–50, p. 2.

[38] Soloway, *Demography and Degeneration*, p. 336.

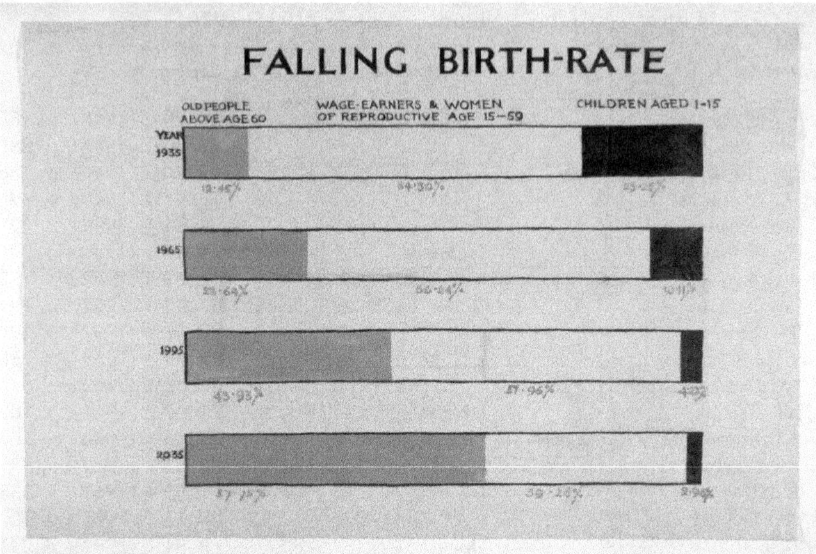

Fig. 5.2 From 1934 there was concern about the falling birth rate (reproduced from the Eugenics Society Archive, Wellcome Library, London)

sixty years until 1934.[39] There had then been a slight increase until 1939, followed by fluctuations during the war, before what was later termed the 'baby boom' started in 1946. Some of these developments took eugenists by surprise. They had become natalist in the 1930s under the impression that the population was declining and did not foresee that fertility would begin rising during the war. It was underway even before the Commission was appointed in 1944.[40] C. P. Blacker said the two influential articles which appeared in *The Times* on 28 and 29 September 1936 overstated the case, but nevertheless made the provincial press take up the 'depopulation' issue.[41] However, when the tide turned, it brought with it new global challenges (Fig. 5.3).

[39] Frank W. Notestein, 'The Report of the Royal Commission on Population: A Review', *Population Studies* 3:3 (1949), pp. 232–233.

[40] Soloway, *Demography and Degeneration*, p. 355.

[41] Blacker, *Eugenics*, p. 148.

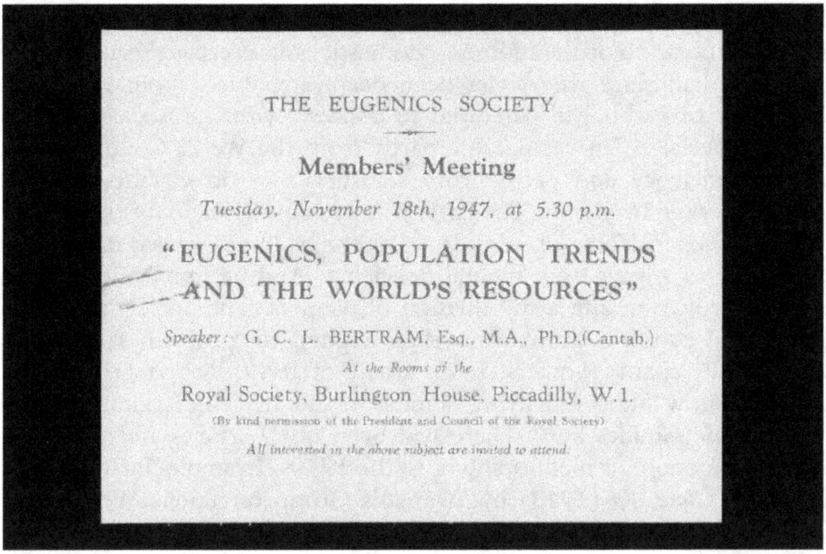

THE EUGENICS SOCIETY

Members' Meeting

Tuesday, November 18th, 1947, at 5.30 p.m.

"EUGENICS, POPULATION TRENDS AND THE WORLD'S RESOURCES"

Speaker: G. C. L. BERTRAM, Esq., M.A., Ph.D.(Cantab.)

At the Rooms of the

Royal Society, Burlington House, Piccadilly, W.1.

(By kind permission of the President and Council of the Royal Society)

All interested in the above subject are invited to attend.

Fig. 5.3 Colin Bertram's presentation on global population problems (reproduced from the Eugenics Society Archive, Wellcome Library, London)

Global Population Concerns and Controversial Issues

A new topic within eugenics was arising. It was introduced by an influential figure in the movement after 1945, the biologist Colin Bertram. A fellow of St John's College, Cambridge, he would serve as general secretary between 1957 and 1963. The novelty related to the effects human populations were having on the environment and how they could be 'managed' in order to limit the destruction of ecosystems. Bertram gave the 1951 Galton Lecture on 'Eugenics and Human Ecology'.[42] Four years earlier, he had spoken on 'Eugenics, population trends and the world's resources' at a members' meeting. Galton had speculated

[42] SA/EUG/A43. Annual Report 1950–51, p. 2.

that mankind might in the future be tended 'as the sheep on a well-ordered moor'.[43] For the moment, domestic issues received priority. The leadership had since the outset been deeply troubled about degeneration. Over time this had narrowed to concern with the 'social problem group' especially. The term came partly from the Wood Committee on mental deficiency and partly from statistician David Caradog Jones's *Social Survey of Merseyside*. The latter had been undertaken for the Society and published in 1934. Caradog Jones's three volumes viewed many cases of poverty as arising from mental deficiency. And he noted higher than average fertility among a sub-division of paupers. The Society's description 'social problem group' was then aligned to 'problem families' as described by charity workers. In an article of 1949, Robert Oxon Black explained how improved living standards and hygienic conditions had passed some families by.[44] There had been other articles and books on the phenomenon, or mentioning it, in the 1940s. Examples included *Our Towns: A Close Up* (1943) by Women's Group on Public Welfare and *Problem Families* (1946) by Robert Wofinden.

The Society began inquires in six districts in 1947 to be led by a dedicated Problem Family Committee under Blacker. Illustrating that eugenics relied on scientific methods though perhaps with unwarranted inferences, the procedure was copied from the Wood Committee. E. O. Lewis had estimated the incidence of mental deficiency by investigating six localities which together were meant to be statistically representative of the nation.[45] Problem families were defined as consisting of the intractably ineducable and being assisted or monitored by multiple social agencies. The Society wanted the Ministry of Health to intervene in an unspecified manner once the survey was complete. Lack of responses from one area (Warwickshire) meant it went ahead with five areas only. The self-appointed task resembled the field work undertaken by the Galton Laboratory before the war, but without the use of a dedicated staff. The Society instead relied on charities and the authorities filling in prepared forms. When the final report reached the office in 1950, the committee could collate information from Bristol, North Kensington,

[43] Francis Galton, *Hereditary Genius. An Inquiry into its Laws and Consequences* (London: Macmillan, 1892), p. 343.

[44] SA/EUG/D169. Problem Families by Robert Oxon Black.

[45] SA/EUG/D169. Problem Families. Proposed Pilot Inquiries, p. 3.

Luton, Rotherham and the West Riding of Yorkshire. In 1952 Blacker published the book *Problem Families: Five Inquiries*. In contrast to the approach before the 1930s, he allowed for environmental effects while placing emphasis upon mental deficiency as a crucial factor.[46]

This project was the clearest example of 'negative' eugenics in the first decades after the war. It risked confirming unpleasant stereotypes about the Society. The families in question suffered from social deprivation and might be seen as victims rather than instigators. The Poor Law was abolished in 1948 and the welfare state was, at least initially, thought to be a cure for social deviance.[47] The critics were mindful of even worse forms of eugenics. In December 1951 Blacker spoke to members on '"Eugenic" experiments conducted by the Nazis on human subjects'. He had been contacted by a government lawyer to comment on the 'eugenic' aspects of these.[48] Unsurprisingly, he saw no connection to the doctrine at all and gave other explanations for what had occurred. The following year he published the book *Eugenics. Galton and After*. It was an attempt to reconfigure the ideology as rational and reasonable. Class prejudice on the part of the members had hamstrung eugenics. He regretted that the poor had sometimes been called 'lower classes', 'riff-raff' and 'dregs'.[49] But he did not concede a resemblance to Nazi views. That was a little disingenuous. In November 1951 he, the lawyer Cecil Binney and social worker Moya Woodside had held a one-day symposium on 'voluntary' sterilization. Binney seemed to hold the door open to sterilization occurring without the consent of a person deemed incapable of agreeing.[50] Blacker too was willing to accept compulsion if 'after prolonged trial and patient effort, persuasion had been found wanting'.[51] The event sought to regain the initiative on the signature policy of the Society in the 1930s.

[46] Patrick T. Merricks, 'Should Such a Faith Offend? Bishop Barnes and the British Eugenics Movement, c. 1924–1953', Ph. D. thesis, Oxford Brookes University, 2014, p. 306.

[47] Mazumdar, *Eugenics, Human Genetics and Human Failings*, p. 142.

[48] C. P. Blacker, '"Eugenic" Experiments Conducted by the Nazis on Human Subjects', *Eugenics Review* 44:1 (1952), p. 9.

[49] Blacker, *Eugenics*, pp. 139, 141.

[50] Cecil Binney, 'Legal Problems Raised by Modern Discoveries about Sex', *The Medical-Legal Journal* 21: 3 (1953), p. 96.

[51] Blacker, *Eugenics*, p. 232.

To counterbalance Blacker's report on problem families, the Society initiated an investigation into 'promising families' too.[52] Why were some families so successful at raising bright, healthy and happy children? They did not necessarily belong to the highest social classes, though the Society was still partial to them. Kenneth Hutton gave a talk on the intelligence, social origins and fertility of pupils and ex-pupils of Winchester College, where he taught chemistry. They were much cleverer than the general population, especially the scholarship boys, and had slightly lower fertility than average.[53] The method chosen by the Promising Families Committee was to obtain recommendations from headmasters and then circulate a questionnaire to the parents of the child. The aim was to discover what factors limited further childbearing by them. The 81 families of grammar school pupils in a southern English county constituted the sample. Meeting four times in 1954, the committee hoped to move on to primary schools later. By 1957 it could report that it had succeeded in this.

Bertram's attentiveness to the environment and other species made his attendance along with Blacker and Dr Margaret Hadley Jackson at the week-long Third International Conference on Planned Parenthood in Bombay in 1952 apposite. At this conference the International Planned Parenthood Federation was founded, receiving three yearly grants of £ 250 from the Eugenics Society. Blacker became influential in the new organization. The focus on the natural environment was perhaps a vicarious argument for why people in developing countries needed to restrict their breeding. It could be a global equivalent of its targeted propagation of contraception at home. The Society had long maintained a connection with the North Kensington Marriage Welfare Centre, making another renewable grant of £ 750 to it in 1953.[54] It covered birth control and other eugenic functions such as advice on marriage or reproduction, premarital health examinations and the treatment of infertility. The type of advice can be seen from a 1954 pamphlet called 'Eugenic Guidance' drawn up specifically for this centre.[55] It focused on anxieties relating to

[52] SA/EUG/A45. Annual Report 1952–53, p. 2.

[53] Kenneth Hutton 'Intelligence Quotients and Differential Fertility. Some Observations from Winchester College', *Eugenics Review* 44:4 (1953), p. 215.

[54] SA/EUG/A46. Annual Report 1953–54, p. 2.

[55] Hanson, *Eugenics, Literature and Culture in Post-War Britain*, p. 45.

weakness in body and mind, which the pamphlet said would be transmitted. It was not written to be reassuring. For the emphasis on global population, the Society worked with Political and Economic Planning again. The latter published a report on world population and resources in 1955. It was an expansion of joint work undertaken with the Society, which had resulted in two broadsheets on the same theme. The intention was to distribute them to delegates at the United Nations World Population Conference in Rome in September 1954.[56] Ernest Simon, formerly a Liberal MP, led the efforts on the Society's side (Fig. 5.4).

Connections forged at the Bombay conference are likely to have spawned an interest in developing countries. It led to a talk on 'Infant mortality in India, 1901–1951: A demographic revolution' by Professor Sripati Chandrasekhar in 1954. He was made an honorary fellow that year and the North Kensington Marriage Welfare Centre joined collectively at the same time.[57] In 1955 the Galton Lecture was given by Professor Sir Cyril Burt on 'The Meaning and Assessment of Intelligence'. Member of the council Hans Eysenck had been one of his students. Later that year, Eysenck spoke on the incipient field of personality differences in a lecture entitled 'The nature and inheritance of extraversion'.[58] There were often family or friendship ties between members. Judith Hubback spoke on 'The fertility of educated women' in February 1955. The topic would form part of her book *Wives Who Went to College* (1957).[59] She was the daughter-in-law of adult education principal the late Eva Hubback, who had been on the Society's council in the 1930s and 1940s.

The Society had a penchant for science with political implications. As immigration from what was later called the New Commonwealth gathered pace, race rose in importance. The anthropologist Geoffrey Ainsworth Harrison addressed a members' meeting in March 1957 on the measurement and inheritance of skin colour.[60] This time it was invested with practical applications. He later used a spectrometer to measure the complexion of mixed-race people and others in Liverpool, for a project

[56] Lord Simon of Wythenshawe, 'Some Aspects of World Population and Food Resources', *Eugenics Review* 46:2 (1954), p. 91.

[57] *Eugenics Review* 46:1 (1954), p. 50.

[58] SA/EUG/A48. Annual Report 1955–56, p. 6.

[59] Hanson, *Eugenics, Literature and Culture in Post-War Britain*, p. 30.

[60] SA/EUG/A49. Annual Report 1956–57, p. 7.

Fig. 5.4 C. P. Blacker accepting the Galton Medal from Sir Charles Galton Darwin (reproduced from the Eugenics Society Archive, Wellcome Library, London)

funded by the Society.[61] It had echoes of Rachel Fleming's research in the 1920s, which had also been anthropometrical and partly conducted in Liverpool. Harrison investigated the medical histories of residents of a particular district with a view to studying intermarriage between black and white people especially (Fig. 5.5).

In October 1957 Bertram became the new general secretary, taking over from Cedric Carter, who had held the post for five years. Blacker, by contrast, had defined an era through serving four times as long. At the jubilee dinner of the Society, held in December that year to mark fifty years of its existence, Blacker was awarded the Galton Medal.[62] On this occasion, it received fair coverage in the *Manchester Guardian*, which saw the Society's purpose as humanizing the struggle for survival. The newspaper wrote it had succeeded in implementing mental deficiency legislation and tax breaks for parents.[63] For its anniversary in 1957, the Society published a new statement of aims.

Light on specific policies, it continued to support 'voluntary' sterilization and suggested that eugenically 'inferior' couples should have just one or two children, in contrast to 'valuable' ones who should aim for at least four.[64] It supported birth control in all forms and the termination of pregnancy when requested by the prospective parents. Abortion was especially pertinent in preventing hereditary afflictions on similar grounds to the factors justifying 'voluntary' sterilization. Having by now had experience of the welfare state, it went back to recommending a graded system of child benefit, proportional to income and the number of children. On immigration and race mixture, it confined itself to stating that greater knowledge was needed. The new general secretary Bertram wrote a broadsheet on West Indian immigration, which appeared in 1958. Its recommendations comprised studying to what extent miscegenation had occurred and the possible consequences, educating people on the laws of heredity, restricting immigration by barring less resourceful types and introducing an annual quota.[65] Although the broadsheets were not as crude as Sir Oswald Mosley's agitation in the 1959 general election, the

[61] *Eugenics Review* 51:4 (1960), p. 205.

[62] SA/EUG/A50. Annual Report 1957–58, p. 6.

[63] *Manchester Guardian*, 4 December 1957, p. 6.

[64] *Eugenics Review* 50:1 (1958), pp. 27, 29.

[65] *Daily Telegraph*, 2 October 1958.

Daily Mail

SCIENTISTS ON POLAR FLIGHT

RAF Intensifying Research into Roof-of-the-World Navigation

ICELAND TO THE YUKON

Cambridge Expert Taking Secret List Equipment

Three co-pilots

Ice-cap Party to Get Supplies

Fig. 5.5 Colin Bertram on the front page of the *Daily Mail* (reproduced from the Eugenics Society Archive, Wellcome Library, London)

message was a warning about the dangers inherent in vast movement of peoples.

Mosley had agreed with 'mainline' eugenic assumptions in the interwar period, though the Society chose not to engage with the British Union of Fascists.[66] He had repeated such ideas after 1945.[67] When he stood for Parliament in the North Kensington constituency in 1959, a year after race riots targeting West Indian immigrants in that part of London, he did not mention eugenics. He did, however, use the standard argument of the birth rate. There would soon be '5 × as many Blacks to compete with jobs when post-war boom over'.[68] He concentrated on the facts on the ground, a mixture of bemoaning immigration for allegedly reducing native wage rates and sensationalist material about how cultures and cleanliness marked the new arrivals out from local people.[69] By the time of Enoch Powell's infamous 'Rivers of Blood' speech in Birmingham in 1968, support for restrictions on immigration had grown. Workers in several cities went on strike in support of Powell after he was sacked from the shadow cabinet and regular politicians had begun listening to sceptics in the main destination areas for immigrants.[70] Harold Wilson's governments pursued a two-pronged strategy of restricting immigration combined with anti-discrimination legislation. They passed the Race Relations Acts of 1965, 1968 and 1976, which satisfied eugenists too. By the 1960s many in the Society were on the tolerant side of the spectrum. The reaction to the broadsheets in 1958 from members and others who had received them brought support and criticism in about equal measure.[71]

The statement of aims had described miscegenation as of 'profound importance' and the distribution of the broadsheets was extensive. All MPs and Privy Councillors received a copy, as did some peers and bishops,

[66] Oswald Mosley, *Fascism. 100 Questions Asked and Answered* (London: B U F Publications, 1936), note 76. SA/EUG/C191 Blacker to Archibald James 10 Oct 1933.

[67] Oswald Mosley, *The Alternative* (Ramsbury: Mosley Publications, 1947), p. 287.

[68] Queen Mary Archive, London. Chesworth Papers. PP 2/46, Manuscript about the meetings, pp. 7–8 (handwritten section).

[69] D. S. Lewis, *Illusions of Grandeur. Mosley, Fascism and British Society 1931–81* (Manchester: Manchester University Press, 1987), p. 243.

[70] Tony Benn, *Office Without Power. Diaries 1968–72* (London: Arrow Books, 1989), p. 60.

[71] SA/EUG/D103. The critical letters were from P. H. G. Scott (4 Nov 1958), Thelma Mendoza (10 Dec 1958) and Charles Brunton (21 Oct 1958).

the headmasters of the main independent schools and a number of trade unions.[72] The Society decided to repeat what it felt was a success by producing a second broadsheet, dealing with both 'problem' and 'promising' families. Before publication, the broadsheet was halted.This was mostly because the Society wanted to wait before conducting further propaganda. The official reason given was that its findings, now solely relating to 'promising' families, were better suited for an article in the *Eugenics Review*.[73] Its questionnaires to such parents showed that increased financial assistance would encourage promising families to increase their fertility. Although affluent families were overrepresented among the children studied, less well-off families showed energy and ability in their leisure activities.[74]

Originally regarded as a means to 'correct' the differential birth rate, in the post-war period the Society accepted contraception across the board. At a members' meeting in 1959, Blacker spoke in favour of a proposition that the British population should be stabilized at 40 million. No vote was taken on the motion, which had the virtue of addressing both recent concerns over environmental degradation and immigration. That year witnessed the Society gaining the freehold of 108 Whitfield Street, Fitzrovia under a legacy from its famous member, Marie Stopes, who had died in 1958. She had shared similar goals for eugenics as Blacker and was friendly to him.[75] A pioneer of contraception, Stopes had established Britain's first birth control clinic in Holloway Road, London in 1921.[76] Four years later it moved to her own house, becoming the clinic she hoped the Society would run. This tallied well with its aims and ambitions. Although it would keep the Marie Stopes Memorial Foundation separate from itself, in practice it belonged to the Society. Already funding and guiding a birth control clinic in North Kensington, it was gaining the opportunity to manage one according to its own preferences, while taking little risk since the capital came from a bequest. It was a new venture for

[72] SA/EUG/A51. Annual Report 1958–59, p. 7.

[73] SA/EUG/A52. Annual Report 1959–60, p. 8.

[74] C. O. Carter, 'Promising Families: Some Conclusions', *Eugenics Review* 52:4 (1961), pp. 197–198.

[75] SA/EUG/K1. Stopes to Blacker 20 May 1952.

[76] Jeffrey Weeks, *Sex, Politics and Society. The Regulation of Sexuality since 1800* (London: Longman, 1981), p. 189.

a new age. Previously it had used the same means of propaganda as the Malthusian League, an earlier birth control association: the public lecture, the tract and the press.[77] The strongest link was Charles Vickery Drysdale, who had a background in the League and gravitated towards the Society when the former became defunct in the aftermath of the First World War.[78]

A Crisis of Eugenics?

The Mental Deficiency Act of 1913 was the eugenics movement's greatest achievement. It was superseded by the 1959 Mental Health Act, which updated the definitions of subnormality. It came with a shift of emphasis towards treatment rather than solely institutionalization.[79] In 1959 the Society had its standing enhanced by the appointment of Sir Julian Huxley as president. The presidency was an honorific position, but having a leading scientific figure as its frontman was beneficial, especially since Huxley was known to be progressive. He had helped to move biology away from an undue focus on race and been consulted by the anthropologist Ashley Montagu when the latter wrote the two UNESCO Statements on Race. These were published in 1950 and 1951. The first received criticism for being moralistic rather than scientific, for instance the assertion that humans felt an urge towards brotherhood. Privately, Blacker opposed 'the egalitarian nonsense put out by Unesco and the left wing'.[80] The second statement contained commentary which included a variety of views. Huxley had been the first director-general of UNESCO between 1946 and 1948. He took advantage of the position to promote scientific humanism or a type of progressive eugenics, running into opposition from conservatives and Catholics.[81] He succeeded the physicist Sir Charles Galton Darwin, the grandson of the naturalist, as president of the Society.

[77] F. D'Arcy, 'The Malthusian League and the Resistance to Birth Control Propaganda in Late Victorian Britain', *Population Studies* 31:3 (1977), p. 430.

[78] *Bulletin* 10:1 (1978), p. 27.

[79] Hanson, *Eugenics, Literature and Culture in Post-War Britain,* pp. 39, 95.

[80] Wellcome Library. Blacker papers. PP/CPB/B21/1. Blacker to Sir Charles Darwin 17 Sep 1960.

[81] Stefan Kühl, *For the Betterment of the Race. The Rise and Fall of the International Movement for Eugenics and Racial Hygiene* (Basingstoke: Palgrave Macmillan, 2013), p. 139.

Huxley's short presidency lasting until 1962 was nevertheless not an outstanding success. The 1961 Galton Lecture, due to be given by the president, had to be postponed due to his indisposition and the pressure of other commitments. Attendance at meetings was low, with fewer than a dozen members present at the annual general meeting. As a reaction to neither of the two members' meetings held in 1961 generating much interest, the Society discontinued this regular feature of its calendar.[82] At the same time, Blacker, a great asset, stood down as honorary secretary. The *Eugenics Review* changed publishers from Cassell, an imprint of Hamish Hamilton, to the lesser known Pergamon Press (Table 5.2).

Blacker had been aware of the negative spiral on which the Society found itself. Membership was down to 456 in 1956 from its highest point of 768 in 1932.[83] On 17 February 1957 he wrote a memorandum about what could be done to reinvigorate the Society. He noted American 'reform' eugenist Frederick Osborn's Galton Lecture the year before, which had included suggestions for how eugenic aims could be promoted without using the name. He did not recommend this course of action, as it would discourage the membership further. Nor did he suggest becoming a Trust or Foundation rather than a society for members. Instead, he proposed increased cooperation with other bodies, such as the British Social Biology Council (the former British Social Hygiene Council, which Sybil Gotto had founded). And he recommended focus on emigration and immigration, with the ultimate goal of reducing the population in 2000 to what it had been in 1900, which was only 38 million. That would allow a campaign against West Indian immigration without dwelling on questions of colour. A cognate opinion was offered

Table 5.2 Publishers of the *Eugenics Review* 1945–1968 (including imprints)

Date	*Issue*	*Publisher*
January 1945	36:4	Macmillan
April 1946	38:1	Hamish Hamilton
April 1949	41:1	Cassell
January 1961	52:4	Pergamon Press
March 1965	57:1	Oliver & Boyd

[82] SA/EUG/A54. Annual Report 1961–62, pp. 3–4, 7.

[83] SA/EUG/J7. The Eugenics Society's Future by C. P. Blacker, p. 1.

by Colin Bertram, who took the latter suggestion forward, except that he did write about race. He too wanted to work against overpopulation globally. The broadsheet idea was a substitute for publicity, as he did not feel the Society could directly influence public opinion yet.[84] Bertram sketched ideas for a council meeting on 9 October 1957, convened to consider future policy, and then wrote a memorandum in November 1957. Blacker's proposals, meanwhile, were considered by the council on 1 May 1957 without achieving traction.

Others also vouchsafed a sense of direction for the Society. A different faction consisting of Alan Parkes, Alan Stevenson, James Tanner and Conrad Waddington forced an extraordinary general meeting on 23 November 1960. Parkes had circulated a memorandum in January that year with the proposition of changing its name to the 'Galton Society' and limiting the consecutive term for all serving honorary officers to six years.[85] He was defeated over the change of name and had to concede that the honorary treasurer should be exempt from the retirement rule. However, it would still mean the enforced departure of Blacker as honorary secretary, as he had served in that capacity since 1952. Blacker suspected that Parkes, who would in fact receive funding for research on artificial insemination in the 1960s, had designs on the Society's capital of more than £ 100, 000.[86] Removing Blacker, and originally his friend the honorary treasurer Aird Whyte, were meant to facilitate this. At the extraordinary general meeting, Parkes agreed not put the motion. The process of change nevertheless moved forward because in the run-up to the extraordinary general meeting, Blacker had said he would stand down at the next annual general meeting in 1961. And the executive meeting taking place earlier on 23 November 1960 had rubber stamped reintroducing the Darwin research scholarships, lapsed during the Second World War, and renewing the grant to the International Planned Parenthood Federation.[87]

Both these policies were passed, pivoting the Society towards research and cooperation with likeminded organizations. The International Planned Parenthood Federation had expanded rapidly with 32

[84] SA/EUG/J7. The Policy of the Eugenics Society by Colin Bertram, p. 5.

[85] PP/CPB/B21/3. Memorandum by Certain Members of the Council of the Society.

[86] PP/CPB/B21/2. Blacker to Aird Whyte 14 Oct 1960.

[87] PP/CPB/B21/2. Agenda of 23 November 1960.

chapters across four regions in 1961.[88] It had been founded by American birth control campaigner Margaret Sanger, with Blacker acting as administrative chairman.[89] The Galton Lecture, showcasing serious academic content, could still attract 60 members. Two Darwin fellows had in fact already been appointed, one on the cellular basis of inheritance and one researching variations in childbirth.[90] This of course cost money with the annual subscription doubling from the old rate of one guinea a year.

At about the same time, the Galton Laboratory at University College London was boosted by a significant donation of £ 200, 000 from the Wolfson Foundation. It was scarcely a centre of eugenics anymore, but the medical and other types of genetics researched by Penrose and Hans Grüneberg, a former Jewish refugee from Nazi Germany, was ongoing. Grüneberg held a chair, though the professorship of biometry had stood vacant since the departure of Haldane. The staff's work had suffered from the cramped conditions at the main quadrangle of the College.[91] This funding would enable what was still called the Department of Eugenics, Biometry and Genetics to be rehoused in a refurbished building near the Euston Road, about five minutes' walk from the main campus. It became known as Wolfson House.[92] The federal University of London was still involved as it leased the building to the College. By 1965 all mention of Galton's doctrine would be gone from the department and the chair he founded. Similarly to the traditional emphasis of the French version, it had become a branch of medicine.[93] That year Harry Harris took up what had become the Galton chair of human genetics. Wolfson House opened in May 1967.[94]

[88] Hanson, *Eugenics, Literature and Culture in Post-War Britain*, p. 123.

[89] PP/CPB/B21/1. Blacker to Julian Huxley 11 Mar 1960.

[90] SA/EUG/A53. Annual Report 1960–61, pp. 3, 7.

[91] *Eugenics Review* 53:3 (1961), p. 133. Also reported in *The Times* on 31 May 1961.

[92] Adam Rutherford, *Control. The Dark History and Troubling Present of Eugenics* (London: Weidenfeld & Nicolson, 2022), p. 136.

[93] Jean-Paul Gaudillière, 'Le Syndrome nataliste: Hérédité, médecine et eugénisme en France et en Grande-Bretagne, 1920–1965' in Jean Gayon, Daniel Jacobi and Marie-Claude Lorne (eds.), *L'Eternel retour de l'eugénisme* (Paris: Presses Universitaires de France, 2006), pp. 179, 197.

[94] D. A. Hopkinson, 'Harry Harris. 30 September 1919–17 July 1994', *Biographical Memoirs of Fellows of the Royal Society* 42 (1996), p. 162.

The Galton Laboratory had taken on a new role as a centre for human genetics. The Eugenics Society's renewed focus on research encompassed contraception, a direct way of continuing to pursue its objectives. The council gave £250 to its Medical Committee for trialling a foam tablet to be used as a spermicide. On 2 May 1961 the Marie Stopes Clinic was officially opened by Sir Russell Brain, a neurologist, and began prescribing the contraceptive pill to some of its patients. One of the pill's inventors, Gregory Pincus, initiated a conference in London that year called 'Man and his Future'.[95] Julian Huxley gave the opening address with other eugenists in attendance. These developments were accompanied by another look at what could be done on 'voluntary' sterilization.[96] Blacker had returned to the issue in articles and had written the book *Voluntary Sterilization. The Last Sixty Years* as well as the pamphlet 'Voluntary Sterilization. The Need for Reassessment'. Both were published in 1960. The general climate might be more auspicious in the 1960s, being affected by the ongoing sexual revolution. In 1963, however, the ad hoc committee for the issue concluded that the Society should not do anything active on 'voluntary' sterilization. Due to the ill-fated campaign in the 1930s, it was felt that any pressure applied would be counterproductive.[97] Writer and former R. A. F. pilot Richard Rumbold had left a bequest to the Society, and it was decided not to spend it on sterilization propaganda.

It was the sustained efforts to introduce 'voluntary' sterilization in the 1930s which put the Eugenics Society at odds with many politicians. When vasectomy was introduced in 1972, the context had shifted towards the operation being genuinely desired by the patient. Patterns of greater openness about sexuality, the prevalence of contraception, abortion being available and less deference to authority made it a safer choice. The decade in which a particular case of sterilization took place is indicative of whether it was voluntary or not. With vulnerable patients, forced sterilizations might have happened as late as the 1970s, especially in the United States. Even in Britain, doctors might take less care with 'problem

[95] Paul Weindling, 'Julian Huxley and the Continuity of Eugenics in Twentieth Century Britain', *Journal of Modern European History* 10:4 (2012), p. 494.

[96] SA/EUG/A55. Annual Report 1962–63, pp. 7–8.

[97] SA/EUG/A56. Annual Report 1963–64, p. 8.

families' and immigrants. For instance, the injectable contraceptive Depo-Provera, with known serious side effects, tended to be prescribed to these groups.[98]

When the Society decided not to get involved on sterilization, it was for legal as well as practical reasons. On 21 November 1963 another extraordinary general meeting took place, removing the propaganda dimension in the Memorandum of Association. This allowed the Society to update its registration as a charity under the Charities Act of 1960.[99] In 1960 it had been decided to conduct a social survey of attitudes to eugenics, pausing its broadsheets until it had received enough feedback. Reflecting its elite origins, a thousand people were initially chosen at random from *Who's Who* and sent a questionnaire.[100] Four hundred of them replied in time for their responses to be collated. Eighty-nine per cent understood broadly what eugenics meant, 40% were aware of the latest development in encouraging promising families, 55% supported making 'voluntary' sterilization easier, 31% supported legalizing abortion and 44% thought that world population was too high for the available resources.[101]

A New Organizational Structure

The survey was intended to point a way for future work. In 1965 the Society withdrew its statement of aims, with an abbreviated update becoming available the next year.[102] The deliberation on what its precise goals should be ended in further emphasis on being a learned association. The introduction of a symposium, a yearly conference with numerous sessions spread over two days, was one way this manifested itself. The first of these, on 'Biological Aspects of Social Problems', took place on 1 and 2 October 1964 at University College London. It aimed at combining genetics and social science. The proceedings were published and, though a smaller event, resembled the First and Second International Congresses of Eugenics in this respect. The four sessions covered

[98] Hanson, *Eugenics, Literature and Culture in Post-War Britain*, pp. 118–119.

[99] SA/EUG/A56. Annual Report 1963–64, p. 4.

[100] *Eugenics Review* 54:2 (1962), p. 57.

[101] SA/EUG/D75. Gallup Poll on Eugenics.

[102] SA/EUG/A58. Annual Report 1965–66, p. 5. SA/EUG/A59. Annual Report 1966–67, p. 3.

demography, social mobility, genetics and fertility control. The last of these sessions, chaired by Julian Huxley, included two papers on 'problem families' and two on immigrants or inhabitants of 'underdeveloped countries' in the contemporary parlance. Lionel Penrose chaired the session on genetics. He criticized the early assumptions of eugenics and stressed that he was not a member of the Society.[103] The table below gives details of the symposia held by the Society and shows the range of its concerns during this period (Table 5.3).

The second symposium, in 1965, covered the topic of human ability, another mainstay of eugenics. It ran along parallel lines to the first and helped establish a format. The fourth session covered differential fertility in relation to intelligence. Cedric Carter, a geneticist, spoke directly on the relationship, while the statistician Bernard Benjamin concentrated on the socio-economic aspects of fertility.[104] Declining average intelligence caused by differential fertility was a eugenic discourse going back decades. It had been given a wider audience through immunologist Sir Peter Medawar's 1959 Reith Lecture on the BBC.[105] Perhaps its elements were on the verge of change though. An analysis of the 1961 census returns showed that unskilled workers were no longer having larger families than professionals. In fact, the largest families were to be found among those with jobs in the highest echelons.[106]

In 1964 Colin Bertram was succeeded as general secretary by Faith Schenk, who became the first woman in the post (although Cora Hodson had served as business secretary 1920–1931, before the current role had come into being).[107] Schenk would oversee major organizational changes connected to the academization of the eugenics movement. The field would be widened with new thinking. She had started out as business secretary in 1946 as the successor to Gwendolyne Collyer.[108] In 1971 she married fellow member the demographer Peter Cox.

[103] J. E. Meade and A. S. Parkes (eds.), *Biological Aspects of Social Problems* (Edinburgh: Oliver & Boyd, 1965), p. 111.

[104] J. E. Meade and A. S. Parkes (eds.), *Genetic and Environmental Factors in Human Ability* (Edinburgh: Oliver & Boyd, 1966), p. viii.

[105] Hanson, *Eugenics, Literature and Culture in Post-War Britain*, p. 57.

[106] *Eugenics Review* 59:1 (1967), p. 1.

[107] *Eugenics Review* 45:2 (1953), p. 78.

[108] PP/CPB/B15. Blacker to Schenk 1 May 1946.

Table 5.3 List of the Society's symposia 1964–1990

Year	*Title*	*Venue*	*Reference*
1964	Biological Aspects of Social Problems	University College London	Annual Report 1964–65, p. 8
1965	Genetic and Environmental Factors in Human Ability	University College London	Annual Report 1965–66, p. 7
1966	Social and Genetic Influences on Life and Death	University College London	Annual Report 1966–67, p. 5
1967	Genetic and Environmental Influences on Behaviour	Zoological Society of London	Annual Report 1967–68, p. 6
1968	Biological Aspects of Race	Zoological Society of London	Annual Report 1968–69, p. 7
1969	Biological Aspects of Sex	Zoological Society of London	Annual Report 1969–70, p. 6
1970	Biological Aspects of Human Fertility	Zoological Society of London	*Bulletin* 2:4 (1970), p. 69
1971	Population and Pollution	Zoological Society of London	*Bulletin* 3:4 (1971), p. 74
1972	Resources and Population	Zoological Society of London	*Bulletin* 4:4 (1972), p. 86
1973	Population and the New Biology	Zoological Society of London	*Bulletin* 5:1 (1973), p. 11
1974	Equalities and Inequalities in Education	Zoological Society of London	*Bulletin* 6:1 (1974), p. 13
1975	Equalities and Inequalities in Health	Zoological Society of London	*Bulletin* 7:2 (1975), p. 44
1976	Equalities and Inequalities in Family Life	Zoological Society of London	*Bulletin* 8:4 (1976), p. 108
1977	Perimeters of Social Repair	Zoological Society of London	*Bulletin* 9:1 (1977), p. 20
1978	Changing Patterns of Sexual Behaviour	Zoological Society of London	*Bulletin* 10:1 (1978), p. 19
1979	Changing Patterns of Conception and Fertility	Zoological Society of London	*Bulletin* 11:1 (1979), p. 19
1980	Changing Patterns of Child-Bearing and Child Rearing	Zoological Society of London	*Bulletin* 12:2 (1980), p. 53
1981	Demography of Immigrants and Minority Groups in the United Kingdom	Zoological Society of London	*Bulletin* 13:2 (1981), p. 55
1982	Developments in Human Reproduction and Their Eugenic and Ethical Implications	Royal Geographical Society	*Bulletin* 14:4 (1982), p. 117

(continued)

Table 5.3 (continued)

Year	*Title*	*Venue*	*Reference*
1983	The Biology of Human Intelligence	Zoological Society of London	*Bulletin* 15:2 (1983), p. 63
1984	The Health and Development of Children	Zoological Society of London	SA/EUG/P32
1985	The Political Economy of Health and Welfare	Zoological Society of London	SA/EUG/P34
1986	Natural Human Fertility: Social and Biological Mechanisms	Zoological Society of London	SA/EUG/P34
1987	Evolutionary Studies: A Centenary Celebration of the Life of Julian Huxley	Zoological Society of London	SA/EUG/P34
1988	Controlling Legal Addictions	Zoological Society of London	SA/EUG/P35
1989	Applications of Molecular Genetics in Medicine: Some Ethical Implications	Zoological Society of London	Annual Report 1989, folio 5 recto
1990	The Genetics, Demography and Health of Minority Populations	Zoological Society of London	SA/EUG/P36

The Galton Foundation, a charitable trust set up by the Society, came to fruition in August 1968.A product of the fiftieth anniversary in 1967 and the idea of promoting eugenics under a different name, it enabled propaganda to continue after this had been curtailed in the Society itself. Its remit to advance 'biosocial science' included the conservation, evolution and progress of mankind.[109] Interconnected with the changes, the *Eugenics Review* would be replaced by the new *Journal of Biosocial Science* after December 1968. The periodical would be an academic journal, rather than a house magazine for eugenics. News and discussion would be the preserve of a regular newsletter.

Professor Sir Alan Parkes took over as president in 1968 and a revised statement of the Society's aims was approved. It was not widely publicized, meaning it had become relatively unimportant after the ideological functions were transferred to the Galton Foundation. For the first time, the Galton Lecture was disassociated from his 16 February birthday and

[109] SA/EUG/A60. Annual Report 1967–68, pp. 5–6.

integrated instead into the Society's symposium. That autumn's symposium considered 'Biosocial Aspects of Race', a timely topic given the Commonwealth Immigration Act of 1968, which had replaced the earlier legislation from 1962.[110] In 1969 a government report *The Problem of Coloured School Leavers* recommended remedial action for this group, arguing that fostering equality did not always imply equal treatment.[111] Although possibly stigmatizing, this was an early example of anti-racism and 'positive discrimination'.

Another separate entity was the Marie Stopes Memorial Foundation. It managed the activities relating to birth control at 106 and 108 Whitfield Street. It was linked to the Society for Constructive Birth Control, responsible for the finances and premises of the clinic. Both were founded in 1960 as separate companies under the chairmanship of Colin Bertram and with Faith Schenk as secretary.[112] The name of the latter deliberately echoed Stopes's own 'Society for Constructive Birth Control and Racial Progress', which had included eugenists Carr-Saunders, Huxley and John Maynard Keynes as vice-presidents in 1927.[113] Since Stopes had bequeathed her assets to the Eugenics Society, it sold her house to the Society for Constructive Birth Control for £ 2, 600. It worked to reduce the birth rate among mothers considered 'unsuitable'. This happened through advice on contraception, including pioneering the contraceptive pill and training overseas nurses on how to give it to patients. There was a big jump in the number of women benefiting in 1964.[114]

The same year, though, the first stirrings towards winding up the new organization occurred. The economist James Meade, honorary treasurer of the Eugenics Society and later a Nobel laureate, thought doing so would save time and resources.[115] For instance, the two societies had needed to work jointly on repairing the façade of the building. In 1967 it was decided to try to transfer the patients to the boroughs of Camden and Westminster, the two local authorities responsible for the district

[110] SA/EUG/A61. Annual Report 1968–69, pp. 2, 4, 6–7.

[111] Hanson, *Eugenics, Literature and Culture in Post-War Britain*, p. 113.

[112] SA/EUG/K38 The minute book, pp. 1, 4. SA/EUG/K39 Minutes of 10 February 1960.

[113] SA/EUG/K1. Stopes to Hodson 10 Jun 1927.

[114] SA/EUG/A57. Annual Report 1964–65, p. 9.

[115] SA/EUG/K38 The minute book, pp. 9, 16.

which the clinic served.[116] The process was only completed in 1976. In January the clinic was handed over to the private company Population Service (Family Planning), which gained a short lease of the building.[117] The Marie Stopes Memorial Foundation and the Society for Constructive Birth Control were liquidated later in the year.[118] Their substantial assets of nearly £ 70, 000 and approximately £ 80, 000 respectively were channelled to the Eugenics Society.[119] The sums easily counterbalanced general deficits above £ 8, 000 in 1976 and £ 9, 000 in 1977.

As a replacement, the Marie Stopes Research Fund was set up, facilitating studies on birth control, demography, eugenics and sexology from 1978.[120] The associated research committee could suggest topics in need of inquiry, and therefore worthy of funding, to the council. The number of applications rose over time, with six proposals under consideration in 1979, eight in 1981 and fourteen in 1985.[121]

The Quarrel Over Abortion

In 1967 abortion was legalized for the first time in Britain. The Society welcomed it as a social reform. This particular issue was nevertheless at the root of a serious quarrel within it. The altercation was triggered by the appointment of a research fellow in the person of Madeleine Simms to study the effects of the Abortion Act.[122]

At the 1970 annual general meeting the Cambridge zoologist Charles Goodhart referred to the Research Advisory Committee, which he said consisted entirely of people who were pro-abortion. He suggested that it broaden its base by including members from the other side as well.[123] The council regretted that partiality had been imputed and agreed to consider the proposal. In reality, the leading figures were all in favour of abortion,

[116] SA/EUG/K39 Minutes of 10 July 1967.

[117] SA/EUG/K39 Minutes of 29 January 1976.

[118] SA/EUG/A69. Annual Report 1976, p. 8.

[119] SA/EUG/A70. Annual Report 1977, pp. 7–8.

[120] SA/EUG/P46. Minutes of the council 26 February 1976, p. 3.

[121] SA/EUG/K48. Folios 1–6.

[122] SA/EUG/A70. Annual Report 1969–70, p. 6.

[123] SA/EUG/A71. Annual Report 1970, p. 2.

with Huxley acting as vice-president of the Abortion Law Reform Association 1969–1970.[124] Goodhart had written a recent article in which he claimed that the commonly repeated statistic of 50–100, 000 illegal abortions a year in Britain was too high.[125] A High Anglican, he was an opponent of legalized abortion and therefore did not wish to exaggerate the problem of back-street abortions.

In 1973, however, there was talk of disbanding the committee, a move which Goodhart criticized at that year's annual general meeting held in July.[126] His intervention again showed his misgivings about the research of Madeleine Simms. Clearly the hope was that the reconstituted committee might control or restrict her work on abortion. John Peel, a geographer and veteran member of the council, explained that Mrs Simms's research conducted with funding from the Society had now ended. Instead, the Social Science Research Council had awarded her a grant to study governmental committees and abortion. She was still housed within Eugenics Society premises, which was paying her overheads and administering the grant. Goodhart would in the future cast doubt on Mrs Simms's independent standing. That her work was liked by the council is evident. *Abortion Law Reformed* (1971), the book which she co-authored with Keith Hindell, was reviewed with satisfaction by Barbara Thomson in the *Bulletin*. It deserved 'to be widely read' and would appeal to both specialists and readers without prior knowledge.[127] It featured a foreword by David Steel, the Liberal MP who is most closely associated with the passage of the Abortion Act in 1967.

At the annual general meeting in July 1974, Goodhart twice intervened on this topic. First, he asked about the Research Advisory Committee and was told that it had been disbanded since the Society did not employ a research fellow. He went on to query the truthfulness of this. John Peel responded on behalf of the council that it had a research officer, Mrs Simms, and that the Society provided office space for her and 'greatly facilitated' her employment.[128] It was right to do

[124] Weindling, 'Julian Huxley and the Continuity of Eugenics', p. 493.

[125] C. B. Goodhart, 'The Frequency of Illegal Abortion', *Eugenics Review* 55:4 (1964), p. 200.

[126] SA/EUG/A66. Annual Report 1973, p. 1.

[127] *Bulletin* 3:2 (1971), p. 36.

[128] SA/EUG/A67. Annual Report 1974, pp. 1-2.

so, he claimed, because her project had begun under its auspices. Goodhart had persistently followed up his dissatisfaction with the research fellowship. At a council meeting earlier in 1974, Cedric Carter and John Thoday both reported having received a letter from him. In response, they conceded that Madeleine Simms should describe herself as affiliated to the Social Science Research Council and use her private address when writing articles of a political nature.[129]

This was turning into the most serious quarrel for some time, caused by differing views on abortion. Although draped in phraseology relating to the existence of committees or the employment status of an individual, what Goodhart really wanted was that the Society should diminish its support for abortion. He was right on the technical issues, as the accounts for 1973 and 1974 showed that £ 160 and £ 107 respectively had been spent on Madeleine Simms's 'research fellowship'. Using different sources, five years after his previous article, he published a paper claiming that 100, 000 illegal abortions per annum before 1967 was five times too high an estimate.[130] He may have influenced Thoday, his colleague who held Fisher's former chair at Cambridge, because at the 1976 annual general meeting the latter started asking about Madeleine Simms's research fellowship in 1974 and 1975.[131] Carter repeated that it was the Social Science Research Council which had funded her. The Society had merely provided accommodation, stationery, the use of a telephone and incidental expenses for her. Madeleine Simms's Research Council funding had lasted until 30 April 1975. Later records showed that the Society had disbursed the considerable sum of £ 13, 275 to Mrs Simms for her research fellowship between 1970 and 1976.[132]

The Bulletin of the Eugenics Society

In 1969 the new *Bulletin* became available. Edited by Faith Schenk, its purpose was to bring news, book reviews and short articles to the membership, taking over from the former *Eugenics Review*, which had

[129] SA/EUG/P45. Minutes of 20 February 1974, p. 4.

[130] C. B. Goodhart, 'Estimation of Illegal Abortions', *Journal of Biosocial Science* 1:3 (1969), pp. 235–245.

[131] SA/EUG/A69. Annual Report 1976, p. 1.

[132] SA/EUG/K46. Some grants given by the Eugenics Society, p. 2.

additionally printed scientific articles. In April a letter was circulated to medical libraries and departments offering a free year-long subscription. Twenty-six of these took up the offer.[133] Various organizations in London and the United States also received the *Bulletin* free of charge, sometimes in return for their own journal. The academic papers would henceforth go into the *Journal of Biosocial Science*. Although officially independent, it published the proceedings of some of the Society's symposia.[134] The journal did, however, cast the net wider by carrying articles by scholars unaffiliated to eugenics.

An old-style eugenist, the psychologist Richard Lynn, warned against 'New Environmentalism' in the *Bulletin* in 1969. He was worried about the brain drain, fuelling his concern with statistics showing that the emigration of graduates amounted to 20% of the annual output of new degree holders.[135] It was furthermore increasing by 5% each year. Lynn regarded increased funding to schools to counteract the tendency, suggested in the previous issue, as a case of 'tax and spend' and essentially futile. He reminded readers that Galton had explained the collapse of civilizations as occurring when the ablest disappeared. Moreover, Britain could be caught in a vicious cycle whereby a comparatively low standard of living pushed talented individuals to leave, further reducing economic success. This was so similar to what Sir Keith Joseph, the Conservative MP, would claim in a notorious speech in 1974 that it is possible he was a reader:

> 'If we do nothing, the nation moves towards degeneration, however much resources we pour into preventative work and the over-burdened educational system. It is all the more serious when we think of the loss of people with talent and initiative through emigration as our semi-socialism deprives them of adequate opportunities, rewards and satisfactions'.[136]

Joseph was one of the few politicians in the post-war period who evidently believed in eugenics. Like Enoch Powell, he made a fateful

[133] SA/EUG/A70. Annual Report 1969–70, p. 5.

[134] David Pilgrim, 'The Eugenics Legacy in Psychology and Psychiatry', *International Journal of Social Psychiatry* 54:3 (2008), p. 278.

[135] *Bulletin* 1:3 (1969), pp. 2–3.

[136] Brian MacArthur (ed.), *The Penguin Book of Modern Speeches* (London: Penguin, 2012), p. 409.

speech to Conservatives in Birmingham.[137] Joseph's speech within a week of announcing that he would challenge Edward Heath for the leadership of the party, suggested making contraception more widely available. Mothers at the bottom of the social scale were having too many children and they were by definition ill-fitted for the responsibility of parenthood. He considered the consequences of this to be exacerbated by talented people emigrating due to lack of opportunities.

Such a speech inevitably had a negative impact in the progressive 1970s and it cost Joseph the chance to compete for the top job. Seemingly oblivious to the furore which the remarks had caused, he took a whole month to withdrew from the leadership contest.[138] On the day that he did so, Margaret Thatcher, who shared his political outlook, announced her candidacy. Joseph was the minister responsible for the contraceptive pill being introduced as a prescription drug, causing an increase in health expenditure which he regretted. The consolation was that at least the pill carried a higher charge than 'life-saving' drugs.[139] It had been passed by the smallest possible margin.

An article in the *Bulletin* entitled 'Ten Years of Change' in June 1970 provided a useful summary of recent trends. The Society's principles remained as before. They had not so easily crystallized themselves into set objectives, however, as with the 'social problem group' in the 1930s and 1940s or with promising families in the 1950s. There were no longer clear criteria to decide who exactly was 'best' and discourse thereon was frowned upon. Medical science had advanced with techniques such as sex determination and organ transplantation coming into use. These might have eugenic implications. It was certain genetic engineering, possibly the greatest development in the last decade, would.[140] Scientific advances in biochemistry and molecular biology had enormous potential to change lives.

The issue of migration had periodically been considered by eugenists, with restrictions both ways being advocated in the 1920s. Another article in June 1970 entitled 'Immigrant Births and Mixed Parentage' looked at

[137] Hanson, *Eugenics, Literature and Culture in Post-War Britain*, p. 106.

[138] Hugo Young, *One of Us. A Biography of Margaret Thatcher* (London: Pan Books, 1989), p. 93.

[139] *Bulletin* 10:2 (1978), p. 49.

[140] *Bulletin* 2:2 (1970), pp. 23, 29–30.

the latest statistics on births. Noting that there had been 405, 000 births in a recent period of six months, it explained that 18, 000 of these were to couples from Africa, the Caribbean or the Indian subcontinent. There were another 2, 000 births where one half of the couple was from these places. Although these facts were not a fool-proof indication of race or origins, it was likely that 4 or 5% of births were to 'coloured' people. Although the non-white adult population stood at 2% of the total, it did not mean that it was growing twice as fast as the indigenous population. It had to be borne in mind that immigrants were younger on average than native Britons. The writer did, however, believe that completed family size was one child higher among immigrants than among the latter.

This was less prejudiced than what Enoch Powell and other Conservative right-wingers were claiming in the 1960s and 1970s, namely that non-whites made up only 2.5 to 3% of the population, but that 20 to 30% of all births were to them.[141] On the eugenic side of the issue, the article went on to consider miscegenation. Human geneticists agreed that there were no adverse effects from such mating. There were, however, sociological consequences, including the psychological difficulties of having parents from two separate cultures, which they should take into account when planning how many children to have. Even so, miscegenation was the single most effectual way of integrating coloured people. While the Society contained a variety of views, the general thrust of opinion among the membership was probably that integration, not repatriation, represented the preferred outcome of immigration.

There was scope for discussion and disagreement with the revival of the members' meetings in 1971. They had been a regular feature of the calendar until 1962, when lack of interest forced their abandonment. The first revived meeting was on whether the authorities should be encouraging a two-child policy.[142] When it took place, on 24 March 1971, attendance was above forty.[143] This was back to the normal level in bygone days and augured well for their reintroduction. Nigel Priestley nevertheless later complained that the meeting had not been genuine

[141] Soloway, *Demography and Degeneration*, pp. 360–361.

[142] *Bulletin* 2:4 (1970), p. 77.

[143] *Bulletin* 3: 2 (1971), p. 29.

because Cedric Carter, an opponent of any limitation, had been set up against a straw man.[144]

Eliot Slater, a psychiatrist who took a genetic approach to mental illness and who had trained in Germany in the 1930s, gave a paper at the Society's 1971 symposium. That year he published a textbook with Valerie Cowie called *The Genetics of Mental Disorders*, which would become a classic.[145] Between 1959 and his retirement in 1969, Slater had headed a centre for psychiatric genetics under the Medical Research Council. His contribution at the session on 'Ethics and Population Increase' argued than humans had no right to destroy other species even if it proved beneficial to a portion of us. The present rates of reproduction were too high. He would prioritize the survival of the Blue Whale over the lives of the potentially 3 million humans who subsisted from it, he said.[146]This was ecological radicalism at a time when awareness about environmental degradation was rising. It came with the sting in the tail that it was especially poorer countries who should be decreasing their population growth. Just as it had been argued in the early twentieth century that social problems were *caused* by the existence of hereditarily afflicted people, modern environmental problems were to some extent laid at the door of the global poor. Stefan Kühl sees this strand as the contemporary version of left-wing eugenics, while the right-wing encapsulation equates to racism and using heredity to oppose social reform.

The increased focus on the environment was reflected in new additions to the Society library such as *Battle for the Environment* and *The Environmental Revolution*.[147] These books, by Tony Aldous and Max Nicholson respectively, had both been published early in the 1970s. Better-known books which stimulated the same concerns were Paul Ehrlich's *The Population Bomb* (1968) and the Club of Rome's *Limits to Growth* (1972).[148] And there was a direct connection to eugenics through sterilization

[144] SA/EUG/P34. Priestley to Benjamin 21 May 1987.

[145] Roelcke, 'Eugenic Concerns, Scientific Practices', p. 28.

[146] *Bulletin* 3:4 (1971), p. 74.

[147] *Bulletin* 5:1 (1973), p. 14.

[148] Jane Cottingham, 'Sexual and Reproductive Connections and Disconnections in Public Health' in Ignatz Sanger (ed.), *Advances in Reproductive Technology Research* (New York: Nova Science Publishers, 2013), pp. 145–146.

campaigns in Latin America and India starting up due to overpopulation concerns.

Sterilization and the opposition it encountered was at the heart of why eugenics became despised. The movement sought to reinvent itself as an academic enterprise to overcome its insalubrious reputation. The American Eugenics Society chose to rename itself the Society for the Study of Social Biology from 1972 onwards. The British society had established the Galton Foundation for its propaganda work and for managing the *Journal of Biosocial Science*, while retaining its old name for everything else. In 1962 it too had debated whether a complete change of name would help. A memorandum on the issue had included the argument that eugenics was seen as 'an evil word associated with Nazi doctrines'.[149] There was little appreciation of the Society's 'kind' approach and research-led activities. The document listed argument and counterarguments on each point. What probably clinched the decision in favour of the status quo was that alterations would require a 2/3 majority and the largely elderly membership were emotionally attached to the name.[150] A need had been identified, to which no solution yet existed. The default option also won out with the statement of aims, which was kept the same in 1962 as in 1957.

In the 1970s the Society was officially functioning as a learned association. Where there had previously been discussion of policies and campaigns, the council's emphasis had by now shifted to academic matters. These included the symposia, other lectures, publications and grants for individuals conducting studies pertaining to eugenics. In 1974 Helen Macbeth of Oxford University's Anthropology Department received a grant of £ 420 for a study of Orkney Islanders who had settled in Aberdeen.[151] After the establishment of the Marie Stopes Research Fund, the disbursements could be substantial. In 1977 Moira Gallen of the Family Planning Association applied with a project on abortion in the slums of Manila, operating with a budget of more than £ 6, 000.[152] The Society's leadership was almost acting as a research council with lengthy forms available for those who sought funding. Potential members were

[149] SA/EUG/J10. Folio 1.

[150] SA/EUG/J10. Folios 2–3.

[151] SA/EUG/P45. Minutes of Meeting 16 October 1974, p. 2.

[152] SA/EUG/P46. Special Projects. Shanty Town Abortions, p. 8.

asked about their fields of interest and approved according to these and their qualifications. Although the application process was a formality, the process accorded with being a learned association. Members' meetings and other activities were advertised in the *Bulletin*. In the immediate post-war period, the Society had actively been recruiting members. No longer campaigning, it did not necessarily feel the need to increase the membership much. When it was still a pressure group, in 1938, 13% of the membership had been white-collar workers.[153]

As mentioned, the Society would choose cooperation with other organizations to advance its objectives. An example was the new Caradog Jones Lecture, an annual event in conjunction with the Royal Statistical Society and Pembroke College, Cambridge. Named after statistician and eugenist David Caradog Jones, it built upon an earlier essay competition which he had initiated.[154] In 1973 the lecture was given by the civil servant F. E. Whitehead on 'Statistics and Social Policy' on 27 March at Middlesex Hospital Medical School.[155] Another example was Professor Nelson Morton of the University of Hawaii giving the Darwin Lecture in 1974 on interracial crosses in man.[156] The Darwin Lecture was the other main event in the Society's calendar. A joint undertaking with the Institute of Biology, it was held that year at the Royal Geographical Society. It led to a debate in which the neuroscientist Steven Rose, not a member of the Eugenics Society, inveighed against scientific racism and in particular the notion that certain races were more intelligent than others.[157] He believed that such racist views were unscientific and that they originated, in their contemporary form, from difficulties in the United States which were proving intractable.

At the end of 1974 Faith Schenk stepped down as general secretary. She had been involved as a member for 28 years. Her tenure had seen the founding of the *Bulletin*, which she had continuously edited. Schenk's successor was Eileen Walters, the second female general secretary appointed. The post constituted ordinary paid employment and Miss

[153] Ian Brown, 'Who were the Eugenicists? A Study of the Formation of an Early Twentieth-century Pressure Group', *History of Education* 17:4 (1988), p. 302.

[154] *Eugenics Review* 56:1 (1964), p. 35.

[155] *Bulletin* 5: 1 (1973), p. 11.

[156] *Bulletin* 6:2 (1974), p. 41.

[157] *Bulletin* 6:4 (1974),pp. 73, 84.

Walters would therefore serve a year on probation.[158] Cedric Carter had been promoted to a professorship and remained the president.

In 1974 the American psychologist Leo Kamin noted that Cyril Burt's studies of identical twins reared apart had jumped from just under twenty to 53 subjects in his later publications.[159] Suspiciously, the correlation for their measured intelligence nevertheless remained the same to three decimal places. Burt's data had strengthened the case for heredity as its determining factor. Dying in 1971, he was an influential figure within the Society for more than half a century. But the coincidence of the correlation not moving at all despite almost trebling the number of participants bordered on the impossible. In 1976 Oliver Gillie, a journalist on the *Sunday Times*, examined Burt's later scientific papers closely. He revealed that Margaret Howard and J. Conway, who had supposedly collected the data, either did not exist or at least could not be traced.

The charge of fraud was levelled at Burt in articles in the *Sunday Times* and its stablemate *The Times*. This naturally exercised members of the Society, such as Colin Bertram who hoped that the council was preparing a response.[160] By chance, on the same day John Fraser Roberts had a letter published in *The Times* defending Burt. His own data corroborated Burt's on the high heritability of intelligence, which he estimated at not less than 60%. (This was lower than Burt's contention of close to 80%.) In meetings with Burt, he had been impressed by the painstaking and accurate way in which the other man had treated his data.

An issue which animated the Society greatly was defending the right to abortion. In early 1977 Lady Vera Houghton wrote a round-robin letter to members asking for help in opposing Conservative MP William Benyon's private member's bill. It would compromise the availability of abortion through sowing doubts about whether the gynaecologist was performing an illegal operation.[161] She dismissed Benyon's arguments and claimed that the number of abortions had fallen between 1973 and 1976 anyway. Her main suggestion was to encourage the members to write to their MP on the matter. The Society answered the minister of health's request for information to be used at the committee stage of

[158] SA/EUG/P45. Minutes of meeting 20 February 1974, p. 3.

[159] Gould, *The Mismeasure of Man*, p. 235.

[160] SA/EUG/P46. Bertram to Carter 30 Nov 1976.

[161] SA/EUG/P46. Letter from Vera Houghton 15 Feb 1977.

the bill by submitting a memorandum.[162] After parliamentary scrutiny, the *Bulletin*, now edited by the geneticist Derek Roberts, described it as 'encouraging' that Benyon's bill had received less support than similar anti-abortion initiatives in the past.[163] It regarded the 1967 law as a landmark and said it would be a pity if it were eroded by poorly informed proposals for change. In 1979 the Society went a step further. The December issue of the *Bulletin* was sent out with a note drawing attention to the leading article urging members to write to their MP in order to oppose Conservative John Corrie's private member's bill on abortion.[164] The note stressed the urgency of the situation, and the editorial gave full details about the bill and the background factors. Again, the Society produced and submitted a memorandum on the issue, this time to the Chief Medical Officer after being contacted by him.[165]

Choosing which births should go ahead was a eugenic activity. Abortion allowed foetuses with genetic conditions and disabilities to be weeded out. What the Society implicitly had desired since the 1930s had become a reality. Janet Chance, wife of Society member Clinton Chance and chairman of the Abortion Law Reform Association, argued in 1938 that one of the benefits of abortion was to prevent the birth of disabled children.[166] Furthering the rights of the disabled was regarded by the Society as a stratagem for attacking the principle of selective termination of pregnancies.[167] It might be the thin edge of the wedge in opposing abortion generally. Attempts would be forthcoming to lower the limit before which a foetus could legally be terminated. The alternative, which it recommended, was to support research which permitted diagnosis of abnormalities as early as possible during pregnancy. It would be a step forward if such diagnosis became possible before twenty weeks. The contrary view is that 'search and destroy' is an affront to the dignity of

[162] SA/EUG/P46. Minutes of meeting 19 April 1977, p. 3.

[163] *Bulletin* 9:2 (1977), p. 37.

[164] SA/EUG/P21. Note following back page of December 1979 issue.

[165] *Bulletin* 11:4 (1979), p. 110.

[166] Calum MacKellar and Christopher Bechtel (eds.), *The Ethics of the New Eugenics* (New York: Berghahn, 2014), p. 26.

[167] *Bulletin* 14:2 (1982), p. 39.

disabled people. It sends out a clear message that society does not value them, according to their rights advocates.[168]

In a sense, the opposite of abortion was creating life which otherwise would not come into being. Worldwide, the development of in vitro fertilization in the 1970s resulted in the birth of several million people who otherwise would not have existed.[169] An older technique of *ab extra* reproduction was artificial insemination. In 1982 its study into Artificial Insemination by Donor was proceeding. Professor Malcolm Symonds and Sue Teper, both at the University of Nottingham, were granted £ 7, 651 from the Marie Stopes Research Fund for a pilot study of A. I. D. subjects.[170] They gave a paper about their research at that year's symposium. It was linked to other reproductive policies through the number of children available for adoption falling after the passing of the 1967 Abortion Act. Success with A. I. D. depended on maternal age, ovulatory function and the husband's level of sperm production. A completely azoospermic male made success more likely than if he had a low sperm count. This was because the donor's sperm would then not be competing with the husband's.[171] Similar reasons governed why the rate of success rose with maternal age. However, Symonds and Teper typified the recent approach of not promoting eugenics consistently, stating that A. I. D. was not for this purpose.[172]

Three Important Symposia

The Society's symposium in 1981 was on the 'Demography of Immigrants and Minority Groups in the United Kingdom'. This event is suited for understanding how immigration and its consequences were viewed by members and invited speakers. An introductory paper by Colin Holmes, an economic and social historian at the University of Sheffield, gave a

[168] Mervyn Ekberg'Eugenics: Past, Present and Future' in Marius Turda (ed.), *Crafting Humans. From Genesis to Eugenics and Beyond* (Goetingen and Taipei: V & R Unipress and National Taiwan University Press, 2013), p. 96.

[169] Rutherford, *Control*, p. 4.

[170] *Bulletin* 14:2 (1982), p. 53.

[171] *Bulletin* 14:4 (1982), p. 118.

[172] Sue Teper and E. M. Symonds, 'Artificial Insemination by Donor' in C. O. Carter (ed.), *Developments in Human Reproduction and their Eugenic, Ethical Implications* (London: Academic Press, 1983), p. 35.

long-term overview of the phenomenon of immigration. Jewish immigrants who had arrived in the period 1880–1914 were fleeing persecution and therefore the host country had sympathy for them. Their integration was aided by the presence of a pre-existing Anglified Jewish community, which recent arrivals from the New Commonwealth lacked.[173] Why the latter had come to Britain was nebulous, being tied up with colonialism and therefore less obvious to the locals. Both these factors had worked to the advantage of Jews in comparison with Caribbean and subcontinental newcomers. Eugenists had in the post-war period supported social mobility and therefore assimilation to middle-class standards. It is probable they wanted the same for minority groups too, and it may have been an easier transition to make for Jews, less visibly different than black and Asian people. In a later paper Barry Kosmin, of the Board of Deputies of British Jews, shed light on demographic trends among this group from 1850 to the present day. He found that Jews were characterized by low synagogue attendance, high usage of contraception, an aging population with a high death rate and low birth rate. This meant that there was no increase in size at all.[174] It matched the subtext that Jews were no threat whatsoever from a demographic or eugenic point of view. What he really was seeking to explain though, was why there had been a shift from seventy years ago when the opposite situation had obtained. He found this in Jews adopting English marriage patterns. In other words, they had integrated well.

It seems a hierarchy was informally being built up about the desirability of various groups. John Hollis gave a paper on New Commonwealth immigration in Greater London. Across the metropolis Asians were more likely to be owner-occupiers (62%) than the white population (45%), whereas 50% of West Indians lived in council housing.[175] Both Asians and West Indians in London had a socio-economic group distribution which resembled that of the indigenous population, except that among the latter

[173] Colin Holmes, 'The Promised Land? Immigration into Britain 1870–1980' in D. A. Coleman (ed.), *Demography of Immigrants and Minority Groups in the United Kingdom* (London: Academic Press, 1982), pp. 14–15.

[174] Barry A. Kosmin, 'Nuptiality and Fertility Patterns of British Jewry 1850–1980: An Immigrant Transition' in Coleman (ed.), *Demography of Immigrants and Minority Groups*, p. 245.

[175] John Hollis, 'New Commonwealth Ethnic Group Populations in Greater London' in Coleman (ed.), *Demography of Immigrants and Minority Groups*, pp. 132, 137.

it was rare to be a professional or a manager. The Asian population in the Midlands and the North had a much higher incidence of manual workers, due to their employment in the textile industry. In a book review two years earlier, council member Elizabeth Still had claimed that since both female Asians and poor West Indians had low literacy rates, there was a tendency for these two groups not to be knowledgeable about contraception. The result was potentially a higher birth rate within these groups than was beneficial in a modern industrial economy.[176] Hollis found that overcrowding was more prevalent among Asians than the indigenous population, West Indians forming an intermediate group.

The Galton Lecture within the symposium, delivered by Bernard Benjamin, professor of statistics at City University in London, showed that Irish immigrants were more affected by suicide and admission to psychiatric hospitals than the general population.[177] Since by this stage such facts hardly had any eugenic implications, the symposium as a whole largely confirmed that the Society had become an academic endeavour. The only findings which may have fed into political narratives was the higher birth rates of Asians and blacks, though was being adjusted downwards towards the average for the latter. Benjamin would became president of the Society in 1983.

1982 was the seventy-fifth anniversary of the founding of the Eugenics Society. It was celebrated with a small reception at Imperial College during that year's symposium. The great-niece of Francis Galton, Dorothy Middleton, was the guest of honour and she contributed some memories of her distinguished ancestor. So much time had passed that it had been forgotten that Galton offered only moral support in the founding of the Society. Her memories of him included that Galton had established it and that he had endowed University College London with a chair of human genetics. This was not entirely accurate either, as the chair had only been renamed in 1963 and was in fact intended to be based at the University of London. She was right, however, that when Galton discussed 'race improvement', it had the meaning of 'mankind' and not any particular ethnic group as in contemporary parlance.[178] Moreover, the emphasis

[176] *Bulletin* 11:1 (1979), p. 27.

[177] Bernard Benjamin, 'Variation in Mortality in the United Kingdom with Special Reference to Immigrants and Minority Groups' in Coleman (ed.), *Demography of Immigrants and Minority Groups*, pp. 65–66.

[178] *Bulletin* 14:4 (1982), pp. 109–110.

had been on the gathering of data and the facilitation of study, with a goal of persuading rather than compelling. She also shared a few family memories of Galton, including how a Siamese-twin rat had run up his leg during a formal tea party at his home. The celebration was much more muted than for the fiftieth anniversary in 1957. On that occasion there was coverage over a dozen pages in the *Eugenics Review*, a jubilee meal at the Savoy Hotel on the Embankment, numerous speeches and greetings from supportive organizations like Political and Economic Planning and the Family Planning Association.[179]

In recent years there had been a downgrading of the ideology. There was often focus on related matters with presentations by regular scholars, who ignored eugenics. When it had been a pressure group, there were members who saw it as a social club only. But they were at least interested in discussing eugenics theoretically. In the 1980s there were applicants wishing to join because of the funding and networking opportunities. In 1983, however, the symposium returned to the heart of eugenics as it had developed over decades with 'The Biology of Human Intelligence' as its theme. Academic standards were maintained with two of the papers given at the event failing peer review for inclusion in the proceedings.[180] They would probably have been adequate for the *Eugenics Review*, which had printed speculative pieces and biased work. Even the controversial professor Hans Eysenck added to the store of knowledge in his Galton Lecture at the symposium. On the centenary of the publication of Galton's *Inquiries into Human Faculty and Its Development*, which had included his studies of individual differences in psychology as well as launched the term 'eugenics', Eysenck argued that Galton had been right after all. Galton had seen intelligence as a biological function connected to learning and problem solving. It was inherited and could be probed through tasks involving inspection time, motor-sensory aptitude and comparisons. Because one particular investigation in 1901 had reported no correlation at all between these and general intelligence, Galton's approach had been discontinued.[181] In actual fact, Eysenck

[179] *Eugenics Review* 50:1 (1958), pp. 9–22.

[180] SA/EUG/P34. Minutes of 22 November 1984, p. 4.

[181] H. J. Eysenck, 'The Galton Lecture 1983. Intelligence: New Wine in Old Bottles', in C. J. Turner and H. B. Miles (eds.), *The Biology of Human Intelligence. Proceedings of the Twentieth Annual Symposium of the Eugenics Society London 1983* (Nafferton: Nafferton Books, 1984), p. 40.

claimed, inspection time was closely connected to intelligence. Galton had succeeded in isolating an essential trait and his concept was preferable to that of Alfred Binet, who had included education and specialized abilities.[182]

Galton had also been a pioneer in studying twins. John Hewitt and Krystyna Last used his paper from 1875 as the introduction to their own. The main utility of twin studies was to help distinguish between heredity and the environment. Unsurprisingly, monozygotic twins had more in common than their dizygotic counterparts, who were ordinary siblings born at the same time. Based on a plethora of research, the two speakers working in the departments of psychology and genetics respectively at Birmingham University believed intelligence to be 45% heritable.[183] This was a lower estimate than the 80% which traditionalists such as Society members the late Sir Cyril Burt and the American Arthur Jensen assumed. Burt had only reluctantly acknowledged slight environmental influence after Fisher's equation showed a way to apportion it.[184] Fisher had also doubted that 'feeblemindedness' was caused by a single Mendelian recessive gene.[185]

A paper by Nick Mascie-Taylor examined intelligence in relation to a myriad of factors. It may be seen as indicative of where the debate stood in 1983. The findings regarding these had fluctuated during the long history of eugenics. Mascie-Taylor presented results which showed that the cognitive ability of children correlated to parental social class. The family size of the most intelligent tended to be just over two children and declined from that point. An only child was on average more intelligent than those in families of three children. Birth order mattered, as each child born was on average less intelligent than his or her next older sibling.[186] With research into such topics becoming less biased, it is likely that these results were an improvement on, for instance, Pearson's early twentieth-century results about the handicapping of the first born.

[182] Eysenck, 'The Galton Lecture 1983', p. 35.

[183] J. K. Hewitt and K. A. Last, 'Twin Studies of Intelligence: Recent Data, New Questions' in Turner and Miles (eds.), *The Biology of Human* Intelligence, p. 82.

[184] Pilgrim, 'The Eugenic Legacy in Psychology and Psychiatry', p. 279.

[185] Mazumdar, *Eugenics, Human Genetics and Human Failings*, p. 94.

[186] G. C. N. Mascie-Taylor, 'Biosocial Correlates of IQ' in Turner and Miles (eds.), *The Biology of Human* Intelligence, pp. 101–105.

December 1983 saw the final issue of the *Bulletin*, which was replaced by equivalent publication *Biology and Society*. The change of title was part of the process once again of deemphasizing eugenics as an ideology. As the subtitle was *The Journal of the Eugenics Society*, it was still clear for what purpose it existed. The main title gave the impression of a professional or learned association's newsletter or magazine. This was by now quite accurate anyway. In its final years as the 'Eugenics Society', the restructuring occurred along the lines agreed upon in the 1960s and 1970s.

Cedric Carter died in March 1984. The proceedings of that year's symposium were dedicated to his memory. A new Cedric Carter Memorial Medal was minted to honour him, with the first recipient being Colin Bertram in 1985.[187] In 1984 the Warnock Committee on Human Fertilization and Embryology reported and recommended that research on foetuses 14 days or older should be a criminal offence. Enoch Powell, who had been elected as a Ulster Unionist MP, attempted to surpass the Warnock Committee through introducing his own private member's bill to ban any research on human embryos. The Society opposed this bill. Apart from preventing research on genetic conditions such spina bifida and hindering better forms of contraception emerging, it would give legal status to embryos. This would affect foetuses too and therefore facilitate amendments to the 1967 Abortion Act. *Biology and Society* for March 1985 carried a loose-leaf insert encouraging members to take action, especially writing to their MP on the matter.[188] Although the bill passed its second reading, it did not progress further down the parliamentary path.

Still occasionally active as a pressure group, the Society was restructuring. In 1985 it sold 106–108 Whitfield Street, to some extent an iconic property although it had belonged to a subsidiary organization. The £ 170, 000 raised was put towards the purchase of another house.[189] It agreed a contract with Macmillan to publish symposia proceedings under the joint imprint of it and the Society.[190] This led to the standardization of material to be included. Although previous proceedings had also had a professional layout, the onus was placed on contributors

[187] SA/EUG/P34. Minutes of 22 November 1984, p. 1.

[188] SA/EUG/P34. Minutes of 12 March 1985, p. 5.

[189] SA/EUG/P34. Minutes of 2 July 1985, p. 2.

[190] SA/EUG/P34. Minutes of 26 March 1986, pp. 5–6.

to a greater degree. In the past, some had submitted handwritten papers and one Galton lecturer had not prepared anything for publication at all. In 1986 the statement of aims was being rewritten with the mention only of 'responsible parenthood' and none explicitly of eugenics.[191] The same year a new general secretary, Linda Brooks, took over.

The Marie Stopes Research Fund became a core activity of the Society and helped in the transition from pressure group to learned association. It had been led by the geneticist Professor John Beardmore since its inception in 1976.[192] Reviewers were consulted and references taken up, with the forms in use being similar to those of national research councils. It is nevertheless clear that a coterie of influential members were the prime beneficiaries. Derek Roberts was awarded £ 42, 383 between 1979 and 1987, Nick Mascie-Taylor £ 13, 720 between 1983 and 1986, Beardmore himself £ 8, 200 in 1983 and Bernard Benjamin £ 12, 332 in 1987.[193] However, if the Society really liked a proposal, it might be funded with five-figure sums, even if the applicant was not a member. This was the case for two separate proposals on sex therapy in 1979 and 1983. They may have been favoured because of Stopes's interest in sexuality. Smaller sums were available for individual projects.

The main business for Mrs Brooks consisted in moving out of 69 Ecclestone Square and beginning preparations for officially letting go of eugenics. It was agreed that the Society's presentational brochure should not contain any mention of the word. It therefore removed the concluding sentence, 'This is precisely the aim of Eugenics', from the Galton quote explaining what it stood for. The new name chosen for the organization was the Galton Institute. Its symbol would be the flower named after him, *Galtonia candicans* also known as the summer hyacinth. The final formalities of the name change for the Society occurred in 1989. In 1988 it moved to 19 Northfields Prospect, in Wandsworth, South London.

As the Society was transitioning, the general secretary sent out a membership roster to the fellows and members.[194] The information

[191] SA/EUG/P34. Minutes of 26 March 1986, p. 2.

[192] SA/EUG/P46. Minutes of 26 November 1976, p. 3.

[193] SA/EUG/P35. Submission to DTI for change of name The Galton Institute, appendix D.

[194] SA/EUG/P35. Letter from Linda Brooks c. 1 September 1988.

contained therein, provides a general picture of who belonged at this late stage (Table 5.4).

One matter for discussion was whether to pursue the appellation of 'Institute'. The Department of Trade and Industry, responsible for the Register of Companies, informed the potential Galton Institute that the name could only be used for organizations carrying out advanced research or professional bodies of the highest standing.[195] There was also a requirement to show that it did not duplicate already existing institutes in its activities. The bar had in other words been set high, but the Society had been transforming itself into a learned association over decades. The title of 'fellow' had long been reserved for persons making a contribution to biosocial science.[196] Others, with only an interest in areas covered by the term, were admitted as members.

Instead of opting for a less prestigious title, a submission was prepared showing its high standards of research, lecturing and admission. It explained that the word 'eugenics' had fallen into disrepute. The new name honoured Sir Francis Galton FRS rather than the doctrine. It was also, wrongly, claimed that he had been the instigator of the original

Table 5.4 Membership of the Eugenics Society 1988

Total membership	380
Fellows	240
Members	137
Unknown status or institutional member	3
Resident in the United Kingdom	300
Foreign or Commonwealth resident	80
Title of professor	59
Title of doctor	132
Title of the honourable	1
Title of knight or lord (if no academic title)	7
Title of lady	3
Title of mister or military title	93
Title of mistress	83
No title given	2

[195] SA/EUG/P35. S. Lewis to Linda Brooks 5 August 1988.

[196] This had been a policy since 1946, when it was decided that fellowship would be restricted to those making a contribution to eugenics or the Society. SA/EUG/A39. Annual Report 1946–47, p. 2.

Society.[197] *The Eugenics Review* had appeared in 1909 and not at the founding in 1907, as was stated in the submission.[198] This mild misdating nevertheless supported the contention that the Society was a learned association. It admitted that it had originally been a pressure group. But the Galton Lecture had been given by some of the century's most distinguished scholars. Among the topics, subsequently realized, were family allowances, a Social Science Research Council, family planning within the National Health Service and a national system of genetic screening. Much of this information was printed verbatim in *The Lancet* after the name change had gone through.[199]

The annual general meeting for 1987 was belatedly held on 5 July 1988 in Ecclestone Square, but in a hotel as the Society's own premises had been sold for £ 700, 000 in December 1987.[200] The sale had been triggered by extensive refurbishment being required to the outer building. As the upgrade would cost in excess of £ 100, 000, the Society decided that disposing of it would be a less time-consuming option.[201] The new premises were worth less than the old at £ 150, 000 and the balance would be invested in the Society's activities. The main item on the meeting's agenda was to consider the extraordinary motion of the name change to the 'Galton Institute' . Charles Goodhart, who had tenaciously opposed the stance on abortion almost two decades earlier, fought to keep the old name. There were also two attempts to adopt a longer name by adding either 'Human Ecology' or 'Biology and Society' to the name. There was a modicum of support for the latter, but it was clear that the overwhelming majority preferred the simpler title. Although it did not become part of the official name, the notepaper of the new Institute mentioned that it was for the study of biology and society.

Early in 1989 approval was granted by the Department of Trade and Industry for the name change. It observed that the object clauses of the 'company' still referred to eugenics and suggested that this be changed,

[197] SA/EUG/P35. Submission to DTI for change of name The Galton Institute, pp.1–2, 4.

[198] Francis Galton, 'Foreword', *Eugenics Review* 1:1 (1909), p. 1.

[199] *The Lancet*, 19 August 1989, p. 461.

[200] SA/EUG/P35. Report of 1987 Annual General Meeting, pp. 2–3.

[201] SA/EUG/P35. Report for inclusion in the 1988 annual report, p. 2.

owing to the lack of esteem in which the term was held.[202] There now only remained gaining the consent of the annual general meeting for the change of name and new articles of association. This was forthcoming without dissent on 4 July 1989, with the meeting being held at the new premises.[203] In 1990, the Institute could report that its principal activity in the past year had been to promote 'biosocial science'.[204]

Although the Institute would continue to have a role in supporting the *Journal of Biosocial Science*, this publication was not officially connected with it anymore.[205] There was no report in it about the change of name. In 1989, the Institute's last year as the Eugenics Society, all the articles in the journal were written by outside scholars, except that Richard Lynn and Nick Mascie-Taylor contributed one co-authored article each and Geoffrey Ainsworth Harrison a book review. By chance, all three pieces appeared in issue 4, of October 1989, after the name change had occurred. The last time the journal printed something by a profiled member of the Eugenics Society was thus in July 1988, when Madeleine Simms had a book review published.[206] This presented a vivid contrast to the old *Eugenics Review* and epitomized how there had been a change of scene.

Conclusion

The last years of the Society in its traditional incarnation coincided with the return of old-fashioned attitudes in the guise of Thatcherism. Margaret Thatcher hailed Victorian values, which, extraneously, had been the starting point for popular eugenics. In 1912 Leonard Darwin stated with minor caveats that the poorer classes on average contained more of the naturally 'unfit' than did the richer classes.[207] Later, he wished aloud that the birth rate of those who were not skilled workers or professional

[202] SA/EUG/P35. J. D. M. Stephens to R. A. Peel 27 Feb 1989.

[203] SA/EUG/A95. Annual Report 1989, folio 4 r.

[204] SA/EUG/A96. Annual Report 1990, folio 3 r.

[205] It subsidized it by £ 6, 425 in 1991 and £ 6, 600 in 1992. SA/EUG/A98. Annual Report 1992, folio 5 r.

[206] Madeleine Simms, 'People Count. A History of the General Register Office', *Journal of Biosocial Science* 20:3 (1988), pp. 377–378.

[207] SA/EUG/A4. Annual Report 1911–12, p. 7.

men could be cut.[208] In 1926 he warned against miscegenation too. Though he was concerned about the 'shiftless' rather than 'respectable' trade unionists and artisans, his prejudice differed little from the later Thatcherites. The Conservative governments in the 1980s may have deliberately allowed unemployment to rise to weaken the trade unions and tried to instil market discipline on those with low incomes.

Margaret Thatcher herself once described Britain as being 'swamped by people with a different culture'.[209] In the same interview in 1978, she predicted there would be four million ethnic minority inhabitants by 2000. Eugenics was passé by 1979, as her mentor Sir Keith Joseph discovered, but right-wingers launched attacks on sections of the poor for being 'workshy' and cut funding for community projects. Especially the Monday Club supported the repatriation of coloured immigrants. Thatcher and her confidants disapproved of non-patriarchal families. Section 28 of the Local Government Act 1988 banned the promotion of homosexuality and criticism of single mothers became a mantra in parts of the tabloid press. It was as if workers and the marginalized had taken on a similar role as the 'residuum' held for 'mainline' eugenists. The discourses were not identical. Eugenists had not concerned themselves with the varieties of family type or promoted socially conservative policies. They first accepted and then argued for abortion and wanted to make child support easier to obtain in the 1930s. Its attitudes on social policy did not undergo change. At the 1976 symposium Robert Chester gave a paper which was understanding of the position of single mothers in society.

Common ground between Thatcherism and eugenics is more evident in a professed belief in meritocracy. Even Galton had been in favour of social mobility. 1980s Conservative policies such as the sale of council houses to their tenants at low cost, provided a ladder upwards for ambitious working-class people. In 1980, 1/8 of the age group went to university, compared to 1/5 in 1990.[210] These changes accorded in principle with the goals of previous generations of eugenists. They, at least officially, believed in giving opportunities to poor but respectable families from the 1930s onwards. After the war, eugenists thought that selection

[208] Leonard Darwin, 'Some Birth Rate Problems', *Eugenics Review* 12:3 (1920), p. 151.

[209] Young, *One of Us*, p. 111.

[210] Peter Mandler, 'Educating the Nation II. Universities', *Transactions of the Royal Historical Society* 25 (2015), p. 10.

for schools by the 11 + would enable talented youngsters to progress in life. Such attitudes had the ring of 'positive' eugenics. As for 'negative' eugenics, its future might lie with contraception, selective termination of pregnancy and efforts to halt population growth worldwide.

There were major changes in eugenics occurring after 1945. Academically, the discipline morphed into human genetics. It would continue with the same methods and many of the objectives, bereft of the ideological ballast. In 1963 the chair of eugenics at University College London was renamed the chair of human genetics. The subtitle *Annals of Eugenics* was dropped from its publication *Annals of Human Genetics* in 1965. Human genetics was recognized and in tune with the concerns of contemporary society. Class snobbery, prejudice based on science and the odious embrace of the Nazis were making eugenics infamous as an ideology.

The Eugenics Society had above all been an interest group for professionals and others who wanted to support the doctrine's application to social problems. It toned down its mission after it halted the campaign for 'voluntary' sterilization in 1938. The intention was that this would only be a temporary lull before the inquiry of the Royal College of Physicians had provided fresh impetus. The war intervened, the Royal College did not recommend it when it reported in 1946 and sterilization, already the most controversial eugenic policy, never got going again. The Society even decided to keep quiet when it heard that there was support for the measure as a facet of reproductive choice in the 1960s. This summed up its less strident approach after 1945 and its acknowledgement that eugenics was toxic to outsiders. Its eugenic goals would need to be implemented on an individual level rather than for groups of people.

In the post-war period, there were a few novel concerns raised, while others followed on from what had been established in the early days or the interwar years. There was a focus on intelligence, sometimes a replacement for social class as a desideratum. It began to be linked to the differential birth rate and the risk of society moving in a dysgenic direction. Intelligence as a vital characteristic allowed the Society to continue holding the torch for an efficient, well-planned and modern society, without offending large swathes of its fellow citizens. It never doubted that middle-class people were more intelligent than workers, and this had been shown to its satisfaction in the 1930s and earlier. Just like 'people healthy in body and mind' was a coded way of saying families of eugenic worth, an 'intelligent population' meant the same. Regulating the birth rate would occur through contraception, financial inducements and, later, abortion and the treatment of infertility.

It was less common to link race to intelligence, although this was possible too. A few Society members, including Hans Eysenck, Richard Lynn and Arthur Jensen did take up the cudgel for scientific racism in this manner. There were others, such as Colin Bertram and the biologist John Randal Baker, author of *Race* (1974), who were opponents of miscegenation. It was black people among the ethnic minorities who had to bear the brunt of scientific racism, while Asians might be criticized on a cultural level. Jews were not criticized at all after 1945. In past decades there had been almost exclusive focus on social class. In an early post-war opinion on immigration, it was suggested that Caribbean postgraduate students might not be a problem, unlike unskilled labourers.[211]

Race gained in importance to become almost equal to class once Britain had opened up to New Commonwealth immigration from the late 1950s. The Society raised the perceived problems involved in West Indian immigration in the first of its broadsheets in 1958. Although it argued politely, the aim was to reduce the influx from the Caribbean through barriers to entry. It sought the attention of interested parties, such as trade unions, politicians and headmasters, perhaps hoping they would formulate their own arguments and take the issue forward. Once Britain had become a multi-racial society in the early 1970s, attention shifted to the birth rates of immigrant groups and how integration could be effected. But many eugenists had a matter-of-fact approach to race and there were even anti-racists among them. The latter included Julian Huxley, president of the Society from 1959 to 1962.

In the first ten years of the post-war period, the term 'problem families' was a significant aspect of eugenic discourse. It was a truncated version of the 'residuum' of Victorian and Edwardian Britain. Known initially as the 'social problem group', they were thought to share traits such as mental deficiency, ineducability, criminality, poverty and poor hygiene. The Society conducted studies to identify them with greater certainty and to reveal whether this 'condition' was hereditary. From the mid-1950s, the Society drew back as it realized it was stepping into a political minefield. It preferred switching attention towards promising families. Headmasters were contacted, followed by the parents of the

[211] *Eugenics Review* 46:4 (1954), p. 200.

'model pupils', in order to discover how some families were so successful at raising their children. It was inquired into and discussed how such families might be persuaded to increase their fertility. Conversely, there was a focus on allying with groups who were seen as undesirable through offering them advice and contraception. This included the pill after it had become available in 1961. The Society had established a connection with a birth control clinic in North Kensington and in 1959 it was bequeathed Marie Stopes's own clinic, which it ran into the 1970s.

Female left-wing eugenics supporters in the 1930s and 1940s, such as Frida Laski and Janet Chance, had championed abortion rights. This became Society policy too for a limited range of cases. The 1957 statement of aims included termination of pregnancy at the request of the parents.[212] The fact that it appointed two female general secretaries, Faith Schenk in 1965 and Eileen Walters in 1975, may have contributed to its increased focus on abortion in those decades. It resisted any weakening of the Abortion Act of 1967 and hoped eventually to extend the time limit during which a termination could be carried out. It supported the use, extension and refinement of prenatal diagnostic tests, agreeing with couples who decided on a termination if the foetus was malformed or afflicted with a condition. It is clear that it considered Down's syndrome to be serious enough to justify an abortion. Just as contraception had not been accepted by all members in the 1920s, so there was a minority who disagreed with abortion in the 1970s. Both types failed to derail the Society's strong support.

The *Eugenics Review* came to an end in 1968. In 1961 Cassell had decided against continuing to publish it, as had Macmillan in 1946. The name in itself seemed a problem, so from 1969 it was replaced by the innocuous-sounding *Journal of Biosocial Science*. Exclusively academic in content, it featured articles by eugenists as well as writers on sociology, criminology and medical science. A new Galton Foundation was established to manage it. To make up for the loss of news and to facilitate shorter articles, the *Bulletin of the Eugenics Society* was founded.

Some classical points of view were allowed to stand, but the Society also developed novel arguments for eugenics. Members such as C. P. Blacker and Colin Bertram were in favour of reducing the population, either in Britain or globally, which either genuinely or vicariously could

212 *Eugenics Review* 50:1 (1958), p. 29.

be framed as supporting green policies. Blacker participated in the Population Trust of Lord Simon of Wythenshawe, while Eliot Slater showed compassion for other species and argued for the end of anthropocentrism. Sir Alan Parkes, a medical scientist, became president in 1968. He had received generous funding for research on artificial insemination in that decade, including making actual medical discoveries. The use of assisted reproductive technology would be consensual, as with the majority of eugenic policies. The Society was transitioning into a learned association and attempting to increase its impact in the contemporary era.

CHAPTER 6

Eugenics and New Technologies of Reproduction

When Francis Galton announced his plans for breeding more people with desirable qualities to raise the standards of society, he feared being ridiculed for proposing a human stud farm. In Victorian times the only respectable method of attaining Galton's ambition was earlier marriage.[1] He advocated this in *Hereditary Genius* as an initial eugenic policy. Recognizing that it would take time before his doctrine found acceptance, he did not openly propose anything contentious. When 'negative' eugenics became prevalent in the 1900s, he suggested that the right to marry could be taken away from lunatics, the 'feebleminded' and sufferers from hereditary afflictions. The main argument in his second public lecture to the Sociological Society, 'Restrictions in marriage' on 14 February 1905, was that marriage customs were regulated by what was socially expedient and not based on immutable traditions. Indeed, they varied by civilization with some old practices now regarded as taboo.

As Galton wanted eugenics to become the guiding light of the polity, he hoped that in time what served that ideology would be approved. Just as he thought mankind would create a better society in the future, he had

[1] Francis Galton, *Hereditary Genius. An Inquiry into its Laws and Consequences* (London: Macmillan, 1892), pp. 352–356. Francis Galton, *Inquiries into Human Faculty and its Development* (New York: Macmillan, 1883), pp. 320–323.

D. Redvaldsen, *A History of British Eugenics since 1865*, Medicine and Biomedical Sciences in Modern History,
https://doi.org/10.1007/978-3-031-72290-5_6

ideas for other forms of reproduction too. In a note written on 28 June 1894 and sent to Cambridge scientist Walter Heape, who had successfully transferred embryos between rabbits in 1890, he suggested that artificial insemination could be used for humans. Marked 'private for the present', it explained that this would take place in a 'different form of civilization to our own'.[2] Institutions would be established in which female surgeons artificially fertilized unmarried women, so that they might bear their first child for the state. Prudery dictated that no men should be admitted and as much as possible of the hymen would be preserved during the operations. Galton's utopia Kantsaywhere was redacted after his death by his niece, Millicent Lethbridge. She destroyed parts of the story, especially 'love episodes'. May some of these have involved unmarried people?[3] In a letter to Karl Pearson, Galton noted that George Bernard Shaw would be speaking to the Eugenics Education Society and hoped 'that he will be under self-control & not be too extravagant'.[4] But the disagreement with Shaw was over tactics rather than deeply held beliefs. Shaw said openly what Galton distanced himself from in public (Fig. 6.1).

Early Artificial Insemination and Eutelegenesis

There might be valid reasons why reproduction could take place outside marriage. The Australian educator Marion Piddington recommended artificial insemination for war widows and women unable to find a husband due to the scarcity of men in the wartime generation. She first proposed this in her short story *Via Nuova, or Science and Maternity* (1916), published under a pseudonym due to its controversial nature.[5] In 1929 she promoted classic eugenics in a Sydney newspaper without mentioning her earlier idea.[6] With Galton's original views ceding to prejudice, it was not until the 1930s that 'positive' eugenics again achieved traction. In

[2] University College London, Galton Archive. Galton/2/5/2/1. A plea to experiment on the artificial fertilisation of mammalia by F.G, folio 1.

[3] Debbie Challis, *The Archaeology of Race. The Eugenic Ideas of Francis Galton and Flinders Petrie* (London: Bloomsbury Academic, 2013), p. 12.

[4] Galton/3/3/7/20 folios 718–719. Galton to Pearson 26 Feb 1910.

[5] Ann Curthoys, 'Eugenics, Feminism and Birth Control: The Case of Marion Piddington', *Hecate* 15:1 (1989), p. 75.

[6] Marion Piddington, 'For Whom are we Building the Bridge? Task of Eugenics', *The Land*, 18 October 1929.

Fig. 6.1 Galton's niece, Millicent Lethbridge (reproduced from the Eugenics Society Archive, Wellcome Library, London)

1934 the Ministry of Health's Brock Committee studying proposed sterilization legislation estimated that carriers outnumbered sufferers by at least ten to one on mental deficiency, with even higher ratios for the rarer genetic diseases.[7] Its conclusion was that the 'voluntary' sterilization of both the afflicted and carriers should be legalized. However, a need to rethink what eugenics should encompass was beginning to appear. Even if sterilization were implemented, it might only be a piecemeal success.

A possible way forward for 'positive' eugenics appeared in the *Eugenics Review* in 1935 and from an unlikely source. Herbert Brewer's article was entitled 'Eutelegenesis', a term he had coined himself to mean 'good birth from a distance'. Unlike the typical Society member, Brewer was an autodidact and a socialist. In the post office where he worked, he would take copies of the *British Medical Journal* out of their wrappers overnight, returning them in the morning after perusing them.[8] He had privately nurtured eutelegenesis as an idea for many years. On 21 March 1934 he overcame his inhibitions and sent C. P. Blacker a paper about his project. In today's parlance it was a sperm bank. Eutelegenesis involved the semen of 'eugenically valuable men' being harvested and stored in a laboratory, where it would be available for women who wished to have a child. Blacker was enthused by the notion, though also worried about the objections which would have to be overcome.[9] Except for the sexual aspect, this really was stud breeding because the progeny would be born outside marriage with no connection between the parents. Moreover, in the context of the 1930s, it was controversial how the semen would be collected, either by masturbation, through married men using a condom while having intercourse with their wives, nocturnal emissions or even electrical stimulation of the nerve's reflexes.[10] When Brewer's piece was published in the July 1935 issue, it carried a rider stating that the Eugenics Society took no responsibility for the views expressed.

As Brewer was unable to advance eutelegenesis much in practice, he pleaded for a professional scientist to take it up on his behalf. In 1935 only about a third of attempts at artificial insemination were successful.

[7] Herbert Brewer, 'Eutelegenesis', *Eugenics Review* 27:2 (1935), p. 123.

[8] Daniel J. Kevles, *In the Name of Eugenics. Genetics and the Uses of Human Heredity* (Berkeley CA: University of California Press, 1986), p. 352.

[9] Wellcome Library, London. Eugenics Society Archive. SA/EUG/C42. Blacker to Brewer 5 Apr 1934.

[10] SA/EUG/C42. Brewer to Blacker 25 Apr 1934.

Lack of knowledge about the hormonal and chemical predispositions in the woman required for pregnancy to occur accounted for this, as did the obvious problem of how to store spermatozoa so that it remained vital outside the male body.[11] Nobody seemed ready for the task in Britain, but coincidentally an American had thought of the same idea. Hermann Muller was a geneticist who wrote the book *Out of the Night. A Biologist's View of the Future* in 1936. Its seventh section dealt with eutelegenesis, for which Brewer was given equal credit in the British edition.[12] Like Brewer, Muller was a socialist. Both would later argue that their proposals amounted to a kind of nationalization of the gene pool. Everyone would be given the right to be born with the finest heredity at their disposal.[13] Eutelegenesis tended to be accepted by progressives ranging from the Liberal C. P. Snow to the communist J. B. S. Haldane, who wrote in the *Daily Worker* about it.[14]

Muller believed that only in a socialist society would eugenics succeed in accomplishing its tasks. Under capitalism it was simply a tool of class oppression. Even eutelegenesis might not work because women were liable to choose semen from undesirable celebrities. Brewer was willing to accommodate women's ideals and wanted to make a start in the here and now. His preferred list of donors comprised 'saints, philosophers, scientists, poets, artists, musicians and athletes'.[15] In other words, he was looking for men with profound cognitive, aesthetic, moral and somatic ability. All would be screened for hereditary diseases and deformities, while vitality, longevity and intelligence were characteristics which they should possess. Muller was especially interested in making sure that intelligence and traits leading to a cooperative disposition were included.[16] Working for the Soviet Union's genetics research programme at the time,

[11] Kevles, *In the Name of Eugenics*, p. 189.

[12] H. J. Muller, *Out of the Night. A Biologist's View of the Future* (London: Victor Gollancz, 1936), p. 9.

[13] SA/EUG/C43. 'Eugenics and Politics. A Word to the Political Left', p. 12.

[14] Kevles, *In the Name of Eugenics*, p. 191.

[15] Herbert Brewer, 'Eutelegenesis', *Eugenics Review* 27:2 (1935), p. 124.

[16] Laurence Karp, 'Eutelegenesis: A Historical Perspective', *American Journal of Medical Genetics* 5:4 (1980), pp. 327–328.

it is likely that he saw his contribution as adding biology to the revolution.[17] Muller was applying his own method, in which gaining control of something large was predicated on first controlling a smaller world.[18]

Apart from the difficulties over conception, eutelegenesis was a realistic proposal. As artificial insemination was already in use for stock breeding, it was merely a matter of applying it to human beings and choosing them carefully. New technical solutions were only required for the safe storage of semen without the sperm deteriorating. Other writers had imagined scenarios whereby breeding was separated from sexual love. Such arrangements usually involved the best specimens being on hand to continue families. Curbing the fertility of the 'unfit' while inviting a few highly selected men to artificially inseminate women worthy of preserving the race was the vision of French eugenist Georges Vacher de Lapouge. Haldane wrote about ectogenesis in *Daedalus* (1923), impersonal pregnancies where eggs were fertilized inside an ovary taken from a woman who had died.[19] Although the book was critical of the eugenics movement, saying it had tried to breed supermen before developing its thought well enough, such impersonal pregnancies offered tremendous opportunities for improving the race.[20] As soon as the natural sexual connection was cut, a different motivation was needed for having a child. Making it as intrinsically wellborn as possible therefore made sense.

A wedge policy towards normalizing eutelegenesis was formulated later. If, Brewer argued, a couple could not have children naturally, artificial insemination represented their only way of conceiving. Ten per cent of marriages in the 'civilized world' were barren.[21] If the spouses had decided on artificial insemination anyway, they might as well use

[17] Mark B. Adams, 'Eugenics in Russia 1900–1940' in *idem*, *The Wellborn Science. Eugenics in Germany, France, Brazil and Russia* (New York: Oxford University Press, 1990), p. 194.

[18] Susan Squier, 'Negotiating Boundaries. From Assisted Reproduction to Assisted Replication' in E. Ann Kaplan and Susan Squier (eds.), *Playing Dolly. Technocultural Formations, Fantasies, and Fictions of Assisted Reproduction* (New Brunswick NJ: Rutgers University Press, 1999), p. 107.

[19] Martin Richards, 'Artificial Insemination and Eugenics: Celibate Motherhood, Eutelegenesis and Germinal Choice', *Studies in the History and Philosophy of Biological and Biomedical Sciences* 39:2 (2008), pp. 212–213.

[20] J. B. S. Haldane, *Daedalus or Science & the Future. A Paper Read to the Heretics, Cambridge on February 4th, 1923* (New York: E. P. Dutton & Company, 1924), p. 57.

[21] Herbert Brewer, 'Eutelegenesis and Sterility', *Eugenics Review* 28:4 (1937), p. 343.

the best sperm available. The same arguments were relevant to couples who desired additional children, but who were unable to have them by natural means. Beyond a couple's own personal reasons, society might benefit too if they aimed to make their baby the best possible in the circumstances. This point of view was presented by Brewer in his second article on eutelegenesis in the *Eugenics Review* in 1937. Artificial insemination by Donor had always had a eugenic flavour. When carried out on a woman in Philadelphia in 1884 by Dr William Pancoast, he chose the best looking of his medical students as the donor.[22] Artificial insemination by Husband had been carried out for longer. In 1785 a Scottish surgeon named John Hunter used this technique on the wife of a London cloth merchant who could not fall pregnant naturally due to the opening of her husband's penis being too far down the shaft. Hunter gathered the semen and used a warm syringe to administer it to the wife. The American gynaecologist James Marion Sims described another successful attempt in a work published in 1866. This was probably artificial insemination by Husband.[23] A. I. H. can only have a eugenic effect through facilitating or hindering the procedure based on whether the couple is considered potentially worthy parents or not. In recent decades, it has become possible to select the best sperm cells produced by the husband, increasing the chances of conception.

In 1939 eutelegenesis was covered in the well-known medical journal *The Lancet* and the Eugenics Society began listing it under its aims and objectives as a subject worth further investigation. But the latter admitted that eutelegenesis presented problems of legality and morality in a society where monogamous marriage was the norm.[24] Its assessment was based on discussions in an ad hoc committee consisting of Basil Bramwell, Clinton Chance, Cecil Binney, Herbert Brewer and C. P. Blacker.[25] At least it complemented recent developments in the doctrine. There had

[22] Judith Daar, *The New Eugenics. Selective Breeding in an Era of Reproductive Technologies* (New Haven: Yale University Press, 2017), p. 8.

[23] J. Marion Sims, *Clinical Notes on Uterine Surgery. With Special Reference to the Management of the Sterile Condition* (New York: William Wood & Co, 1866), pp. 365–369.

[24] SA/EUG/A31. Annual Report 1938–39, p. 4.

[25] British Library of Political and Economic Science, London. Carr-Saunders papers. B/3/10. Eugenics Society. Interim Report covering the period from May 16th, 1939 to August 15th, 1939, p. 3.

been an upgrading of 'positive' eugenics recently. The Population Investigation Committee, which tried to find means of increasing the birth rate, was established in 1936.[26]

Brewer had been aware that a physician might be prosecuted for practising A. I. D. Until the 1970s it happened haphazardly with few, if any, records kept. A husband who knowingly put himself down as the father on the child's birth certificate was committing perjury. The barrister Cecil Binney investigated these matters for Brewer in 1936. Binney may have been the authority for Brewer's suggestion that the husband should adopt the child born of A. I. D.[27] Brewer had one last innovation to his cherished idea. It was what he called 'penectogenesis' (meaning close to Haldane's ectogenesis), or what is known today as in vitro fertilization.[28] A woman's ova would be fertilized outside her body and implanted in her or another woman. The reason he saw this as a step forward was that it enabled the selection of ova from a highly capable female donor, making the eugenic effects twice as strong. In 1943 a unanimous council of the Society thought this was going too far.[29] There were technical obstacles as well. Capturing the early ova was one unsolved problem, although progress was being made here. In 1934 Gregory Pincus at Harvard University had succeeded in transferring female monkey eggs and he later achieved the same with rabbits. Pincus's academic career was subsequently cut short, but he was involved in creating the contraceptive pill which became available in Britain in 1961. Brewer would always argue that 'convention' and 'morality' were separate entities and that the highest morality consisted of examining how new practices meshed with objective truth.[30] Because he was debating with eugenists, he liked to remind them how Galton had not felt bound by convention. It was not known at the time that Galton had proposed something similar.

[26] Naomi Pfeffer, *The Stork and the Syringe. A Political History of Reproductive Medicine* (Cambridge: Polity Press, 1993), p. 15.

[27] SA/EUG/C43. Brewer to Blacker 26 Mar 1936.

[28] SA/EUG/C44. Brewer to Blacker 19 May 1943.

[29] SA/EUG/C44. Blacker to Brewer 30 Jun 1943.

[30] John McMillan, 'The Return of the Inseminator: Eutelegenesis in Past and Contemporary Reproductive Ethics', *Studies in the History and Philosophy of Biological and Biomedical Sciences* 38:2 (2007), p. 406.

Post-War Developments on Artificial Insemination

A. I. D. would receive publicity in 1945 when the obstetrician Mary Barton, her husband Bertold Wiesner and urologist Kenneth Walker wrote about the practice in the *British Medical Journal*.[31] Walker argued that since the Anglican Church had accepted birth control within marriage, involving a separation of sexuality and reproduction, logically it should accept artificial insemination too.[32] In December 1945, due to wider media coverage, the archbishop of Canterbury's Commission on artificial insemination was established. When it reported in 1948, it recommended prohibition.[33] The practice constituted adultery, the birth certificates were falsified and the physician carrying out the procedure acted as both the instigator and the accessory to a crime. These conclusions were in harmony with legal practice. The verdict in the English law case Russell vs Russell in 1924 was that *ab extra* insemination constituted adultery.[34] In 1945 a physician was sent to prison and struck off the Medical Register for falsely registering two children as being his and his wife's.

Artificial insemination was described in the Society's revised statement of aims in 1950 as having obvious eugenic possibilities.[35] It nevertheless urged caution and wished to overcome the prevailing laws and sentiments against it first. The risk existed of genetic half-siblings marrying if a donor had a lot of children, marriage would be weakened by use of the technique and children might feel alienated if told that their biological father was unknown to their mother. The first scenario was not an exaggerated danger, particularly if the donor lived in a small town where his offspring might well meet each other.[36] However, after freezing sperm became possible in the 1950s, families of full siblings could be created

[31] Pfeffer, *The Stork and the Syringe*, p. 115.

[32] Angus McLaren, *Reproduction by Design. Sex, Robots, Trees, and Test-Tube Babies in Interwar Britain* (Chicago: University of Chicago Press, 2012), p. 126.

[33] *Artificial Human Insemination. The Report of a Commission Appointed by His Grace the Archbishop of Canterbury* (London: S. P. C. K, 1948), pp. 37, 40, 42.

[34] A. C. M. Schellen, *Artificial Insemination in the Human* (Amsterdam: Elsevier, 1957), pp. 292, 323.

[35] Richards, 'Artificial Insemination and Eugenics', p. 217.

[36] Schellen, *Artificial Insemination*, p. 165.

with the same donor being used several times.[37] The technique took a great leap forward in 1953 with reports of A. I. D. going ahead with frozen sperm.

In 1956 the Society convened a symposium on artificial insemination by Donor with C. P. Blacker, Cecil Binney, Margaret Hadley Jackson, Philip Bloom and Alan Parkes speaking. Jackson was a specialist at Exeter Family Planning and Infertility Clinic, and active in the Family Planning Association.[38] Bloom was a psychiatrist and A. I. D. practitioner. Parkes is credited with the discovery in 1949 that using glycerol when freezing sperm ensures that it is not damaged.[39] Binney as usual clarified the legal aspects of the technique. He said it was not a crime for a married woman to choose it behind her husband's back since ordinary adultery was not illegal.[40] Nor did he consider impregnation in this way to be adultery, reasoning that adultery was sexual in nature while artificial insemination was not.[41] If an A. I. D. child were used to defeat someone's claim to a title or annuity, the legal position would be the same as for an ordinary illegitimate child doing so. In 1944 the Medical Defence Union, an organization protecting physicians from prosecution, had produced a consent form for A. I. D. patients. The patient and her husband had to declare that the birth of a child through the method would not defeat any other person's claim to a title, estate, interest or funds.[42] The form, however, was probably not legally valid as no such assurance could reasonably be given. The physician could in any case be sued for performing the operation unskilfully or negligently.

It was Alan Parkes's team at Rothamsted Experimental Station in Hertfordshire which had made the discovery about glycerol. He hoped to make further innovations, stating that cryo-technology of cells and tissues seemed promising.[43] Describing her own experience, Jackson said the

[37] Daar, *The New Eugenics*, p. 8.

[38] McLaren, *Reproduction by Design*, p. 132.

[39] Jon Turney, *Frankenstein's Footsteps. Science, Genetics and Popular Culture* (New Haven: Yale University Press, 1998), p. 163.

[40] Schellen, *Artificial Insemination*, p. 322.

[41] Cecil Binney, 'Legal and Social Implications of Artificial Insemination', *Eugenics Review* 40:4 (1949), pp. 201, 203.

[42] Pfeffer, *The Stork and the Syringe*, p. 119.

[43] 'Artificial Insemination (Donor)', *Eugenics Review* 48:4 (1957), pp. 204, 207, 209–211.

oldest semen sample she had succeeded with was 15 ½ hours old. She thought deep-freeze methods would be improved in the future. Bloom counselled that to avoid psychological damage, it was best not to tell the child how he or she had been conceived. In fact the couple should forget about it themselves. Beyond the practical aspects, it was left to Blacker to focus on how the method could be used in eugenics. He also related a story of how he had been approached by someone who wished to set up a sperm bank with celebrity donors and was hoping to profit financially.

Conversely, writing in 1958, Blacker suggested that being asked to donate gametes to a National Eugenics Corporation could be an honour for eminent men and women similar to being included in the *Dictionary of National Biography*.[44] Cold War considerations were affecting his thinking, such as the need to maintain the sperm banks below ground to avoid the risk of damage by radiation. There was already an element of designer babies, since the physician involved tried to satisfy the wishes of the couple. There might be other eugenic uses for artificial insemination too, such as correcting a demographic curve.[45] Having already discussed eutelegenesis in 1939, the council of the Society deliberated on artificial insemination in 1958. In March that year it set up an A. I. D. Investigation Council consisting of Parkes, Jackson and Binney.[46] This body gathered evidence from almost all the twenty doctors practising artificial insemination in Britain. It prepared a memorandum for the Home Office's Committee on Artificial Insemination. The Home Office needed clearer guidance on the subject after a Scottish legal ruling by a judge in Edinburgh in 1958 found that A. I. D. without the husband's consent did not constitute adultery.[47]

The Church of England's Lambeth Conference debated the issue in 1958. Its conclusion that A. I. D. equalled adultery contributed to making the road ahead a winding one. It was probably in line with the public's view. When the *News Chronicle* published two articles on A. I. D. in February 1958, nine out of ten letters from readers condemned

[44] C. P. Blacker, 'Artificial Insemination: The Society's Position', *Eugenics Review* 50:1 (1958), pp. 51, 53.

[45] G. W. Bartholomew, 'The Development of Artificial Insemination', *Eugenics Review* 49:4 (1958), p. 193.

[46] Pfeffer, *The Stork and the Syringe*, pp. 118–119, 212.

[47] *Report of the Departmental Committee on Human Artificial Insemination* (London: H. M. Stationery Office, 1960), p. 2.

the practice.[48] Cedric Carter, the Society's new general secretary, was not as enthusiastic about eutelegenesis as Blacker. He nevertheless disagreed with the archbishop's commission that artificial insemination was a form of adultery.[49] The Society convened a members' meeting on the impact of the Lamberth Conference with the Rev Dr Sherwin Bailey, a liberal theologian, as the speaker.[50]

The Society's memorandum to the Home Office's Committee on Artificial Insemination was submitted in May 1959.[51] Alan Parkes and Philip Bloom additionally gave oral evidence to the committee. Herbert Brewer was not directly involved in the 1950s. He was glad that the Society was continuing to discuss and research it.[52] Still working on the idea, in an article for the American magazine *Balanced Living* in March 1961, he quoted the French biologist Jean Rostand that sperm deeply frozen could be used for donating long after the individual himself had died.[53]

The Departmental Committee, led by the Earl of Feversham, reported in July 1960. Among its recommendations were that single women should not be inseminated in this way, while in a marriage both husband and wife should consent in writing before A. I. D. was carried out.[54] The Committee regarded A. I. D. as a practice which it hoped would cease or become less frequent in the interests of society. It did not wish to institute a ban as the practice was difficult to detect and not prevalent, with only about a hundred births a year resulting from A. I. D. In order to avoid legal recognition, neither would it regulate the practice. As for eugenics, the Departmental Committee's report mentioned the role of A. I. D. both as a fertility treatment and for the sake of improving the race. It acknowledged but did not engage with Brewer's work. The archbishop's commission had briefly dealt with the eugenic aspect of A. I. D.[55] Both Blacker and another member of the Society, the biologist Peter Medawar, had given evidence to the commission. Medawar's BBC Reith Lectures

[48] Pfeffer, *The Stork and the Syringe*, pp. 120–121.

[49] SA/EUG/C44. Carter to Brewer 4 Jan 1955.

[50] SA/EUG/A51. Annual Report 1958–59, p. 6.

[51] *Report of the Departmental Committee*, p. 99.

[52] SA/EUG/C44. Brewer to Blacker 3 Nov 1958.

[53] SA/EUG/C44. Herbert Brewer: Ethics, Parenthood and Conception, p. 4.

[54] *Report of the Departmental Committee*, pp. 6, 34, 69, 72, 78, 81.

[55] *Artificial Human Insemination*, pp. 9, 17, 31.

on the 'Future of Man' took place in 1959 and he won the Nobel prize for physiology or medicine the following year.[56]

After the A. I. D. Investigation Council had submitted the memorandum, its focus changed towards better techniques of storing spermatozoa over time. In July 1961 it appointed the biologist Agnes Onions to test spermatozoa routinely and make calculations about the exact temperatures required to preserve it intact.[57] She had experience of testing spermicides and access to a laboratory. Succeeding her in 1962 was D. W. Richardson, who continued the work until 1966, when he left before the completion of the project.[58] Meanwhile, Hermann Muller wrote an article in *Science* in 1961 calling for 'germinal choice'.[59] The concept implied that families could have children the natural way or they could choose eutelegenesis. Muller thought that many of them would choose eutelegenesis when it became clear how superior those children would be to the average.[60] In the United States the connection to eugenics was maintained and in 1980, thirteen years after his death, the 'H. J. Muller Repository for Germinal Choice' was unveiled in California. The brainchild of businessman Robert Graham, the inventor of shatter-proof spectacles, it was nicknamed the 'Nobel Prize Sperm Bank' because it was known that some Nobel laureates in science had donated samples.[61] Graham saw it both as an offering to women and a means of 'positive' eugenics.[62] Unlike an earlier notion of an American sperm bank based on eugenics, it became a reality. During the Second World War, the eugenist Frances Seymour had proposed opening a clinic in London, where British women would be inseminated with the sperm of American men.[63]

[56] 'The Reith Lectures 1959. The Future of Man', *Eugenics Review* 52:3 (1960), p. 145.

[57] SA/EUG/A54. Annual Report 1961–62, pp. 9–10.

[58] Faith Schenk and A. S. Parkes, 'The Activities of the Eugenics Society', *Eugenics Review* 60:3 (1968), p. 153. SA/EUG/A59. Annual Report 1966–67, p. 5.

[59] David Plotz, *The Genius Factory. Unravelling the Mysteries of the Nobel Prize Sperm Bank* (London: Simon & Schuster, 2005), p. 31.

[60] Karp, 'Eutelegenesis', p. 328.

[61] Plotz, *The Genius Factory*, p. 4.

[62] C. O. Carter, 'Eugenic Implications of New Technology' in *idem*, *Developments in Human Reproduction and their Eugenic, Ethical Implications* (London: Academic Press, 1983), p. 207.

[63] Pfeffer, *The Stork and the Syringe*, p. 210.

Although the report of the Feversham Committee had been dismissive of A. I. D., by the early 1970s the British Medical Association noticed an increased demand for it. It decided to establish a sub-panel with Sir John Peel, a former president of the Royal College of Obstetricians, as chairman to investigate the practice and formulate a policy. In 1973 the panel recommended that A. I. D. children should be deemed legitimate in the law, provided that the husband consented and had taken steps to register himself as the father on the birth certificate. It wanted A. I. D. to be made available on the National Health Service subject to certain provisos. This in fact happened soon afterwards with Margaret Hadley Jackson as the pioneer.[64] Her view was that it appealed to unconventional people less affected by religious, moral or legal scruples.[65]

The Society commended the technique as facilitating an improvement of the population, even if used on a small scale. A. I. D. was eugenic due to donors being chosen carefully. They tended to be highly fertile and have good personal characteristics.[66] It was only if marital breakdown subsequently occurred or if there was a dispute about inheritance that conceptions of this kind posed a problem. To ward off these dangers, the Society recommended that children born using A. I. D. should also be legally adopted by the husband of the mother. It wanted a revision of the law so that all children born in a marriage had equal status, whether conceived naturally or artificially. Eugenics was not a significant factor when A. I. D. became a recognized option. Most sperm banks in Britain today do not select for particular characteristics. Only being free of congenital diseases and matching the recipient family's physical appearance are taken into account. In the late 1960s and in the 1970s, however, a new reproductive technique was being invented in Britain which had a connection to eugenics. That aspect of its history is not common knowledge.

[64] Sue Teper and E. Malcolm Symonds, 'Artificial Insemination by Donor: Problems and Perspectives' in Carter (ed.), *Developments in Human Reproduction*, p. 41.

[65] William J. Finegold, *Artificial Insemination* (Springfield IL: Charles C. Thomas, 1976), p. 21.

[66] *Bulletin* 9:2 (1977), pp. 51–52.

In Vitro Fertilization

In 1982 the Galton Lecture was given by Robert Edwards, the pioneer within in vitro fertilization.[67] He was a member of the Society. In 1999 his remarks about how in the future it would be considered a sin to allow a child with a genetic disease to be born, raised hackles. Thinking about the 'quality' of children to be born would be necessary.[68] Edwards had begun collaborating with the gynaecologist Patrick Steptoe in 1968. One of Steptoe's innovations was the use of laparoscopy, allowing the harvesting of ripe ova through a small incision in the abdomen.[69] It helped overcome some of the difficulties they were experiencing. In 1969 they managed to implant a fertilized egg in the womb of a woman, though it did not develop normally there. Edwards and Steptoe would induce ovulation by hormone treatment, which led to oestrogen levels rising so high that implantation of the embryo failed. After some years without success, they decided to insert the embryo at the best time of the natural menstrual cycle. This led to conception and in 1978 Louise Brown, the first test-tube baby, was born at Oldham General Hospital. The fact that a baby had been conceived in a laboratory surpassed the feat of artificially induced conception in the womb.

Though not quite Haldane's ectogenesis, the opportunities offered by in vitro fertilization were stupendous. Lesley Brown had one of her own eggs fertilized, but medical science was getting closer to allowing someone else's ovum to be used. The first known egg donation was carried out in Australia in 1984.[70] Using sperm and ova from highly selected donors and placing the embryo in the womb of almost any healthy woman is a eugenic dream come true. Brewer's 'penectogenesis' was becoming a reality. And it could be taken even further. The Society was interested in genetic engineering in the late 1970s.[71] Organisms might have a gene added, producing a protein which would modify the characteristics of the host. In 1980 Herbert Boyer and Stanley Cohen began applying the technique devised by Daniel Nathans, who won the

[67] *Bulletin* 14:4 (1982), p. 117.

[68] Daar, *The New Eugenics*, p. 184.

[69] Turney, *Frankenstein's Footsteps*, p. 166.

[70] Erica Haines, 'Egg Donation' in Annette Burfoot (ed.), *Encyclopedia of Reproductive Technologies* (London: Routledge, 2018), p. 269.

[71] *Bulletin* 11:2 (1979), p. 37.

Nobel prize for physiology or medicine in 1978. It involves dislodging a DNA molecule and attaching it to a plasmid, a free-floating piece of DNA found in bacteria. This recombinant DNA molecule can be transferred to another living organism, where it might produce an associated protein.[72]

Genetic Modification and Gene Therapy

In the 1980s and 1990s work on modifying plants genetically made advances. They became known as genetically modified food. Gene therapy was used on humans from 1993, adding genes not present naturally and the absence of which can cause serious diseases. It helped against hypercholesterolemia, a life-threatening condition, by adding the necessary gene to the liver of the patient. Similar treatment worked for adenosine deaminase deficiency, whereby metabolism is difficult, for eight children in the United States and Italy in 1995. In 1996 at the University of Texas Medical School, the mutated gene P-53, present in half of all common cancers, was replaced.[73] The act of removing genes is called biopsy . Gene therapy can be used to treat an individual already having the potential to develop a disease. Gametal cells are the egg and sperm cells, the modifications of which thereby affect future generations. The only type of gametal modification which is currently practised is maternal spindle transfer. It is the same as pronuclear transfer, described below, with the variation that it preserves the mitochondria of the mother instead of a fertilized cell. The aim is to produce healthy ova where they are at risk of disease.[74] The technique is also medically simpler than with pronuclear transfer.

A long-term safety audit of biopsy has not yet been undertaken.[75] Its practice needs to be balanced against the fundamental question of when it is worthwhile to remove genes which may cause a disorder. Genetic conditions vary in their rates of penetrance, meaning whether they will lead to the actual disease or not. Huntington's chorea has 100% penetrance and therefore will always manifest itself during the individual's natural life if the genes for it are present. However, other conditions

[72] Nathan Van Camp, *Redesigning Life. Eugenics, Biopolitics, and the Challenge of the Techno-Human Condition* (Brussels: Peter Lang, 2015), p. 17.

[73] Richard Lynn, *Eugenics: A Reassessment* (Westport CT: Praeger, 2001), p. 254.

[74] Calum MacKellar and Christopher Bechtel (eds.), *The Ethics of the New Eugenics* (New York: Berghahn, 2014), p. 101.

[75] MacKellar and Bechtel (eds.), *The Ethics of the New Eugenics*, pp. 83–85.

exist with much lower penetrance. Is removing the genes in question ultimately worthwhile then? In Britain there is increasing use of preimplantation genetic diagnosis (PGD), previously carried out only to detect the most serious diseases. It indicates that the threshold for the use of PGD is being lowered and that currently biopsy is favoured over the default safety aspect of keeping all genes. On the other hand, the reach of PGD is limited because only a small number of embryos can be created, whereas there are about six thousand possible genetic illnesses. Given a couple's genetic makeup, there may be few or no embryos created which are free of these potential diseases.

Another problem is a philosophical one of identity. Just how many genes can be added or removed before the actual person inherent in the embryo changes? It is a question which the ethicist Nicholas Agar, who supports the new eugenics, has pondered. If the gene therapy occurs in such a way that the potential changes occur late in life or are minor, the individual may be regarded as the same despite the modification. However, a substantial change or one which affects childhood involves the actual replacement of one human being by another.[76] For eugenists this is arguably an advantageous outcome. An individual is seen as the bearer of a particular strand of genetic code, affecting himself or herself during the lifetime and others in subsequent generations. If a better code can be imprinted, it is seen as leading to benefits while the non-appearance of the less fortunate person is fortuitous. Agar nevertheless thinks that there are limits to how much we should improve ourselves, as he argued in *Humanity's End: Why We Should Reject Radical Enhancement* (2010).[77]

Prenatal Diagnosis

With eugenics basing itself on a greater degree of volition after the Second World War, an important tool for perceived human improvement became available in the form of prenatal diagnosis. In the words of a participant in the movement, Richard Lynn:

[76] MacKellar and Bechtel (eds.), *The Ethics of the New Eugenics*, p. 103.

[77] Alison Bashford, 'Julian Huxley's Transhumanism' in Marius Turda (ed.), *Crafting Humans. From Genesis to Eugenics and Beyond* (Goetingen and Taipei: V & R Unipress and National Taiwan University Press, 2013), pp. 153–154.

> The first testing method to be developed was amniocentesis. This consists of inserting a needle through the abdomen of the pregnant woman into the amniotic fluid surrounding the fetus and the withdrawing of a fluid sample. This sample is then analyzed for the presence of defective genes and other genetic anomalies. Amniocentesis was first developed in the 1930s for the diagnosis of erythroblastosis fetalis, a blood disorder. [...] In 1967 it became possible to use amniocentesis for the fetal diagnosis of Down's syndrome; and in 1968 the first abortion of a Down's fetus was reported.[78]

In 1972 David Brock and Roger Sutcliffe of Edinburgh University discovered that it could also be used for detecting anencephaly and spina bifida. Within a few years, almost all the hundred or so chromosomal conditions and 23 defects of metabolism could be traced.[79] A latter addition is chorionic villus sampling, in which the needle only touches the outer membrane of the foetus (the chorion).[80]

Chorionic villus sampling has a 1–3% chance of inducing a miscarriage and for amniocentesis the risk is about a third of that.[81] Maternal serum alpha-feta protein analysis is to check for elevated levels of AFP in a blood sample from the pregnant woman. Higher or lower levels than usual may indicate neural tube diseases or Down's syndrome. Another safe form of screening is an ultrasound scan about which Lynn explains:

> Ultrasound scan involves passing high frequency sound waves through the mother's body and analyzing the echoes reflected from the fetus. A picture of the fetus is built up, and it becomes possible to detect 80 to 90 percent of major structural anomalies of the brain, heart and limbs, including anencephaly and spina bifida. The method cannot be used until 16 weeks into the pregnancy at the earliest. [...] By the early 1990s most pregnant women throughout the Western nations were offered an ultrasound scan, had the test carried out, and had the impaired fetuses aborted.[82]

[78] Lynn, *Eugenics*, p. 248.

[79] Kevles, *In the Name of Eugenics*, p. 257.

[80] MacKellar and Bechtel (eds.), *The Ethics of the New Eugenics*, p. 71.

[81] Richard. T. Hall, 'Prenatal Diagnosis Technologies' in Burfoot (ed.), *Encyclopedia of Reproductive Technologies*, pp. 195–196.

[82] Lynn, *Eugenics*, p. 249.

Under current British law, two physicians must agree that a foetus has a 'serious' disability if a late termination is to occur.[83] Abortion is not by definition a eugenic procedure. It is an article of faith for feminists that a woman should be in control of her own body. In the 1930s supporters of abortion would often highlight the argument that it would diminish maternal mortality.[84] There was 'some overlap' between the Abortion Law Reform Association and the Eugenics Society.[85] The honorary treasurer of the association when it was established in 1936 was Frida Laski, a convinced socialist eugenist.[86] She would explain how a working-class mother could not afford the £ 25 or £ 50 fee payable for an abortion in Harley Street. She had used the same argument for birth control in 1925.[87]

Ectogenesis and Eutelegenesis

The main political reaction in Britain to the arrival of in vitro fertilization was the formation of the Committee of Inquiry into Human Fertilization and Embryology in 1982. Led by the philosopher Mary Warnock, it was tasked with recommending legislation for the new reproductive technology. The remit covered both present and future developments in research and infertility treatment, such as cloning, surrogacy, ectogenesis, parthenogenesis and gene therapy.[88] Its report, issued in 1984, urged the founding of a statutory licensing authority to regulate both research and infertility treatment. There should be legal limits on how science could proceed ethically, infertility services should be expanded through the National Health Service and the law updated to take the children born from such envisaged techniques into consideration. The two issues where unanimity could not be achieved were surrogacy and

[83] MacKellar and Bechtel (eds.), *The Ethics of the New Eugenics*, p. 74.

[84] Jeffrey Weeks, *Sex, Politics and Society. The Regulation of Sexuality since 1800* (London: Longman, 1981), p. 195.

[85] Stephen Brooke, *Sexual Politics. Sexuality, Family Planning, and the British Left from the 1880s to the Present Day* (Oxford: Oxford University Press, 2011), p. 100.

[86] SA/EUG/D1 Abortion Law Reform Association Conference 1936, pp. 1, 34.

[87] Martin Pugh, *Women and the Women's Movement in Britain 1914–1959* (Basingstoke: Macmillan, 1992), p. 136.

[88] Susan Merrill Squier, *Babies in Bottles. Twentieth-century Visions of Reproductive Technology* (New Brunswick NJ: Rutgers University Press, 1994), pp. 20–21, 28.

embryo research. In 1990 the Human Fertilisation and Embryology Bill was passed in Parliament. It permitted research on embryos up to fourteen days after conception, with a statutory body established to regulate such research. The Act would have implications for infertility, genetic disease, miscarriage, contraception and the diagnosis of chromosomal abnormalities.

Over time in vitro fertilization witnessed new techniques. One of these can be used for cloning, deliberately creating twins or quintuplets. The woman begins by taking fertility drugs to increase her production of eggs and stimulate the ripening of them. Spermatozoa is then added in a petri dish with 50, 000 or so to an egg. The fertilized egg is placed in a tissue culture median and there undergoes cell division until it is a blastocyst of eight or sixteen cells, whereupon it is implanted into the womb. Afterwards pregnancy progresses normally. To clone the embryo, its cells are separated at the stage when it consists of four cells which become identical blastocysts.[89] This simulates the natural process of twinning. It is probable that four such blastocysts are the maximum number into which it is possible to split human embryos. It is a separate question whether the woman can bear quintuplets. The splitting often happens naturally with implanted foetuses anyway. The rate of twinning is 23% for such births in Britain.[90]

The Warnock Committee recommended a ban on ectogenesis. It proposed making it a criminal offence to grow a human embryo outside the body for more than 14 days. In *Daedalus*, Haldane had imagined a situation a hundred years hence when less than 30% of births happened naturally. Perhaps not having to bear children anymore would be beneficial to women, thus promoting gender equality. Ectogenesis could potentially have profound eugenic consequences. A test-tube baby can undergo much more design than any child conceived in the womb. In Haldane's vision, the small number of people who were responsible for providing gametes for the new children were highly selected. This was having momentous effects: 'the advance in each generation in any single respect, from the increased output of first-class music to the decreased

[89] National Academy of Sciences, National Academy of Engineering, Institute of Medicine, National Research Council, *Scientific and Medical Aspects of Human Reproductive Cloning* (Washington DC: National Academy Press, 2002), p. 26.

[90] Hrafn Asgeirsson and Salvör Nordal (eds.), *Reproductive Technology and Surrogacy. A Global Perspective* (Copenhagen: Nordic Council of Ministers, 2015), p. 25.

convictions for theft, is very startling'.[91] Civilization would have collapsed had it not been for the new form of reproduction, as the least desirable members of society were outbreeding the virtuous and able. Here was the discontent which Sheila Jasanoff identifies as the motivating factor behind a number of utopias.[92] Haldane's ectogenesis was the escape route from the troubles of contemporary society. Those who were chosen to be ectogenetic mothers underwent an operation to remove their ovaries. This operation was portrayed as neither disfiguring nor dangerous anymore, though not pleasant either. A necessary condition for eugenics enjoying success, it was seen as an honour. In the late 1960s fiction became reality through the invention of laparoscopy. Haldane, who did not think that regular eugenics was advancing, conceded there were a few disadvantages with ectogenesis. Family life had suffered and instinct had become blurred. Women now routinely had lactation brought on by the injection of placentin.

Alexis Carrel, the French biologist, had succeeded in keeping mammalian ovaries alive outside the body in tissue culture in 1935. This research was funded by the American aviator Charles Lindbergh.[93] Scientists had experimented on culturing human placentas outside a woman's body in 1958, 1962, 1968 and 1969 and foetuses had been perfused.[94] Apart from ectogenesis, Haldane agreed with eutelegenesis. He promised to fund Brewer moderately and was willing to donate gametes if his ancestry was found to be satisfactory.[95] However, like with other forms of eugenics, Haldane did not think eutelegenesis would lead to a breakthrough. Very few women were choosing Artificial Insemination by Donor and most conceptions were going to remain natural. A. I. D. would have to become prevalent to herald significant change in the genetic makeup of humans. This would bring forth additional people of

[91] Haldane, *Daedalus*, pp. 65–67.

[92] Sheila Jasanoff, 'Imagined and Invented Worlds' in Sheila Jasanoff and Sang-Hyun Kim (eds.), *Dreamscapes of Modernity. Sociotechnical Imaginaries and the Fabrication of Power* (Chicago: University of Chicago Press, 2015), p. 337.

[93] Daniel J. Kevles, 'Le Spectre de Galton et les nouvelles technologies de la reproduction: Genre et eugénisme, enjeux présents et passés' in Jean Gayon, Daniel Jacobi and Marie-Claude Lorne (eds.), *L'Eternel retour de l'eugénisme* (Paris: Presses Universitaires de France, 2006), p. 217.

[94] Squier, *Babies in Bottles*, p. 71.

[95] SA/EUG/C43. Memorandum, 14 Nov. 1935, p. 2.

high intelligence, but not automatically more 'geniuses'. It was a mystery how men of the stature of Galileo and Newton arose.[96] Galton had assumed that once the general standard of the population was raised, more 'illustrious' men would come into being.

An earlier writer who had criticized Galton's understanding on this was Frank Challice Constable. Efforts to breed a Newton, Shakespeare or Milton would not only fail, but would probably in fact *prevent* such great minds being born.[97] Only the *combination* of parents these men had would work, and it would never occur to a eugenist to select the right ones. And if the superlatively gifted were made into a separate caste, it would rupture the connection which exists between them and the wider genetic pool of which they form part. Take away the spark of talent which exists everywhere, and the ordinary members of society would never again be able to foster someone exceptional. These arguments suggested that the medical breeding of talent would fall short. The very unpredictability of genius meant that it could not be analysed and replicated like a cooking recipe. Galton equated eugenic worth with intelligence, health and morality existing together. However, as pointed out by Cullwick Perrins in the *Socialist Review* in 1931, people of exemplary attainment often were not the most vigorous (and it may be added that all were not so virtuous either):

> Guy de Maupassant, Nietzche (sic) (syphilis), Keats, Sterne, Brontë, R. L. Stevenson, Spinoza, Richelieu, Nelson, Mozart, Descartes, and Francis of Assisi (tuberculosis), Carlyle (chronic dyspepsia), Swift (Meniere's disease), Turner, Bacon and Lincoln (sons of mothers mentally unsound). Newton himself was for some time actually insane. But what society would say these ought not to have lived?[98]

Galton proposed competitions to find the best men and women in 1865 and repeated the notion in 'The Eugenic College of Kantsaywhere'. He sorted individuals who were above average into grades, with each

[96] Diane Paul, 'Eugenics and the Left', *Journal of the History of Ideas* 45:4 (1984), p. 575.

[97] F. C. Constable, *Poverty and Hereditary Genius. A Criticism of Mr Francis Galton's Theory of Hereditary Genius* (London: Arthur C. Fifield, 1905), pp. 129–130.

[98] Cullwick Perrins, 'Eugenics without the Churches', *The Socialist Review* New Series Vol III no. 3 (January 1931), p. 142.

consecutive grade signifying greater achievements. If his eugenic measures succeeded in raising the statistical mean by one grade, it would make Adolphe Quetelet's curve shift to the right, meaning there would be more persons in the highest grades. For instance, he had calculated that there were 233 people in a million whose exploits made them 'eminent'. By current standards, this was one in 4, 300 individuals and the level which Galton classified as 'grade F'. If the average were raised one grade, this level of attainment would instead be reached by 2, 423 in a million (previously 'grade E'). This tenfold increase of 'eminent men' would yield vast benefits to the nation, and the natural increase among the even higher levels of intellect even more so.[99]

If eutelegenesis were practised according to Muller's specifications, he had worked out that the incidence of the 'eminent' would rise from one in four thousand to one in a hundred.[100] It scarcely affects the argument that Pearson, a better mathematician than Galton, felt that Galton's calculations considered ability rarer than it actually was. The 'eminent' judges, statesmen, scientists and, especially, clergy in *Hereditary Genius* were scarcely one in four thousand. Pearson initially reckoned that among the classes who received the education required to perform these roles, the incidence was closer to one in five hundred. By the time he had finished the book, he doubted that they were rarer than one in a hundred among their fellows.[101] Implicitly then, an alternative to 'positive' eugenics was simply to make an education available to all children. Galton agreed with widening education, but did not think that it would produce many more outstanding individuals. The lack of an education did not hold a supremely talented person back, only perhaps someone who was a little above average.[102]

The Warnock Committee discussed parthenogenesis. It may be induced artificially through chemical or physical stimulus, causing an egg to undergo cell division without being fertilized by a spermatozoon. This form of asexual reproduction, which is natural in some fish, birds and reptiles, causes offspring to be born with traits greatly resembling

[99] Galton, *Hereditary Genius*, p. 343.

[100] Elof Axel Carlson, *Genes, Radiation, and Society: The Life of H. J. Muller* (Ithaca NY: Cornell University Press, 1981), p. 400.

[101] Karl Pearson, *The Life, Letters and Labours of Francis Galton*. Volume 2 (Cambridge: Cambridge University Press, 1924), p. 104.

[102] Galton, *Hereditary Genius,* p. 39.

the mother's. Eugenically, parthenogenesis therefore would ensure that offspring born of a particularly talented woman do not regress too much towards the mean. Julian Huxley believed in 1923 that it was only a matter of time before the technical difficulties with it would be overcome with mammals and humans.[103] This assertion may have been based on parthenogenesis occasionally occurring in hypertonic sea water, as communicated by American biologist Thomas Hunt Morgan.[104] Nevertheless, even today, it is more difficult to harvest and store eggs than sperm. The extraction needs to be done carefully as it involves an element of risk. Only a few eggs are produced from the hormone inducement. They are usually extracted just before being fertilized, whereupon the embryo is implanted at once or stored.[105] Parthenogenesis does not require fertilization of course, but the ovum would need to start cell division in a laboratory before being implanted.

Prenatal Sex Selection

Prenatal sex selection may also have eugenic effects. This is because women are carriers of, but men are affected by, certain genetic diseases like haemophilia and muscular dystrophy. There are almost two hundred other inherited diseases and conditions which affect only males. Some are mild like colour blindness. However, for parents with diseases in their bloodlines, it is a distinct advantage to have girls rather than boys. As the sperm containing an X-chromosome, making the baby female, are slightly larger and heavier than sperm with a Y-chromosome, containing 2 or 3% more DNA, by adding a liquid the sperm can be sorted by a process of centrifuge. The chromosomes containing an X-chromosome are spun out of the solution. It is also possible to select by staining chromosomes with a fluorescent dye which can be detected by laser or making the sperm swim through a selective medium. Artificial insemination or in vitro fertilization is then used to give the prospective mother the sperm of the type which is desired. In vitro fertilization can be used to grow several embryos, the sex of each of these determined and one with the chosen sex then implanted.

[103] Squier, *Babies in Bottles*, p. 41 modified.

[104] Jon Turney, *Frankenstein's Footsteps*, p. 68.

[105] MacKellar and Bechtel (eds.), *The Ethics of the New Eugenics*, pp. 58–59, 61, 66.

The process of checking the sex and implanting is so rapid that it can be done in a day.

The main disadvantage with sex selection is that sperm sorting does not always achieve what is wanted, with a success rate of 70–90%. (For in vitro fertilization with embryo selection the success rate is higher than 99%, but as explained below, its use would be illegal for any non-medical reason.) Leaving aside the issue of legality for the moment, if sex selection were permitted, it could lead to disappointment. Having invested in expensive treatment, only to find that a baby of the opposite sex to the one desired is born, parents might subconsciously feel hostile towards it and give it fewer opportunities in life.[106] On the other hand, it may reduce the number of abortions which are motivated by the foetus being of the opposite sex to the one wanted. It is likely that this still goes on in Britain, with the sex perhaps being one of several factors which decide whether a foetus is to be carried to term. An abortion may be motivated by the likelihood of genetic illnesses as described above. The Human Fertilisation and Embryology Act, as amended in 2008, prohibits sex selection for any reason that is non-medical. When it is permitted, strict guidelines must be followed by clinics, which need to be licensed.[107] Both the sperm sorting and preimplantation diagnostic techniques of sex selection were legal into the 1990s, for any reason.

In 1990 the technique of sex selection became the crux in a public debate when Mary Warnock agreed it could be used by families of noble lineage to select male embryos to ensure the continuation of their lines.[108] It caused a furore because Warnock said that the statutory body should prevent frivolous sex selection, but that preserving a hereditary title was a legitimate reason to want a male baby. As this constituted extra help with the view to preserving class differences, the statement was unpopular. The same topic featured in earlier decades with regard to artificial insemination. As the Eugenics Society tended to organize wealthy, leisured people a few of whom had titles, it was a concern to some of the membership. Artificial Insemination by Donor is farther from the natural

[106] Tom Shakespeare, 'Sex Selection Should be Outlawed' in Cindy Muir (ed.), *Reproductive Technology* (Detroit: Thomson Gale, 2005), p. 68.

[107] MacKellar and Bechtel (eds.), *The Ethics of the New Eugenics*, p. 65.

[108] Squier, *Babies in Bottles*, p. 100.

order of things than sex selection because the husband is not the biological father and titles run in the male line. However, other claimants to a peerage or annuity may feel that sex selection would give certain family members an unfair advantage over others.

The use of in vitro fertilization has clear eugenic effects. As the embryos are implanted, they can legally be selected according to a medical diagnosis since they are less than two weeks old. Research conducted outside Britain might not be subject to the same limitations and the findings are in any case shared internationally. Preimplantation genetic screening (PGS) allows the parents of the would-be child to learn much about its prospects within days of fertilization. Cells can be extracted and checked for the correct number of chromosomes or whether a mutation associated with disease is present in them.[109] Thus it is possible to detect Down's syndrome and other chromosomal conditions at a very early stage. Preimplantation genetic diagnosis (PGD) is similar but examines the DNA-material directly.[110] It is usually performed when it is suspected that a couple may harbour genetic diseases. At present, it is only possible to detect diseases which are caused by a single gene. Polygenic diseases are not picked up. Preimplantation genetic haplotyping (PGH) resembles PGD. Instead of straightforwardly finding a dysfunctional gene in an embryo, it looks for suspicious 'markers' near a faulty gene which may betoken a disease caused by a mutation. The advantage of this method is that it is better at finding rare mutations, whereas the common ones are often uncovered by PGD.

All these methods provide the opportunity to select among embryos, as not every fertilized egg is used. While this allows couples to select embryos which are free of the genetic taint causing the diseases to which they may be susceptible, it opens up the possibility of creating designer babies too. The Human Genome Project was completed in 2003. Although understanding the genomic sequence does not give information about individual genes, such advances will eventually permit selection.[111] The bioethicist John Harris writes in *Enhancing Evolution. The Ethical Case for Making People Better* (2010) that it is a moral duty to

[109] Daar, *The New Eugenics,* p. 138 modified.

[110] MacKellar and Bechtel (eds.), *The Ethics of the New Eugenics,* pp. 78–79.

[111] Nils Roll-Hansen, 'Eugenics and the Science of Genetics' in Alison Bashford and Philippa Levine (eds.), *The Oxford Handbook of the History of Eugenics* (Oxford: Oxford University Press, 2010), p. 93.

use the new technologies to improve health, intelligence and longevity.[112] To what extent such vistas are *possible* is an unanswered question, but the new eugenics can be as millennial as the old. Traditional eugenics did not rely on complete knowledge of genetics either and much of what it believed turned out to be incorrect. Since selection is the cornerstone of designer babies, embryos will be left over, which could lead to a new practice of embryo adoption. The technology of preserving ova has traditionally been more complicated than that of preserving spermatozoa.[113] A rejected embryo could instead be implanted in another woman and brought to gestation.[114] It is furthermore possible to delay pregnancy after conception. A case was reported in Britain in 2004 in which a mother gave birth after the embryo had been stored in a deep freezer for eleven years.[115] Subsequently, the law was changed to allow storage for five years only. In May 2003 there were 52, 000 frozen embryos in the United Kingdom.[116]

It might be assumed that Preimplantation Embryonic Selection (PES), occurring as part of all three methods discussed, acts as a harbinger of choosing embryos, to virtually unlimited parental satisfaction. This is based on what might become possible in the future rather than on today's realities. As things stand, the number of embryos available for selection from a single course of treatment is about twelve. The hyperstimulation of a woman's ovaries through fertility drugs is a precondition of this and is not without its dangers.[117] As the procedure is costly and intense, there is moreover a limit on how many couples or individuals who can undergo it.

[112] Daar, *The New Eugenics*, p. 191.

[113] Carlson, *Genes, Radiation, and Society*, p. 436.

[114] MacKellar and Bechtel (eds.), *The Ethics of the New Eugenics*, p. 52.

[115] Estela V. Weldon, 'Why Do You Want to Have a Child?' in Mariam Alizade (ed.), *Motherhood in the Twenty-first Century* (London: Karnac, 2006), p. 70.

[116] Cindy Muir, 'Introduction' in *idem*, *Reproductive Technology*, p. 13.

[117] MacKellar and Bechtel (eds.), *The Ethics of the New Eugenics*, p. 81.

Cloning

Cloning is perhaps the final frontier for the new reproductive technologies. Parthenogenesis comes close as it is a form of asexual reproduction too. Although Gregory Pincus succeeded with a rabbit ovum in 1934, he was unable to replicate the experiment with the same result afterwards.[118] Thus artificial parthenogenesis has not yet been developed successfully, whereas cloning is a reproductive technique that made strides in the 1990s. The great symbolic event was the birth of Dolly the sheep in 1996, announced to the press in 1997. Dolly was cloned with a cell from the frozen tissue of a long-dead sheep by Ian Wilmut and others at the Roslin Institute at Edinburgh University. Scientists had successfully cloned other animals and cloned sheep were in existence before Dolly . What was novel was the use of an adult animal cell to create the clone. The success rate had been low, one birth from 277 attempts, but it did demonstrate what was technically possible. No primates have yet been cloned.

There is one technique, merely on the drawing board at the moment, which resembles cloning and parthenogenesis to some extent. Mitochondria are very small entities inside each cell which produce the energy needed to keep the cell functioning. They are inherited from both parents. Pronuclear transfer replicates the procedure used in somatic cloning. The pronuclei from both parents, in an egg with problematic or diseased mitochondria, are transferred into a second egg from a woman with no mitochondrial disease. The second egg has had its nuclei removed beforehand.[119] It is a combination of cells which would allow the useful elements to survive, while the parental essence of the person or persons replicated remains intact. Pronuclear mitochondrial transfer was permitted by Parliament in February 2015 in an amendment to the Human Fertilisation and Embryology Act of 2008. Human clinical trial has commenced, but only one successful transfer has taken place, which was reported for an abnormally fertilized egg.[120]

Unlike how the matter is presented in science fiction, clones are not always identical to each other. Through epigenetics it has become known

[118] Turney, *Frankenstein's Footsteps*, pp. 162, 213.

[119] MacKellar and Bechtel (eds.), *The Ethics of the New Eugenics*, p. 92.

[120] A. S. Reznichenko, C. Huyser, M. S. Pepper, 'Mitochondrial Transfer: Implications for Assisted Reproductive Technologies', *Applied and Translational Genomics* 11 (2016), p. 45.

that certain chromosomes are switched on, while others are present but have no apparent effect. If a female mammal has two X-chromosomes, one is functional in each cell, while the other is inactive.[121] Many human traits are polygenetic, which means that making the trait appear is more complicated than replicating a single or just a few genes. The belief that 'feeblemindedness' resulted from just one gene was one of the fallacies of early twentieth-century eugenics. Human clones are not necessarily wanted as living twins. One of the applications of cloning is to create possible sources of replacement organs or for the sake of testing new drugs. Cloning is associated with eugenics as it is a means of artificial reproduction which allows plenty of design.

Eugenists have always sought to create new and better types of human beings. Sexual reproduction cannot avoid regression towards the mean (a law of heredity discovered by Galton). Even though the offspring of parents who are superior in a particular characteristic are also above average, they are not generally *as superior as their parents*. Cloning the parent is therefore a more certain way of preserving his or her characteristics in the future generations. Also, it is likely that exemplary performance in a particular field is caused by a *combination* of characteristics. Such combinations might not be replicated in the offspring, even though a child shares half the genes of a parent. But while cloning prevents regression and may preserve beneficial assortment of characteristics, it is worth noting that the personality of any person is constructed through a mixture of heredity and life experiences (the environment). A cloned person will not share the same life experiences as the original. Formative events, inspiration and significant mentors that are missing can chart a completely different course in an individual's life. Even clones born at the same time are no more similar than identical twins.

From a eugenic point of view, cloning could be used to create masses of near-identical people. Already knowing the physical and mental characteristics of someone, cloning would make it possible to create approximate copies of him or her. Surrogate mothers would have to gestate all the children in question, which is where 'penectogenesis' comes in, reproducing with no contribution of traits from the child bearer. Just as it is a combination of genes which creates favourable traits, the same principle applies in society. An impressive civilization comes into being

[121] Kerry Lynn Macintosh, *Human Cloning. Four Fallacies and their Legal Consequences* (Cambridge: Cambridge University Press, 2013), pp. xv, 35, 89.

through containing a variety of people. The accomplished men investigated in *Hereditary Genius* were judges, statesmen, peers, commanders, writers, scientists, poets, musicians, painters, clerics, classicists, oarsmen and wrestlers. While these vocations generally require the exercise of the intellect, a victorious commander possesses different attributes to an innovative scientist. Cloning would allow greater control of what type of citizens may be in existence by breeding true to type. Although the personality is too complex to be transmitted with certainty by any reproductive technique, sexual reproduction in essence involves another person, whereas the mixing of genes has already played itself out with cloning.

Apart from cloning through twinning by means of in vitro fertilization, the other method is how Dolly the sheep was created. Starting with an egg, the inherent chromosomes of what would have been the mother are removed. By adding a somatic cell from the animal or embryo to be cloned to the unnucleated egg and stimulating it, sometimes the egg begins to divide. The cell divisions form a blastocyst, an embryo which has not yet been implanted. By placing the blastocyst in the womb of an animal of that species, if it continues to develop and is carried to term, a clone is born.[122] The Human Fertilisation and Embryology Authority regulates research in this area in Britain and the cloning of humans is prohibited by the 1990 Act of the same name. In terms of life opportunities, cloning would allow women who cannot reproduce by any other means to become genetic mothers. With 'penectogenesis' the child is biologically not hers, whereas if cloning were allowed, it would be her daughter or younger twin.

The New Eugenics and the Old

The use of the new reproductive techniques could be made to have eugenic effects. Some are eugenic in themselves. Prenatal diagnosis involves making a choice about what sort of babies are worth having. The fact that the decision is made by the parents rather than by society does not cancel out the eugenic aspect. Even before 1945, eugenists claimed to be proceeding through persuasion and attempting to create an 'enlightened eugenic conscience'. Traditional eugenics was typified by

[122] National Academy of Sciences, *Scientific and Medical Aspects of Human Reproductive Cloning*, p. 23.

restrictions on marriage, segregation in institutions to avoid conception and the attempt to introduce sterilization, but it also included selectively targeting contraceptive advice, incentives to parenthood and giving lectures on heredity. Eugenics is about social selection as an alternative to social Darwinism in which nature does the choosing. Galton found the latter cruel and amenable to change.[123]

Many left-wing supporters such as Lancelot Hogben, H. G. Wells, J. B. S. Haldane and Bertrand Russell did not identify with the doctrine because they saw only 'mainline' eugenics as 'eugenics', whereas their socially reformist and inclusive version was not 'eugenics'. As Haldane wrote in 1926: 'The growing science of heredity is being used in this country to support the political opinion of the extreme right' making 'eugenics abhorrent to many democrats'.[124] It would be a logical fallacy to go along with their interpretation. Wells is known for his advocacy of the sterilization of 'failures'. He and Hogben lent their names to the Eugenics Society's efforts to bring in sterilization in 1931. Hogben explained his reasoning like this:

> I sympathise with Major Church's Bill for the following reasons. While remediable defects of social and physical environment undoubtedly play a part in the incidence of mental disorders and defects of a general nature, there is little doubt that inborn genetic factors contribute to the incidence of mental defect, and that in some cases the genetic factor is the significant element in the situation.[125]

Russell went so far as to suggest that 'imbeciles, then epileptics, consumptives, dipsomaniacs, and finally "all who fail the usual school examinations"' might be sterilized.[126] He wrote that a future socialist state would come to grief if it ignored eugenic concerns.[127] Haldane

[123] For a longer formulation of how eugenics and social Darwinism differ from each other, see Andre Pichot, *The Pure Society. From Darwin to Hitler* (London: Verso, 2009), p. 111.

[124] Gary Werskey, *The Visible College* (London: Penguin, 1978), p. 96.

[125] London Metropolitan University. TUC Collection. Box marked 'Contraception Eugenics to 1988' HQ 750–766. The Committee for Legalising Eugenic Sterilization. Better Unborn, p. 10. See p. 12 for Wells's brief comment of support.

[126] McLaren, *Reproduction by Design*, p. 30.

[127] *Eugenics Review* 14:4 (1923), p. 289. Reprinted from the Independent Labour Party's newspaper *The New Leader*.

believed in ectogenesis and eutelegenesis, by means of which society would improve mankind through selective breeding.

Eugenics is severely criticized for the attitudes its supporters held. Extending the gift of fertility to additional people through new technology, though, does not jar with the moral principles of the majority. In a similar way, birth control is accepted by almost everyone in modern societies. On the level of personal choice, contraception is neutral and not eugenic in nature. The same can be said for the reproductive techniques whereby is not known what type of child would be born. As soon as a glimpse of the child to be born appears, eugenics cannot be held outside the equation. Selecting births based on ability, health, morality or social class is intrinsically eugenic. If new reproductive technology is marshalled only for privileged groups or if choice occurs based on such criteria, eugenics is still in play. This can be seen from articles in the *Eugenics Review* or the *Bulletin of the Eugenics Society* after the Second World War in which counselling couples or monitoring immigrants is recommended. The threshold for access to reproductive medicine is a factor in whether these practices are eugenic or not. To make it easy or difficult according to whether a patient is regarded as an asset or a liability to society, is eugenic. One must take care, however, not to label a policy eugenic simply because it is linked to the birth rate. In 1916, writing in the *International Journal of Ethics*, Bertrand Russell proposed payment to women who gave up work in favour of motherhood and to unmarried women.[128] The purpose was probably to increase the birth rate generally and not just among privileged families. It may have been social Darwinist because Russell mentioned Germany, and in the context of the First World War raging, the idea may have been that a larger population would better be able to withstand war.

Apart from the institutionalization of the 'feebleminded' permitted by the Mental Deficiency Act of 1913, 'negative' eugenics did not have much impact in Britain. Two plainly eugenic clauses were removed during the parliamentary process.[129] The 1959 Mental Health Act abolished the distinction between psychiatric and general hospitals. Outpatient care

[128] Stephen Trombley, *The Right to Reproduce. A History of Coercive Sterilization* (London: Weidenfeld & Nicolson, 1988), p. 77.

[129] Jayne Woodhouse, 'Eugenics and the Feeble-minded: The Parliamentary Debates of 1912–14', *History of Education* 11:2 (1982), p. 134.

expanded during the 1950s and 1960s as drug treatments became available. By the early 1970s, there was a steady decline in the number of patients left in asylums.[130] Eugenic ideology now retained only indirect societal influence through the social Darwinist component of the 'mainline' version. That theory supported free markets, competition and individualism, with Herbert Spencer becoming a late nineteenth-century philosophical icon. His preferred policies were enjoying a resurgence, growing during the 1970s, and brandished by right-wing Conservatives brought to power in 1979.

Reproductive Choice

Liberal attitudes that people should be trusted to make decisions for themselves and that outsiders should respect personal choice being exercised were of use within assisted reproduction. Eugenics is about supposedly objective standards. Most of the time the choices made are in conformity with the doctrine anyway, such as foetuses with genetic anomalies being aborted. In Britain, it is not permitted deliberately to select persons or embryos who are likely to have disabilities in preference to those who are normal, under Sect. 14 (4) of the Human Fertilisation and Embryology Act 2008.[131] Such counterintuitive selection could occur because parents might accept a disability like autism in return for possibly concomitant powers of increased concentration and high intelligence.

A case from the United States came to international attention in 2002. A deaf, lesbian couple, Sharon Duchesneau and Candace McCullough, tried to have a child by artificial insemination and specified that the donor should also be deaf. All sperm banks turned the couple down, as they were not willing to attempt to create a disabled child.[132] The couple eventually found a deaf male friend who agreed to act as a direct donor and a child was conceived. Subsequently, in Britain, other deaf couples expressed a preference for a child sharing their disability. While people are of course free to have a child by natural means with whomever they like, infertility

[130] Barbara Taylor, 'The Demise of the Asylum in Late Twentieth-century Britain: A Personal History', *Transactions of the Royal Historical Society* 21 (2011), p. 197.

[131] MacKellar and Bechtel (eds.), *The Ethics of the New Eugenics*, p. 38.

[132] Van Camp, *Redesigning Life*, p. 27. See M. Spriggs, 'Lesbian Couple Create a Child Who is Deaf Like Them', *Journal of Medical Ethics* 28:5 (2002), p. 283.

treatment can be refused under Sect. 13 (5) of the same Act to a woman whose future child's welfare, or the welfare of any child already living, may be adversely affected by such a birth.[133] This includes supportive parenting, meaning that a child likely to be neglected or one who will be living in penurious conditions, may not be conceived as a result of infertility treatment. This could be seen as a eugenic enactment against the 'feckless', although equally as a child protection measure. And the sperm banks in the Duchesneau and McCullough case were also acting eugenically, though hardly in a manner which offends against a sense of justice.

British eugenists in the 1960s and 1970s were hoping to exert influence on couples through medical counselling, including the size of the family for which they should aim. If a general practitioner or fertility specialist leant on a couple heavily to choose a particular course of action, he or she was practising eugenics. In the 1930s the Mental Hospital and Institutional Workers' Union recognized that mental deficiency constituted a social problem. Although it was this trade union which prodded the Trades Union Congress to adopt a negative view of 'voluntary' sterilization, it realized that a solution was needed. A cartoon in its journal from July 1935 described the 300, 000 mental defectives as 'the shadow over England and Wales'. Another cartoon from November 1935 drew attention to one person in ten being 'too dull or sickly to earn a living unaided' and one in 120 being 'feebleminded'.[134] The disparity between the eugenists and the trade union was that the former were trying to prevent mental deficiency through legislation. They claimed it would go hand in hand with inculcating attitudes in the population and the individuals concerned. After 1963, when they decided not to get involved in pushing for sterilization, only persuasion remained as a realistic alternative. This should happen face to face with the person concerned through medical advice and birth control. Daniel Kevles thinks that such consultations had generally lost their eugenic flavour by the 1970s when genetic counselling became available on the National Health Service. The counsellors were usually physicians with an interest in genetics not going

[133] MacKellar and Bechtel (eds.), *The Ethics of the New Eugenics*, p. 51.

[134] University of Warwick. Modern Records Centre. Unmarked volume. Mental Hospital and Institutional Workers' Journal. 24:7 (1935), p. 9. Mental Hospital and Institutional Workers' Journal 24:11 (1935), p. 17.

beyond the most obvious ramifications for the patient.[135] Few of these practitioners in a system of healthcare for everyone would recommend having children or not based on the social status of the couple. Modern views contrasted with the former eugenic ideal, illustrated by a story in the medical journal *The Lancet* in 1939. An involuntary childless couple had no less than one male and sixteen female volunteers to help them reproduce with either of their gametes because the husband met the criteria of 'high intelligence'.[136]

The new eugenics is based on choice. Mary Warnock argued that everyone will need to make their own moral judgements.[137] Selecting gametes, embryos, foetuses and eventually infants (designer babies born with chosen specifications in relevant categories) is the *raison d'être* of the ideology in its contemporary guise. Regulators and the authorities generally take a back seat and do not apply eugenics as they were urged to do in the period when the doctrine was influential. In fact, lessons having been learned, they act as a brake on developments by forbidding medical procedures which they see as unethical or which they fear may reopen the door to official eugenics. Eugenics had deeper consequences on continental Europe and Scandinavia than it ever did on Britain. Accordingly, the European Parliament declared in 1997 that human cloning could not be tolerated 'as it permits a eugenic and racist selection of the human race, it offends against human dignity and it requires experimentation on humans'.[138] It was probably mindful of experiments on inmates in concentration camps during the Second World War and the severe trauma subsequently caused. Because of the Nazi embrace of eugenics and the racial state, the mere word 'eugenics' is odious and the doctrine must be prevented from returning at all costs. The consensus is therefore to permit medical advances which will bring improvements to health and the diminution of suffering, but also to be on guard against unintended side effects whereby the substance of eugenics might return.

Nevertheless, the individual choices people make in a competitive and money-oriented society collectively provide an opportunity for eugenics

[135] Kevles, *In the Name of Eugenics*, pp. 257–258.

[136] *The Lancet*, 29 July 1939, p. 265.

[137] Richard John Neuhaus 'The Return of Eugenics' in *idem*, *Guaranteeing the Good Life: Medicine and the Return of Eugenics* (Grand Rapids MI: William Eerdmans, 1990), p. 25.

[138] Lynn, *Eugenics*, p. 268.

to gain ground. The majority who make a conscious decision probably desire offspring who are healthier and more intelligent. Some will look for exceptional aptitudes in art, music, athletics or chess. Specialized ability of this kind can be the key to major earnings at the top level of performance or at least to kudos and enjoyment if not taken to its peak. Education may potentially pave the way for exciting, high-earning careers and thus it is hardly surprising that many parents would prefer their children to have high educability. It is not yet possible to select for these traits at an embryo level, but heredity implies that it may be done at a gametal level. Sperm and egg donation can be from highly selected persons, meaning that it is already possible to continue one of the two family lines while adding heredity of proven mettle. Since the patriarchal family has declined in recent decades, it carries only a little stigma to be born outside a relationship. Nobody need ever know that a woman has visited a sperm bank, while artificial insemination or surrogacy makes it possible for a man to have a child with a woman of great talents whom he has never met. Giving evidence to the archbishop of Canterbury's Commission on Artificial Insemination, which reported in 1948, two lawyers argued that all members of society have the right to know each other's identity, which is the significance of names.[139] Social life has taken a different turn during the intervening years and anonymous interaction is increasingly common. A stage has already been reached where 'penectogenesis' (considered as surrogacy with separate egg donation) and eutelegenesis are living realities. Marriage having been weakened since 1935 or 1980 makes eutelegenesis a less redoubtable prospect. If same-sex couples are allowed to have children, why should it be a problem that children in a family are only biologically connected to the mother? In other families where the children are from various past and present relationships, why should a father not add another by procreating with an egg donor? The legal and moral problems identified by the Eugenics Society in connection with artificial insemination in the 1950s are all but gone.

The hurdles that remain relate to the countermeasures imposed by the authorities. There is a £ 45 limit on the expenses which can be claimed by a sperm donor (although the limit is £ 986 for an egg donor) precisely to stop a market in advantageous human traits emerging. This stands in contrast to the United States, where sperm banks or egg donor services

[139] *Artificial Human Insemination*, p. 53.

exist which keep records of their donors' results in scholastic examinations and where there is no cap on the payment for them. In Britain, eutelegenesis and eugenic 'penectogenesis' are not practised directly. Public medicine has only sought to find a good match between the appearance of the donor and the putative father in the family. But consider the information available to clients of the London Sperm Bank in January 2011. The records of the donors list height, hobbies or interests, ethnic origin, occupation, race and religion.[140] It asks its clients to choose the donor with the most appealing characteristics. These variables are broad enough to allow a eugenic choice to be made. Occupation along with hobbies or interests go beyond the data which the Eugenics Education Society felt so fortunate to be able to convince the authorities to include in the 1911 census. A revealing picture emerges from knowing what the donor decided on for his career alongside his avocational interests. The choice is as eugenic as the focus on the differential birth rate was in the early twentieth century. The added information available to clients in the United States consists of data which led to the refinement of eugenics in the 1920s and 1930s. They enable fine-tuning, but the recorded details for British purchasers of sperm are enough in themselves to allow eugenics. The personal choice will usually be in conformity with the doctrine because screening has already taken place and few clients will go for a donor who has unwholesome hobbies and is perceived as struggling to make ends meet.

It may be argued that the options women have in choosing a mate are wider and the categories far more numerous in real life. Knowing or meeting a man in reality permits greater discernment of his characteristics and entire personality. That is true, but selection at a sperm bank is nevertheless more eugenic because the woman is guaranteed the opportunity to try to conceive by that particular donor, which is not the case in normal interaction. Eugenists tried to influence spouse selection through lectures, posters and films. It is likely, however, that the vast majority chooses the best they can manage and who will accept them.[141] Hermann Muller did not think that women would select the best men, from his perspective, if

[140] MacKellar and Bechtel (eds.), *The Ethics of the New Eugenics*, p. 109 endnotes 100 and 101.

[141] Anthony M. Ludovici, *The Choice of a Mate* (London: John Lane & The Bodley Head, 1935), p. 17.

the context was American society in the 1930s.[142] Francis Galton was also aware that wealth might trump desirable characteristics in mate selection, which he regretted. Both of course believed in eugenics. The alternatives to marrying money were valuing altruism (Muller), considering the accomplishments of the individual (Galton) or taking into account the health of everyone in the family (Eugenics Education Society). This was a part of what was called creating a 'eugenic conscience' (Fig. 6.2).

The Underclass and Immigrants

Apart from the new eugenics emanating from reproductive technology, there are still remnants of old-style eugenics around. It is seldom associated directly with genetics, nor are people ranked in terms of value. Nevertheless, what the Eugenics Society originally called the 'social problem group' still exists. In the aftermath of the England riots in 2011, Prime Minister David Cameron created a body to investigate 'troubled families', stating that 'a relatively small number of families are the source of a large proportion of the problems in society'.[143] Now perhaps identifiable by anti-social behaviour combined with long-term unemployment and a lack of qualifications, they may in consequence be relying on various benefits for their livelihood. Such outcomes are not currently believed to be caused by genetics, but social inheritance plays a part and it may be difficult to break away from such a background. A discourse exists in some quarters against single mothers, who in reality are victimized for eschewing the patriarchy, but whose alleged transgressions are failing to provide a proper upbringing for their children and for claiming benefits.[144] It may be a watered-down version of early twentieth-century eugenic thought about 'feebleminded' women. If they belonged to the residuum, women like these were at risk of becoming prostitutes, yielding easily to seduction and often having had children outside marriage anyway.[145]

[142] Muller, *Out of the Night*, p. 142.

[143] Patrick T. Merricks, *Religion and Racial Progress in Twentieth-century Britain. Bishop Barnes of Birmingham* (Cham: Palgrave Macmillan, 2017), p. 228.

[144] See Richard Lynn, *Dysgenics. Genetic Deterioration in Modern Populations* (no place of publication: Ulster Institute of Social Research, 2011), pp. 68–69.

[145] *Report of the Royal Commission on the Care and Control of the Feeble-minded.* Volume 8 (London: H. M. Stationery Office, 1908), pp. 23, 25, 120, 372.

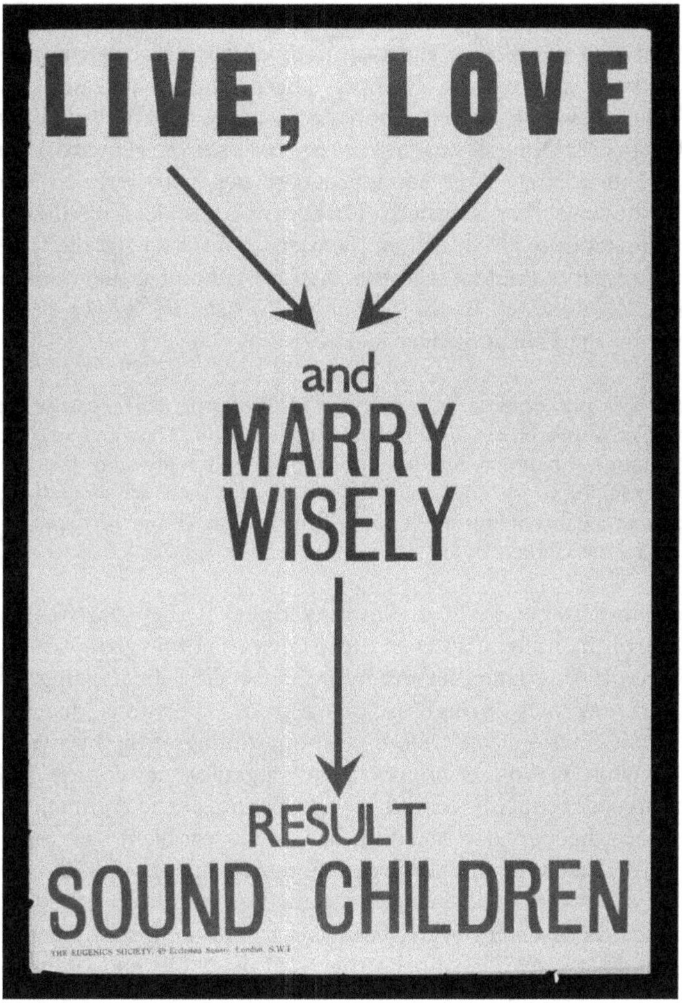

Fig. 6.2 Creating a eugenic conscience (reproduced from the Eugenics Society Archive, Wellcome Library, London)

The United States experienced the breaking down of ethnic barriers in the first two decades of the twentieth century through immigration from southern and eastern Europe. There, and to a lesser extent in Britain, eugenics took on xenophobic connotations. The British psychologist William McDougall was appointed to a chair at Harvard University in 1920. Immediately after emigrating, he began to refer to American research showing that southern Europeans were less intelligent than northern Europeans.[146] In Britain the menace was that the higher fertility of the immigrants could cause the displacement of indigenous people. As Sidney Webb stated in the pamphlet *The Decline in the Birth-Rate*, published by the Fabian Society in 1907:

> Twenty-five per cent of our parents, as Professor Karl Pearson keeps warning us, is producing 50% of the next generation. This can hardly result in anything but national deterioration; or, as an alternative, in this country gradually falling to the Irish and the Jews. Finally, there are signs that even these races are becoming influenced. The ultimate future of these islands may be to the Chinese![147]

Those immigrants could do nothing right. If they started practising birth control themselves, an even more devious ethnic group stood ready to take over. In this pamphlet and in similar articles, the nagging concerns were tied to eugenics through use of the stock expressions 'degeneration' or 'national deterioration'. Doubts about immigration have not gone away and while no longer connected to eugenics, newcomers including refugees are still regularly viewed with suspicion. If the argument is made longitudinal, their greater breeding becomes a problem too. Such arguments were a staple in the heyday of eugenics, but though less common today are still occasionally wheeled out. The only overt supporters of classic eugenics in recent years, besides a few academics working from a theoretical perspective, have been the National Front and other far-right organizations. The fascists bemoan intermarriage between native Britons and foreigners and are homophobic and hostile to disabled people,

[146] William McDougall, *Is America Safe for Democracy?* (New York: Charles Scribner, 1921), pp. 63-65.

[147] Sidney Webb, *The Decline in the Birth- Rate* (London: Fabian Society, 1907), p. 17.

believing that these tendencies and characteristics will destroy the British population over time.[148]

With social attitudes changing, miscegenation has generally not been viewed in a negative light in recent decades. As Commonwealth immigration subsided in the 1970s, members of the Eugenics Society began thinking about how the new arrivals could be integrated. Nothing was as effective as intermarriage between them and locals. They were sure it would not lead to any bad consequences genetically. However, sociological pressures exist of being torn between two cultures, which the parents should pay attention to, among other factors, when deciding how many children to have.[149] There was a time in the late 1950s and in the 1960s when it looked as if the Society might become an anti-immigration pressure group. C. P. Blacker was open to this current and Colin Bertram wrote the broadsheet which was critical of Caribbean immigration in 1958. Blacker, however, was following up his special interest of reducing the population generally.

At Galton's 1904 presentation to the Sociological Society, H. G. Wells suggested in his response to the paper that two healthy parents who belonged to different races might have sickly children.[150] He gave the example of the offspring born to white men and black women in Tanganika. In 1908 eugenist Eugen Fischer showed this was not true for the progeny of Boer or German men by local women in German South-West Africa, whom he had studied.[151] Fischer later fell into line with the German anthropological tradition and lauded racial purity. Jon Alfred Mjøen, a Norwegian member of the Eugenics Education Society, and Charles Davenport, of the Eugenics Record Office on Long Island, also researched racial mixture . Mjøen's approach was to distinguish between

[148] D. S. Lewis, *Illusions of Grandeur. Mosley, Fascism and British Society 1931–81* (Manchester: Manchester University Press, 1987), pp. 250–251.

[149] *Bulletin* 2:2 (1970), p. 31.

[150] University College London, Galton Archive, Galton/2/4/19/9/2. Sociological Society, folio 17.

[151] Paul Weindling, *Health, Race and German Politics between National Unification and Nazism, 1870–1945* (Cambridge: Cambridge University Press, 1993), pp. 143–144, 437.

'harmonic' and 'disharmonic' race crossings and he bred rabbits in an attempt to prove his contentions.[152]

The new eugenics might have a slender effect on the ethnic composition of Britain. Well-integrated groups who trust the medical authorities are probably more likely to seek help in the case of infertility. As with other discretionary medical interventions, ethnic minorities may on average know less about the opportunities available and be less inclined to take advantage of them if they do. This is probably true for socio-economic status in general.[153] Knowledge, confidence and sometimes wealth, if private healthcare is involved, therefore act as gatekeepers for parenthood through assisted reproduction. But the increment is small, since only about 3% of all births occur through such techniques.

Contemporary eugenists who pin their hopes on people choosing the new technologies of reproduction continue to face an overwhelming obstacle: continuing the family line is an instinctive wish. These feelings are held even by those who are not perceived as being talented. As things stand, it is primarily people affected by genetic disease who are likely to turn to gene therapy for purposes of procreation. Competition on a personal level for fame and fortune is not as prevalent as in the United States, where the culture may be more sympathetic to the new eugenics. In 1960, when the technology was decades away, an American eugenist called Calvin Kline opined that the choice of traits for a designer baby exceeded natural parenthood in satisfaction.[154] In the early 1980s about 23 states had laws deeming A. I. D. children to be legitimate.[155]

[152] Jon Alfred Mjøen, 'Harmonic and Disharmonic Racecrossings' in *Eugenics in Race and State*. Volume II. *Scientific Papers of the Second International Congress of Eugenics* (Baltimore: Williams & Wilkins, 1923), pp. 53–58. A version of this paper dealing with humans only was published in the *Eugenics Review* 14:1 (1922), pp. 35–40.

[153] Dion Farquhar, 'Gamete Traffic/Pedestrian Crossings' in E. Ann Kaplan and Susan Squier (eds.), *Playing Dolly. Technological Formations, Fantasies and Fictions of Assisted Reproduction* (New Brunswick NJ: Rutgers University Press, 1999), p. 30.

[154] SA/EUG/C44. Prospectus for the Formation of a Society for the Guidance of Human Evolution by C. W. Kline, p. 20.

[155] Douglas J. Cuisine, 'Legal Implications' in Carter (ed.), *Developments in Human Reproduction*, p. 228.

The Politics of Genetics

Apart from reproductive technology working intrinsically for small-scale eugenics, the doctrine at present benefits from the public's fascination with the unfolding scientific discoveries in genetics. As new genes are found, purportedly responsible for a variety of behaviours, it leads the mind towards a biological world view, as occurred in the aftermath of Darwin's theory gaining credence in late Victorian Britain. The press is quick to seize on the newsworthiness of genes theoretically affecting sexuality, alcoholism, criminality or piety.[156] Is it conceivable that segments of the personality are hardwired in the brain from birth? The psychologist Hans Eysenck was a pioneer of research in this field. He wrote many popular books aiming to disseminate work relating to the genetic input on human characteristics, such as *Crime and Personality* (1964) and *Personality and Individual Differences* (1985). At the time the latter book was published, Eysenck laid claim to being the most cited psychologist in the world. He straddled the divide between popular psychology and academic research. His conclusions came too early to encompass contemporary knowledge of genetics, but he always asserted that he had science behind him.

Eysenck's reputation suffered through his engagement with topics like astrology, paranormal phenomena and, perhaps most damagingly, race. He alleged without many caveats that certain races were more intelligent than others and that social stratification was founded on variations in biological quality.[157] Disadvantaged groups might be incapable of benefitting from social improvements when they occurred. These were echoes of earlier eugenic opinions, but specialist debates had moved on since the early twentieth century. Perhaps not aided by his colourful language, Eysenck became a maverick outside the pale of respectable opinion. His actual results were derivations of the work of Arthur Jensen, who spent two years as a postdoctoral fellow with him at the Maudsley Hospital.[158]

[156] Adam Rutherford, *Control. The Dark History and Troubling Present of Eugenics* (London: Weidenfeld & Nicolson, 2022), p. 109.

[157] H. J. Eysenck, *Race, Intelligence and Education* (London: Temple Smith, 1971), pp. 27, 66, 142.

[158] H. J. Eysenck, *Rebel with a Cause. The Autobiography of Hans Eysenck* (New Brunswick NJ: Transaction Publishers, 1997), p. 153.

It is affiliated to another London institution, the Institute of Psychiatry. Robert Plomin, like Jensen an American, studied twins at the latter from 1995 and reached similar hereditarian conclusions. Jensen's article from 1969 'How Much Can We Boost IQ and Scholastic Achievement', published in the *Harvard Educational Review*, was the catalyst for a spirited debate about social, racial and educational topics, including in Britain.

As has been observed above, the American scene differs from the British. This is true for intellectual discourse as well. Opinions which are relatively common within stateside universities can be rare in the United Kingdom. In the latter country there is little appreciation for research which contends that extreme competition is beneficial or that there are significant inherent differences in capabilities. As early as 1873, Galton argued that excessive competition merely caused suffering and did not lead to any particular benefits. Although Pearson was a social Darwinist, he combined it with socialism and never denied anyone the right to decent material conditions. Indeed, his social views were premised on making use of the talents of all Britons, saying the nation could not afford to waste its human resources. Eugenists like Julian Huxley, C. P. Blacker, J. B. S. Haldane and Lancelot Hogben argued that eugenic worth could not be determined until living standards had been equalized. Otherwise, it would not be possible to tell which observed differences were down to heredity and which were instead the outcome of living in varying social environments.[159] Although there are striking disparities of wealth and income in Britain, these have not been justified intellectually with much vigour since the early days of eugenics. The experiment lasted a few decades, but was overcome from within the movement itself due to a change of tactics, scientific research and the objections posed by the rising tendency of 'reform' eugenics.

'Reform' Eugenics and Transhumanism

In 1943 William Beveridge was able to change the import of eugenics when he put forward that uniform child benefit, and not just the income-gradated scheme in which eugenists believed, would make society better. By giving opportunities to later-born children, it would encourage

[159] 'Eugenics, Socialism and Capitalism. Debate at Members' Meeting', *Eugenics Review* 27:2 (1935), pp. 113–114.

thoughtful parents to increase the size of their family. Such prudent and considerate family units were exactly the ones who should be increasing their birth rate. The welfare state was on the verge of being founded and the partnership of the labour movement with scientific experts and bureaucrats would improve the health and standards of the population.[160] The strand of eugenics which focused on better babies, education for an enlightened citizenry and social mobility was gaining ground. In the 1930s the Society had made working-class women one of the main audiences for its propaganda and they may have influenced each other. Eugenics over time graduated to concentrating on the challenge of the 'social problem group', but its new enemy was less virulent than the old residuum. 'Mainline' eugenics abandoned its discourse from the 1910s and 1920s that unskilled workers lowered the quality of the population. Instead it claimed that people of low intelligence reduced the norms of the citizenry.

In the 1960s and 1970s eugenists devoted attention to race and immigration, while not rejecting a multi-ethnic society out of hand. Some focused on the greater fertility of the immigrant stock, asking what Britain would look like in twenty or thirty years' time. Starting roughly in 1970, there was increasing awareness of the impact of environmental damage, which dovetailed with warnings by a few eugenists in the 1950s about the population explosion. Continuing population growth, especially in Third World countries, would have an impact on other species and nature. It thus constructed a convenient new argument for why there should be checks on who should reproduce. Generally, it was possible to observe popular eugenics transforming itself into the wider notion that scientific thought ought to have an influence on society.

Julian Huxley was among the progressive eugenists. He preserved aspects of the doctrine such as individuals being of varying quality, but did not draw many socio-economic or racial conclusions from them. In 1957 he published a chapter called 'Transhumanism' as part of his anthology *New Bottles for New Wine*. This notion took on a life of its own and became a project of superseding human limits, especially through capacity-enhancing technology. Transhumanism's broader outlook was that mankind now had a cosmic role as the receptacle of a consciousness

[160] Nicholas Timmins, *The Five Giants. A Biography of the Welfare State* (London: Fontana, 1996), pp. 52–53, 55–56, 61–62.

inherent in the universe.[161] The potential was staggering as humans differ as much from single-cell life as the latter do from rocks and stone. The wholesome development of the personality was the highest goal attainable. Thinking also in terms of 'negative' eugenics though, there could be horrendous mismatches of exceptional ability and malevolence. As exploration of the planet was complete, exertions should instead be devoted to mapping the full potential of human beings. Everyone was working far below their true capacity. Science had discovered how to prevent hunger and disease. It offered tremendous vistas of the future. Nobody need accept a life that was unsatisfactory or where he or she was struggling one way or another. However, eugenics still guided Huxley's vision as he stated that he was aiming for quality not quantity, that overpopulation would bring down the curtain on a glorious future and that producing ugliness would be regarded as immoral in the future. Others knew that overpopulation would entail poverty, hunger and violence.[162] Harnessing and teaching spiritual abilities as well as intellectual enhancement was needed. On the one hand, everyone should be improving themselves. On the other, the responsibility to help others also existed.

This prospectus was perhaps an update of what Galton had wanted for eugenics. It would usher in a civilization in which the tone was higher, with marvellous achievements and less suffering. In 1904 he had moreover given imperial ambitions as a justification for 'national eugenics'. In the 1950s with the Empire transforming itself into a 'Commonwealth of Nations', the accent was on universality. Sceptics may have wondered if the poorer part of the globe had any role to play in the betterment of humanity, which Huxley chose to call 'transhumanism'. Developing nations have not been at the forefront of the genetic revolution which has been the clearest marker of transhumanism in recent decades.[163] Huxley, who died in 1975, seems to have regarded educational and self-induced refinement on a scientific basis to be especially important. Technology had a role, particularly since enhancement has been linked to computing, although Huxley was a biologist and concentrated first and foremost on

[161] Julian Huxley, *New Bottles for New Wine* (London: Chatto & Windus, 1959), pp. 13–17.

[162] SA/EUG/C44. Prospectus for the Formation of a Society for the Guidance of Human Evolution by C. W. Kline, p. 28.

[163] Dhavendra Kumar, 'Preface' in *idem, Genomics and Health in the Developing World* (Oxford: Oxford University Press, 2012), p. VII.

evolutionary and medical developments. Huxley's optimism was tied to 'positive' eugenics spreading its wings in combination with the social reform in which he had long believed. Even Galton had believed in social amelioration, while asserting that it was not sufficient in itself. Hereditary measures to enhance human stocks were required, otherwise the improvements in living conditions would quickly be checked and reversed. As the Society claimed in a pamphlet of 1923, slum clearance was merely palliative if nothing was done to tackle the breeding of people gravitating to those areas.[164] Not just 'new wine' (ideas or culture) but 'new bottles' (bodies) had to brought into existence. Huxley was a long-term supporter of Haldane's idea of ectogenesis and wrote the foreword to Brewer's pamphlet about eugenics which addressed left-wing people.[165]

Left-Wing and Right-Wing Eugenics in the Modern Age

After the Second World War the left-wing and right-wing versions of eugenics were reshaped to suit contemporary realities and the experiences of the movement. The left-wing version leant on 'reform' eugenics as formulated in the 1930s and before. Essentially, it offered a positive vision of what science and biological technology could contribute to human progress. Huxley had an important role in crystallizing its values. He was the first director-general of UNESCO between 1946 and 1948, and his views were aligned to its agenda. The other major influence was the Geneticists' Manifesto of 1939. The version was still recognizably eugenic. Huxley updated Galton's aspiration from 1869. If eugenic selection could raise the general level of intelligence just slightly, it would still lead to a significant rise in the number of outstandingly capable people needed to manage complex societies.[166] Concerns about overpopulation constituted a vital component. Huxley, who had been the secretary of the Zoological Society, was one of the founders of the World Wildlife

[164] SA/EUG/C108. Folio 5.

[165] SA/EUG/C43. Foreword by Julian Huxley.

[166] Julian Huxley, 'The Future of Man—Evolutionary Aspects' in Gordon Wolstenholme (ed.), *Man and His Future* (Boston MA: Little Brown and Co., 1963), pp. 16–17.

Fund and remained a Malthusian.[167] He suggested that a share of foreign aid should be directed towards limiting excessive population growth in recipient countries.

Another supporter of reversing population growth worldwide was C. P. Blacker. After 1945, he collaborated with the former Liberal politician Ernest Simon. Simon may have come to eugenics from his association with Eva Hubback. He had an interest in slum clearance and Scandinavian social democracy, about which he wrote books. He joined the Labour Party in the early post-war period. As a noted advocate of contraception, he was active in the International Planned Parenthood Federation. His death in 1960 led to the endowment of a population trust in his name to fund research on global population growth. In 1955 he had written a pamphlet for the Eugenics Society entitled *Some Aspects of World Population and Food Resources*.

The right-wing version of eugenics did not agree with the ravages of the Nazis. It resembled American eugenics, continuing to regard races as inherently unequal. Adherents believed that the differential birth rate was leading to a lowering of average intelligence. Not enthused by the welfare state, they believed it would collapse from the strain of maintaining 'problem families' and immigrants. In 1957 the Society returned to supporting child benefit proportionate to earnings. The right-wing version was hereditarian, with little interest in social influences, and resisted the ongoing process of turning eugenics into human genetics without clear ideological goals. The broadsheets about West Indian immigration in 1958 were part of its agenda. In 1961 the racist journal *Mankind Quarterly* published its first issue in Edinburgh. Reginald Ruggles Gates and Sir Charles Galton Darwin were among the founders.[168] Others who shared these tenets were the psychologists Hans Eysenck and Richard Lynn.[169] It was possible to combine elements from both versions. For instance, Colin Bertram supplemented racism with ecological concerns and C. P. Blacker, who wished to reaffirm eugenics

[167] Alison Bashford, 'Julian Huxley's Transhumanism' in Marius Turda (ed.), *Crafting Humans. From Genesis to Eugenics and Beyond* (Goetingen and Taipei: V & R Unipress and National Taiwan University Press, 2013), p. 156.

[168] Richard Lynn, *The Science of Human Diversity. A History of the Pioneer Fund* (Lanham MD: University Press of America, 2001), p. 458.

[169] Lynn's books *Eugenics. A Reassessment* and *Dysgenics. Genetic Deterioration in Modern Populations* quoted above might be seen as epitomizing right-wing eugenics.

as a valuable tradition, was otherwise closer in spirit to the left-wing version.[170]

Beveridge only became involved in the Eugenics Society in the post-war period, but had given the 1943 Galton Lecture and attended the 1924 Galton Dinner. In 1926 he spoke before the National Council of Public Morals on how to improve the British race. He suggested a mixture of family allowances and birth control, saying that everyone but the poorest were now limiting their families.[171] He agreed that men who rose in social position were generally above average in eugenic worth. During his tenure as director of the London School of Economics 1919–1937, Beveridge established a Department of Social Biology there in partnership with the Rockefeller Foundation. It became a haven for left-wing eugenists such as Lancelot Hogben and David Glass, a statistician who had grown up in the East End. Before it closed down in 1937, the department published research showing that there existed plenty of talent among working-class children. Much of it lay unappreciated in the educational system. Studies by Hogben and Louis Herrman showed lower correlations on intelligence between twins than other researchers had found.[172] This added importance to environmental influences compared to genetic ones. Demography, including differential fertility and population growth, was well covered, including by Hogben's wife Enid Charles.

After 1945 left-wing eugenics emphasized meritocracy and believed that there were working-class youngsters able to distinguish themselves. Blacker wrote approvingly of the Labour government in *Eugenics. Galton and After* (1952).[173] He accepted the criticism of eugenics that it had not recognized fully that the poorest were at a disadvantage in terms of diet, education and housing. The right-wing version became a subculture among traditionalists and Conservatives. It had a certain appeal in protecting long-standing conventions and cultural values in a rapidly changing Britain. Equally, it was provocative to others. In 1971 and 1973

170 SA/EUG/C44. Blacker to Brewer 15 Jan. 1961. He criticized it for becoming mere genetics.

171 *Eugenics Review* 18:2 (1926), p. 159.

172 Lancelot Hogben, *Nature and Nurture* (London: George Allen & Unwin, 1945), p. 94.

173 C. P. Blacker, *Eugenics. Galton and After* (London: Gerald Duckworth, 1952), pp. 139, 144–145.

presentations by Hans Eysenck were disrupted by students at Birmingham University. In 1973 he was also physically attacked while attempting to give a lecture on 'individual differences in brain processes' at the London School of Economics.[174] Students positioned Eysenck as a supporter of fascism and imperialism, which was a focus point in opposing him. Hereditarianism rankled in the context of race, with Eysenck having made himself a standard bearer for scientific racism in 1971. Much of it was imported from the United States, a setting where such thinking has had a platform. The dismantling of Jim Crow produced an outpouring of justification for the earlier system.[175] Even before desegregation, race was a quotidian issue in former slave states. Paul Weindling has explained Huxley's former racism, which the biologist abandoned in the 1930s, as stemming from his first job at Rice University in Texas after 1913.[176] Lancelot Hogben, conversely, became a firm anti-racist while at the University of Cape Town 1927–1930.[177]

The Metamorphosis of Eugenics

Eugenics is not a unified vision or a single practice. There are medical, scientific and popular strands to the ideology. Concurrent with the new technologies of reproduction advancing, the Eugenics Society was transitioning into a learned association for members who liked to discuss a range of topics. In 1984 the *Bulletin* changed its name to the pertinent *Biology and Society*. It folded in 1990, a year after the Society had become the Galton Institute. The *Journal of Biosocial Science* was sent out to members instead.[178] The Society's loyal following supported or acquiesced in the transition to an ostensibly 'biosocial' association, with

[174] Roderick D. Buchanan, *Playing with Fire. The Controversial Career of Hans J. Eysenck* (Oxford: Oxford University Press, 2010), p. 299.

[175] Stefan Kühl, *For the Betterment of the Race. The Rise and Fall of the International Movement for Eugenics and Racial Hygiene* (Basingstoke: Palgrave Macmillan, 2013), p. 3.

[176] Paul Weindling, 'Julian Huxley and the Continuity of Eugenics in Twentieth-Century Britain', *Journal of Modern European History* 10:4 (2012), p. 482.

[177] Saul Dubow, *Scientific Racism in Modern South Africa* (Cambridge: Cambridge University Press, 1995), p. 15.

[178] *The Galton Institute Newsletter* 1 (May 1991), p. 1.

only eight members resigning during 1989.[179] The Institute was scientific with a popular edge in its newsletter, which encouraged debate and short articles, like the *Eugenics Review* had done three decades earlier. The majority of formal publications in its second and third series of occasional papers pertained to medicine. The second series started in 1987 with *A Guide to Cytogenetic Studies in Haematological Disorders* edited by Christine Harrison.

Even before the Galton Institute had become the endorsed name, numerous members were knowledgeable about, or interested in, eugenics and related topics rather than actual supporters of the doctrine. Probably they agreed with elements of it, since it was malleable enough to encompass a variety of observations on how heredity or genetics could be utilized for desirable ends. Hans Eysenck continued to speak before it, contributing to the annual symposium in 1991. Revisiting Galton's *Hereditary Genius*, he argued that the chequered family backgrounds of Newton and Faraday showed that intelligence is genetic.[180] They would not have been able to succeed against the odds, if it were primarily the environment around them which mattered.

The Institute had been a charity since 1960. It was therefore not supposed to become involved in political activities beyond standing up for special concerns. Neither the Marie Stopes Memorial Foundation (wound up in 1976) nor the Galton Foundation crossed that line. The Society for Constructive Birth Control had also been liquidated in 1976. Two years later, it nevertheless reappeared as the Birth Control Trust, wholly under the control of the Eugenics Society.[181] The trust's sole object was to supply poor people in Britain and abroad with free contraception. There was of course an ideological slant to the work. In 2003 it agreed to provide £ 10, 000 per annum for five years to improve access to reproductive health and family planning services in Ethiopia. This could not have profound effects in a country of 75 million people, however, and illustrates the late twentieth-century practice of applying eugenics on a small scale or individual level. The project in Ethiopia was funded through Marie Stopes International, a charity deriving from the new managers of the birth clinic in 106 and 108 Whitfield Street.

[179] SA/EUG/A95. Annual Report 1989, folio 6 v.

[180] *The Galton Institute Newsletter* 3 (December 1991), p. 2.

[181] SA/EUG/A113. Annual Report 2004, folio 9 r., folio 11 r.

In 2007 a proposal was made to establish a closer relationship with the Galton Laboratory. The intended purpose was for the Galton Institute to invest in a programme of human genetics at University College London with medical applications, which would reach out to the wider community. Had this gone ahead, it would have been historic. In 1934 the Eugenics Society had started funding the *Annals of Eugenics* published by the Galton Laboratory. The remits of both had changed since then, becoming respectively 'the science of human heredity and its historical development' or human genetics.[182] It was of course entirely possible to promote eugenic ideas through the means of genetics, particularly with the new opportunities offered by reproductive technology. Robert Edwards and the beginning of successful in vitro fertilization in 1978 proves this. But with eugenics having been significantly diluted over time, there is no reason to assume that geneticists or even members of the Galton Institute sought to do so.

The Galton Institute promoted science through an annual conference, sponsorship of other conferences, research or project funding, its newsletter, website and publications. It is in science that the future of eugenics lies too. In 1976 an editorial in the *Bulletin* was pleased 'experiments in genetic manipulation utilizing artificially prepared DNA recombinant' had been recommended in a departmental report.[183] The probable writer was Derek Roberts, a geneticist at Newcastle University. The research was not without risks, including dangerous pathogens escaping from laboratories. Geneticists today know that a person's characteristics are often acquired on the journey through life. Or the traits are highly polygenetic, with perhaps 709 or 1, 016 genes comprising biological intelligence.[184] It will not be easy to select for this trait.

Our inventive species has never allowed present difficulties to get in the way of seductive visions of the future. In *The World, the Flesh, and the Devil* (1929) Cambridge physicist John Desmond Bernal thought that ectogenesis would set the stage for a three-part process whereby 'perfect man' would be created. It would be a type freed from the sensory, motor and biological constraints of other humans.[185] In 1957 Julian Huxley

[182] SA/EUG/A116. Annual Report 2007, folio 6 r., folio 5 r.

[183] *Bulletin* 8:4 (1976), pp. 101–103.

[184] Rutherford, *Control*, p. 177.

[185] Squier, *Babies in Bottles*, p. 78.

sketched his vision of transhumanism, in which nobody needs accept an unfulfilling life. In 1961 Hermann Muller called for 'germinal choice', believing that the obvious superiority of highly selected individuals would act as a wake-up call to others. In 2010 ethicist John Harris reasoned that it was not just an option, but a duty to apply the genetic technologies to enhance wellness and cognitive skills.

All these visions were powered by, or interacted with, eugenics. Agreement on an exact definition of what it involves may prove elusive. Adam Rutherford, a geneticist who has presented a radio series about the doctrine, does not view medical interventions to reduce suffering as eugenics.[186]There is something to be said for this. The Human Fertilisation and Embryology Act 2008 allows prenatal sex selection only for medical reasons. If motivated by a parent's genetic predisposition, it can reduce suffering because there are many hereditary diseases which only affect men. Other restrictions are in place too, but the Act is still guided by the principle of reproductive choice. And reproductive choice is how eugenics hopes to make its mark in the contemporary era. On the other hand, Galton argued that freeing mankind from suffering was a fundamental reason for embracing the ideology. Informed by the theory of evolution, it was a technical answer, later a political one, to the struggle for existence identified by Darwin.[187] In Edwardian Britain the struggle for existence was visited upon slum dwellers all the time.

Conclusion

Assisted reproduction affords vast opportunities to eugenics. In 1935 the proposal of 'eutelegenesis' was introduced by one of the doctrine's supporters in the *Eugenics Review*. Apart from debating its implications for years, the Eugenics Society undertook research into the practicalities of storing spermatozoa in the 1960s. J. B. S. Haldane had launched a similar vision, called ectogenesis, in *Daedalus* (1923). Pregnancy outside a woman's body did not become a reality in the hundred years that followed. However, in 1978 Robert Edwards and Patrick Steptoe's method of in vitro fertilization succeeded, leading to the birth of a baby.

186 Rutherford, *Control*, p. 189.

187 Charles Darwin, *On the Origin of Species by Means of Natural Selection or The Preservation of Favoured Races in the Struggle for Life* (London: Folio Society, 2006), p. 48.

The technique allows one of several possible embryos to be implanted in the womb, which is momentous in combination with preimplantation genetic diagnosis (PGD). The sex of the foetus can be selected and chromosomal anomalies may be avoided. It is often possible to create twins from the embryos, which is a form of human cloning.

Prenatal screening combined with abortion being available aids eugenics. Gene therapy allows the addition or removal of a gene, which can decrease the likelihood of developing a particular hereditary or congenital disease. It is not yet feasible to choose genetic traits on an embryonic level. Heredity implies that it may be attempted on a gametal level, through the choice of sperm or egg donors for artificial insemination.

The new technologies widen the pool of people who can become parents. Eugenists favour the use of artificial selection and, even more so, social selection as a complement to nature taking its course. They have only a modicum of influence on what type of people who become parents through assisted reproduction. However, when choices are made regarding potential characteristics in the offspring, they tend to accord with eugenic criteria. Even in the 1930s the Eugenics Society asserted that it believed in parenthood as a voluntary role. Contraception would allow couples to decide how large their families should be. The sterilization it recommended was also touted as 'voluntary'. After 1945 it gradually became clear that persuasion would be crucial to the doctrine since support from the authorities was ebbing away. It might have to be exerted on a personal level, which explains why it supported reproductive choice.

Eugenics is a political ideology with an anchoring in science. The same points of view can be made today without needing to refer explicitly to heredity. Only seldom is a version of its vocabulary used, such as 'troubled families' for what eugenists called the 'social problem group' in the 1930s and 'problem families' in the 1950s.

CHAPTER 7

Conclusion

This book has charted a single national case of eugenics over almost a century and a half. Britain was the place of origin of the doctrine. A multitude of other national and regional versions existed. Eugenics presented simple claims about heredity and society. It was malleable enough to be combined with all sorts of social, political and scientific ideas. Even theology could be related to eugenics, as Patrick Merricks has shown. Political issues have been the focus of the present work. This is in line with Chris Renwick's view that social science may have shaped the doctrine to a greater extent than did biology.[1] Eugenics was a justification for right-wing thinking because it explained skewed life chances as a result of heredity. Especially the bourgeoisie owed their good fortune to superior natural endowment. The doctrine was aligned to social Darwinism and social hygiene, dismissing redistribution of resources as exacerbating the 'degeneration' already taking place. It warned that the differential birth rate whereby in the first half of the twentieth century, poorer families had more children than the well-to-do, risked the collapse of civilization. These contentions made up the basic form of eugenics in Britain.

[1] Chris Renwick, 'From Political Economy to Sociology: Francis Galton and the Social-Scientific Origins of Eugenics', *British Journal for the History of Science* 44:3 (2011), p. 345.

D. Redvaldsen, *A History of British Eugenics since 1865*, Medicine and Biomedical Sciences in Modern History,
https://doi.org/10.1007/978-3-031-72290-5_7

In retrospect, after a competing interpretation had arisen, they have been classified as 'mainline' in contrast to the new 'reform' eugenics.

'Reform' eugenics agreed that some individuals were better than others, but what was perceived as 'fitness' under contemporary conditions did not equal eugenic worth. It wanted social reform to accompany eugenic measures and believed that the environment, and not just heredity, had a role in producing personal traits. Every eugenist can be placed somewhere on the spectrum between the 'mainline' and 'reform' versions. Galton, for instance, believed that there was talent to be found outside the bourgeoisie, but still saw characteristics as overwhelmingly determined by heredity. 'Mainline' and 'reform' eugenists were equally supportive of 'voluntary' sterilization.

Neither the Galton Laboratory nor the Eugenics Society were particularly racist, though people who might be described as such formed part of the movement. H. G. Wells, Reginald Ruggles Gates, Hans Eysenck, John Randal Baker and Arthur Jensen (an American member of the British society) went beyond attitudes common in their time to give eugenics a racist slant. The Eugenics Society repeatedly wished to study miscegenation, though it never led to any policies. Ethnicity took on added significance after the Second World War, as immigration from the New Commonwealth became a matter for debate in the 1950s and 1960s. Colin Bertram and C. P. Blacker were opponents of immigration, the former probably to avoid miscegenation as revealed by his authored broadsheet on West Indian immigration in 1958. Reginald Ruggles Gates and Sir Charles Galton Darwin were among the initiators of the journal *Mankind Quarterly*, a mouthpiece for scientific racism first appearing in 1961. Though founded in Edinburgh, it took a greater interest in the United States than in Britain.

The National Front and similar far-right extremists are virtually the only overt supporters of 'mainline' eugenics today. 'Mainliners' in earlier periods were usually social Darwinists, but seldom adherents of the far right. There were Nazi sympathizers among British eugenists in the 1930s, including George Pitt-Rivers and Lord Lymington. They were peripheral in the wider movement. Being Jewish has never been problematic within organized eugenics in Britain. Black and Asian doctors and academics started joining the Eugenics Society no later than the 1940s.

The fact that a doctrine is not especially racist does not mean that it is harmless. Race is just one variable which causes inequality in society. There were people in Britain who had reason to fear eugenics. The Brock

Report of 1934 had unanimously recommended that even carriers of genetic conditions should be 'offered' sterilization. Written work by C. P. Blacker and Cecil Binney makes clear that though touted as 'voluntary', if persuasion over time was to no avail, sterilization could be performed on individuals who had not given their informed consent. Sufferers from hereditary afflictions were at risk, especially mental defectives who had been the target of the Society's 1931 private member's bill. Once the principle had been recognized in the law, there is no telling how it would have developed subsequently. Denmark, like Britain, is a country under the rule of law with democratic traditions. Allowing sterilization there in 1929, with a sunset clause, led to two further laws in 1934 and 1935 extending the practice.[2] Until 1963 the Eugenics Society was considering if it should take up the cudgel for sterilization again.

Artificial insemination was another procedure in which it had a long-term interest. An explicitly eugenic version of it was launched in 1935 as 'eutelegenesis'. Herbert Brewer had coined the term, meaning 'good birth from a distance', in other words impersonal pregnancies with sperm from highly selected donors. The Society realized that legal and moral hindrances stood in the way. Yet the amount of time, effort and money it spent on this proves that it hoped they could be overcome. In the early 1960s, it devoted funds to researching low-temperature semen preservation. In the 1970s it wanted a change in the law to make all children born in a marriage of equal status, whether conceived naturally or with the help of a donor. In the 1980s it gave financial support to its member Sue Teper and her colleague, Malcolm Symonds, for a pilot study of children born from Artificial Insemination by Donor.

Artificial insemination increased its potential range of uses in the 1960s when in vitro fertilization was successfully pioneered by another member, Robert Edwards, and the gynaecologist Patrick Steptoe. Preimplantation genetic diagnosis, cloning, biopsy, genetic enhancement and prenatal sex selection were all made possible or facilitated by it. After the Second World War eugenics increasingly worked on an individual level and these techniques have clear implications for the doctrine. To enable the choice of traits in an embryo raises the prospect of designer babies. It has long been possible on a gametal level, choosing sperm or, later, ova for their

[2] David Redvaldsen, 'Eugenics as a Science and as a Social Movement. The Cases of Denmark and Norway 1900–1950', *Moving the Social. Journal of Social History and the History of Social Movements* 48 (2012), pp. 139, 142–143.

genetic traits. Although the complicated workings of heredity do not guarantee success, neither do current methods of affecting somatic cells.

Important as 'voluntary' sterilization and artificial insemination were to the Society, another practice relating to reproduction had even greater potency. Contraception was discussed from the early 1920s. In 1926 there were clear eugenic indicators for who should be using it. Married women of the lowest social tiers should be encouraged to limit their families, whereas healthy middle-class people should be 'condemned' for thinking along the same lines. In the post-war period it accepted that contraception would be utilized by everybody, but the attitude remained the same. The idea was reformulated in 1957 as the suggestion that 'inferior' couples should have just one or two children, while 'valuable' ones should try for at least four. The following year Marie Stopes, a well-known eugenist, died and left her clinic in Fitzrovia to the Society. The bequest allowed it to run a birth control service, which it did until 1976. Earlier it had funded and forged a connection with the North Kensington Marriage Welfare Centre. On a larger scale, it financed and supported the International Planned Parenthood Federation from its inception in 1952. The Society did not charge the International Planned Parenthood Federation rent for the use of office space at its premises 69 Eccleston Square. In the late 1930s and during the Second World War, neither had it charged the Family Planning Association rent. The latter was a merger of several organizations, including the National Birth Control Association, with which the Society had made contact in 1936.

When organized eugenics began in Edwardian Britain, the intention was that the Society should propagate the doctrine, while the Galton Laboratory researched and refined it. The latter was only a eugenic institution under its first two directors, Karl Pearson and Ronald Fisher. Their consecutive tenures lasted from 1907 to 1943. It abandoned eugenics in favour of human genetics with medical applications immediately after the Second World War. Lionel Penrose held the Galton chair from 1944. The changeover to genetics had been foreshadowed by the founding of the serological laboratory there in 1935. It attempted to make live medical discoveries on the causes of mental disorders and subnormality. Penrose had originally been a eugenist, but during the Colchester survey of mental defect 1930–1938, he had uncovered a variety of genetic, pathological and environmental factors as its causes. This must have made him think again, and in 1949 he announced in a lecture to the Society that he was at variance with its views.

Under Pearson and Fisher, the Galton Laboratory had furthered 'mainline' eugenics through research and publication. Pearson's team continued to study eugenics along the biometric lines which Galton, W. F. R. Weldon and he had pioneered. The topics straddled the Biometric and Galton Laboratories, ranging over a plethora of issues. Investigations were undertaken for governmental and public research bodies, especially during the First World War, when the entire department was commandeered for the war effort. The eugenic work predominantly consisted of statistical studies, based on a great deal of data but nearly always showing hereditary influences as preeminent. Trained field workers gathered raw data and it was categorized according to their interpretations, with many scales in use. The Galton Laboratory also occasionally ran courses for students, teachers, health professionals and others. This was a function established by Galton's will, though subsidiary to the research activities. The Treasury of Human Inheritance and the Eugenics Laboratory Memoirs were the pinnacle of these. The former continued until 1958 and the latter until 1966. They were fact-based medical science literature and therefore not affected by the abandoning of eugenics.

The Galton Laboratory had no popular appeal, which was not a source of regret to Pearson. He sought to influence government and decision makers directly. Take-up was high in the *British Medical Journal*, quality newspapers and the scientific press. The London County Council and the Medical Research Council funded projects and supported the investigations. The journals *Biometrika* and the *Annals of Eugenics*, edited in the Galton Laboratory, were distributed widely by subscription. Diverse institutions in Britain and abroad subscribed, proving the acceptability of academic eugenics. Relations were poor with the Eugenics Society until its influential member Ronald Fisher became director of the Galton Laboratory in 1933. The Society represented popular eugenics, although it too contained leading scientists. At least three members between 1907 and 1990 won Nobel prizes in physiology or medicine and one in economics.

In the 1960s the Society too pivoted towards research and aspired to be a learned association. There were many steps along that road, with its research committee originally being set up in 1912. From 1923 a portion of its membership had been known as 'fellows'. By officially becoming a charity it avoided having to pay income tax, in return for not directly taking part in politics. These pressure-group activities had been crowned with the occasional success, with its influence rising before the 1930s. Its sustained focus on 'voluntary' sterilization in that decade created new opponents, especially the Catholic Church and the Labour Party with its affiliated Trades Union Congress. The class prejudice displayed by eugenists in the recent past, fuelled suspicions about who exactly would be sterilized. The National government and subsequent Conservative governments demurred as sterilization was too controversial. Although the Brock Committee reported in 1934, the time never became right to implement its proposals for 'voluntary' sterilization.

The Society wanted to remain an influence through research, publications, lectures and an annual symposium. Many members had in any case been more interested in discussing eugenics and related topics than in implementing the doctrine. In 1969 the *Eugenics Review* was replaced by the *Journal of Biosocial Science* and the *Bulletin*. The *Journal of Biosocial Science* published academic articles and the *Bulletin* shorter pieces alongside news. The reinvention of eugenics as 'biosocial science' placed it at the intersection of biology and sociology. The name was chosen to avoid negative connotations. For the same reason, the Society changed its name to the Galton Institute in 1989.

A hundred and fifty years ago, Galton described the ideal person as being healthy, intelligent and altruistic. He believed that the progress of society depended on the gifted, who should be encouraged to intermarry and develop a feeling of caste. However, everyone should have their mental and physical characteristics improved. Thirty years later, he listed 'health, energy, ability, manliness and courteous disposition' as the qualities that he valued.[3] The best specimens from every class and sect would be progenitors. From this a civilization would arise, where the tone was higher and its inhabitants better able to fulfil Britain's imperial destiny. Like the Athenians whom he admired, Britons had succeeded in building an empire due to the emigration of well-bred compatriots.[4]

Perhaps the ideal of progressing as a nation still stands, no longer symbolized by painting the map British red. What has changed is that few look to heredity as the means of achieving it. Public health measures, education, enterprise and technology are instead the favoured solutions. Charles Darwin believed that forming communities increased the chances of survival of all their members. Enforceable laws that protect the lives

[3] Francis Galton, 'Eugenics: Its Definition, Scope and Aims' in *idem*, *Essays in Eugenics* (London: Eugenics Education Society), pp. 36–37.

[4] Debbie Challis, *The Archaeology of Race. The Eugenic Ideas of Francis Galton and Flinders Petrie* (London: Bloomsbury Academic, 2013), p. 55.

of fellow citizens, medical help, knowledge sharing and collective defence against outside aggression do just that. Eugenics does not. It sees the future as belonging only to a fixed proportion of the population. How large the proportion should be, what to do about the rest and how to ensure that the best human stocks thrive and multiply were the main questions for eugenists.

Appendix

Holders of the Galton chair at University College.

Period	Holder
1911–1933	Karl Pearson
1933–1943	Ronald Fisher
1944–1965	Lionel Penrose
1965–1975	Harry Harris
1975–1994	Bette Robson
1994–2009	vacant
2009–	Nicholas Wood

Presidents of the Eugenics Society 1908–1990.

Period	President
1908–1909	Sir James Crichton-Browne
1909–1910	Montague Crackenthorpe
1911–1928	Leonard Darwin
1929–1932	Sir Bernard Mallet
1933–1934	Sir Humphry Rolleston
1935–1948	Lord Horder
1949–1952	Sir Alexander Carr-Saunders
1953–1958	Sir Charles Galton Darwin

(continued)

D. Redvaldsen, *A History of British Eugenics since 1865*, Medicine and Biomedical Sciences in Modern History,
https://doi.org/10.1007/978-3-031-72290-5

(continued)

Period	*President*
1959–1961	Sir Julian Huxley
1962–1964	Sir James Gray
1965–1967	Lord Platt
1968–1969	Sir Alan Parkes
1970–1971	Peter Cox
1972–1975	Cedric Carter
1976–1982	Harry Armytage
1983–1987	Bernard Benjamin
1988–1990	Margaret Sutherland

General secretaries of the Eugenics Society 1920–1990.

Period	*General secretary*
1920–1931	Cora Hodson (Business Secretary)
1931–1952	C. P. Blacker
1952–1957	Cedric Carter
1957–1965	Colin Bertram
1965–1973	Faith Schenk
1974–1985	Eileen Walters
1985–1990	Linda Brooks

Membership of the Eugenics Society 1910–1990.

Year	*Members*	*Source*
1910	679	Annual Report 1910–11, pp. 25–35, 38–40, 42, 44, 46–47
1915	975	Annual Report 1914–15, pp. 13–30, 32–33
1920	493	Annual Report 1919–1920, pp. 10–14
1926	561	SA/EUG/A85
1930	713	Annual Report 1930–31, p. 8
1935	673	Annual Report 1935–36, p. 6
1940	586	Annual Report 1940–41, p. 4
1945	495	Annual Report 1944–45, p. 4
1950	544	Annual Report 1950–51, p. 3
1955	515	Annual Report 1955–56, p. 9
1960	498	Annual Report 1960–61, p. 11

(continued)

(continued)

Year	*Members*	*Source*
1965	451	Annual Report 1965–66, p. 10
1970	390	*Bulletin* 3:4 (1971), p. 81
1975	365	*Bulletin* 8:1 (1976), p. 14
1980	408	*Bulletin* 13:1 (1981), p. 15
1988	380	SA/EUG/P35
1990	389	Annual Report 1989, folio 6 verso

Glossary

A. I. D.: artificial insemination by donor.
A. I. H.: artificial insemination by husband.
Alpha-feta-protein: protein emitted by malign or benign childhood tumours.
Anencephaly: lack of hemispheres in the brain.
Biopsy: gene therapy which removes one or more genes.
Blastocyst: an embryo consisting of a few cells only.
Brachydactyly: short stubby fingers.
Cloning: genetic duplication of an individual by asexual reproduction.
Degeneration: a process by which evolution leads to less 'fit' individuals.
Ectogenesis: artificial pregnancy occurring outside the body of a woman.
Embryo: the early stage of a foetus.
'Eutelegenesis': artificial insemination with highly selected semen.
'Feeblemindedness': a supposed condition of 'unfitness' or mental subnormality.
'Fit': individual suited to present conditions of life.
Gamete: germ cell of a male or a female which can fertilize or be fertilized.
Huntington's chorea: hereditary neurone disease in which muscular control becomes difficult.
Hypercholesterolemia: high levels of cholesterol in the blood.
In vitro fertilization: laboratory fertilization of an egg.

D. Redvaldsen, *A History of British Eugenics since 1865*, Medicine and Biomedical Sciences in Modern History,
https://doi.org/10.1007/978-3-031-72290-5

'Mainline' eugenics: standard British version of eugenics emphasizing heredity and social class.

Muscular dystrophy: congenital disease whereby failure of muscles causes incapacitation.

'Negative' eugenics: measures to decrease the breeding of the 'unfit'.

Parthenogenesis: reproduction occurring from an unfertilized egg.

'Penectogenesis': in vitro fertilization.

'Positive' eugenics: measures to increase the breeding of the 'fit'.

Preimplantation embryonic selection: choice of an embryo from several possible ones.

Preimplantation genetic diagnosis: laboratory examination of an embryo for characteristics.

Preimplantation genetic haplotyping: examination of an embryo for traits from one parent.

Pronucleus: nucleus of sperm or egg before fertilization.

'Reform' eugenics: version of eugenics supporting social reform alongside hereditary measures.

Social Darwinism: accepting or applying natural selection to humans in the present.

Spina bifida: congenital condition in which the spinal cord protrudes through the backbone.

Surrogacy: a woman carrying a foetus on behalf of another woman, a couple or a man.

'Unfit': individual not suited to present conditions of life.

Bibliography

Primary sources

Archives

British Library of Political and Economic Science, London. Carr-Saunders papers.
Danish National Archives, Copenhagen. Københavns Universitet.
London Metropolitan University, TUC Collection.
Modern Records Centre, University of Warwick.
National Archives of Norway, Oslo.
Queen Mary Archive, London. Chesworth papers.
University College London. Galton Archive.
University College London. Galton Laboratory.
University College London. Pearson papers.
Wellcome Library, London. Eugenics Society Archive.

Journals

British Medical Journal, The
Bulletin, The Eugenics Society
Daily Telegraph, The
Eugenics Review, The
Galton Institute Newsletter, The
Hansard
Manchester Guardian, The
New Statesman, The

D. Redvaldsen, *A History of British Eugenics since 1865*, Medicine and Biomedical Sciences in Modern History,
https://doi.org/10.1007/978-3-031-72290-5

The Lancet
The Times

Books

Artificial Human Insemination. The Report of a Commission Appointed by His Grace the Archbishop of Canterbury (London: S. P. C. K, 1948)

Benn, Tony. *Office Without Power. Diaries 1968–72* (London: Arrow Books, 1989)

Blacker, C. P. *Eugenics. Galton and After* (London: Gerald Duckworth, 1952)

Cattell, Raymond B. *The Fight for our National Intelligence* (London: P. S. King & Son, 1937)

Constable, F. C. *Poverty and Hereditary Genius. A Criticism of Mr Francis Galton's Theory of Hereditary Genius* (London: Arthur C. Fifield, 1905)

Darwin, Charles. *The Descent of Man and Selection in Relation to Sex* (London: John Murray, 1909)

Darwin, Charles. *On the Origin of Species by Means of Natural Selection or The Preservation of Favoured Races in the Struggle for Life* (London: Folio Society, 2006)

Dickens, Charles. *Bleak House* (London: Penguin, 2012)

Eysenck, H. J. *Race, Intelligence and Education* (London: Temple Smith, 1971)

Eysenck, H. J. *Rebel with a Cause. The Autobiography of Hans Eysenck* (New Brunswick NJ: Transaction Publishers, 1997)

Ferri, Enrico. *Socialism and Positive Science (Darwin-Spencer-Marx)* (London: Independent Labour Party, 1905)

Galton, Francis. *Hereditary Genius. An Inquiry into its Laws and Consequences* (London: Macmillan, 1892)

Galton, Francis. *Inquiries into Human Faculty and its Development* (New York: Macmillan, 1883)

Galton, Francis. *Memories of My Life* (London: Methuen, 1908)

Haldane, J. B. S. *Daedalus or Science & the Future. A Paper Read to the Heretics, Cambridge on February 4th, 1923* (New York: E. P. Dutton & Company, 1924)

Headley, F. W. *Darwinism and Modern Socialism* (London: Methuen, 1909)

Herbert, S. *The First Principles of Evolution* (London: A&C Black, 1915)

Herbert, S. *The First Principles of Heredity* (London: Adam and Charles Black, 1910)

Hogben, Lancelot. *Nature and Nurture* (London: George Allen & Unwin, 1945)

Huxley, Julian. *New Bottles for New Wine* (London: Chatto & Windus, 1959)

Independent Labour Party, *Report of the Coming-of-Age Conference. Held at Bradford, April 1914* (London: Independent Labour Party, 1914)

Laughlin, Harry H. *The Second International Exhibition of Eugenics* (Baltimore: Williams & Wilkins, 1923)
Lombroso, Cesare. *The Man of Genius* (New York: Garland Publishing, 1984)
Ludovici, Anthony M. *The Choice of a Mate* (London: John Lane & The Bodley Head, 1935)
Lynn, Richard. *Dysgenics. Genetic Deterioration in Modern Populations* (no place of publication: Ulster Institute of Social Research, 2011)
Lynn, Richard. *Eugenics: A Reassessment* (Westport CT: Praeger, 2001)
Lynn, Richard. *The Science of Human Diversity. A History of the Pioneer Fund* (Lanham MD: University Press of America, 2001)
MacArthur, Brian (ed.). *The Penguin Book of Modern Speeches* (London: Penguin, 2012)
McDougall, William. *Is America Safe for Democracy?* (New York: Charles Scribner, 1921)
Meade, J. E. and A. S. Parkes (eds.). *Biological Aspects of Social Problems* (Edinburgh: Oliver & Boyd, 1965)
Meade, J. E. and A. S. Parkes (eds.). *Genetic and Environmental Factors in Human Ability* (Edinburgh: Oliver & Boyd, 1966)
Mill, John Stuart. *Considerations on Representative Government* (New York: Prometheus Books, 1991)
Mosley, Oswald. *Fascism. 100 Questions Asked and Answered* (London: B U F Publications, 1936)
Muller, H. J. *Out of the Night. A Biologist's View of the Future* (London: Victor Gollancz, 1936)
National Academy of Sciences, National Academy of Engineering, Institute of Medicine, National Research Council. *Scientific and Medical Aspects of Human Reproductive Cloning* (Washington DC: National Academy Press, 2002)
Pannekoek, Anton. *Marxism and Darwinism* (Chicago: Charles H. Kerr & Co, 1912)
Paul, Eden. *Socialism and Eugenics* (Manchester: National Labour Press, 1911)
Pearson, Karl. *The Ethic of Freethought. A Selection of Essays and Lectures* (London: T. Fisher Unwin, 1888)
Pearson, Karl. *The Grammar of Science* (London: Walter Scott, 1892)
Pearson, Karl. *On the Handicapping of the First-born* (London: Dulau and Co, 1914)
Pearson, Karl. *The Problem of Practical Eugenics* (London: Dulau and Co, 1912)
Professional Classes War Relief Council. 13 & 14 Prince's Gate, S. W. (no place of publication or date)
Punnett, R. C. (ed.). *Proceedings of the Seventh International Genetical Congress* (London: Cambridge University Press, 1941)

Report of the Departmental Committee on Human Artificial Insemination (London: H. M. Stationery Office, 1960)

Report of the Inter-departmental Committee on Physical Deterioration. Volume 1 (London: H. M. Stationery Office, 1904)

Report of the Royal Commission on the Care and Control of the Feeble-minded. Volume 8 (London: H. M. Stationery Office, 1908)

Report of the Thirteenth Annual Conference of The Labour Party (London: Labour Party, n.d.)

Ritchie, David G. *Darwinism and Politics. With Two Additional Essays on Human Evolution* (London: Swan Sonnenschein, 1895)

Schellen, A. C. M. *Artificial Insemination in the Human* (Amsterdam: Elsevier, 1957)

Shaw, Bernard (ed.). *Fabianism and the Empire. A Manifesto by the Fabian Society* (London: Grant Richards, 1900)

Sims, J. Marion. *Clinical Notes on Uterine Surgery. With Special Reference to the Management of the Sterile Condition* (New York: William Wood & Co, 1866)

Small, Laurence. *Darwinism and Socialism* (London: Independent Labour Party, 1907)

Terman, Lewis M. *The Measurement of Intelligence. An Explanation of and a Complete Guide for the Use of the Stanford Revision and Extension of the Binet-Simon Intelligence Scale* (Boston MA: Houghton Mifflin, 1916)

The Problem of the Feeble-Minded. An Abstract of the Report of the Royal Commission on the Care and Control of the Feeble-minded (London: P. S. King, 1909)

Webb, Sidney. *The Decline in the Birth-Rate* (London: Fabian Society, 1907)

Wells, H. G. *A Modern Utopia* (London: Chapman & Hall, 1905)

Wells, H. G. *Anticipations of the Reaction of Mechanical and Scientific Progress upon Human Life and Thought* (Leipzig: Bernard Tauchnitz, 1902)

Whitehead, George. *Socialism and Eugenics* (London: Twentieth Century Press, 1911)

Chapters in Volumes

Benjamin, Bernard. 'Variation in Mortality in the United Kingdom with Special Reference to Immigrants and Minority Groups' in D. A. Coleman (ed.), *Demography of Immigrants and Minority Groups in the United Kingdom* (London: Academic Press, 1982), pp. 43–69.

Carter, C. O. 'Eugenic Implications of New Technology' in *idem*, *Developments in Human Reproduction and their Eugenic, Ethical Implications* (London: Academic Press, 1983), pp. 205-211.

Cusine, Douglas J. 'Legal Implications' in C. O. Carter (ed.), *Developments in Human Reproduction and their Eugenic, Ethical Implications* (London: Academic Press, 1983), pp. 227–236.

Eysenck, H. J. 'The Galton Lecture 1983. Intelligence: New Wine in Old Bottles', in C. J. Turner and H. B. Miles (eds.), *The Biology of Human Intelligence* (Nafferton: Nafferton Books, 1984), pp. 33–52.

Galton, Franics. 'Eugenics: Its Definition, Scope and Aims' in *idem*, *Essays in Eugenics* (London: Eugenics Education Society, 1909), pp. 35–43.

Galton, Francis. 'Local Associations for Promoting Eugenics' in *idem*, *Essays in Eugenics* (London: Eugenics Education Society, 1909), pp. 100–109.

Galton, Francis. 'The Possible Improvement of the Human Breed under Existing Conditions of Law and Sentiment' in *idem*, *Essays in Eugenics* (London: Eugenics Education Society, 1909), pp. 1–34.

Galton, Franics. 'Preface' in *idem*, *Essays in Eugenics* (London: Eugenics Education Society, 1909), not paginated.

Galton, Francis. 'Probability, the Foundation of Eugenics' in *idem*, *Essays in Eugenics* (London: Eugenics Education Society), pp. 73–99.

Galton, Francis. 'Restrictions in Marriage' in *idem*, *Essays in Eugenics* (London: Eugenics Education Society, 1909), pp. 44–59.

Hewitt, J. K. and K. A. Last. 'Twin Studies of Intelligence: Recent Data, New Questions' in C. J. Turner and H. B. Miles (eds.), *The Biology of Human Intelligence* (Nafferton: Nafferton Books, 1984), pp. 65–84.

Hollis, John. 'New Commonwealth Ethnic Group Populations in Greater London' in D. A. Coleman (ed.), *Demography of Immigrants and Minority Groups in the United Kingdom* (London: Academic Press, 1982), pp. 119–141.

Holmes, Colin. 'The Promised Land? Immigration into Britain 1870–1980' in D. A. Coleman (ed.), *Demography of Immigrants and Minority Groups in the United Kingdom* (London: Academic Press, 1982), pp. 1–21.

Huxley, Julian. 'The Future of Man—Evolutionary Aspects' in Gordon Wolstenholme (ed.), *Man and His Future* (Boston MA: Little Brown and Co, 1963), pp. 1–22.

Kosmin, Barry A. 'Nuptiality and Fertility Patterns of British Jewry 1850–1980: An Immigrant Transition' in D. A. Coleman (ed.), *Demography of Immigrants and Minority Groups in the United Kingdom* (London: Academic Press, 1982), pp. 245–261.

Kumar, Dhavendra. 'Preface' in *idem, Genomics and Health in the Developing World* (Oxford: Oxford University Press, 2012), pp. VII-IX.

Mascie-Taylor, G. C. N. 'Biosocial Correlates of IQ' in C. J. Turner and H. B. Miles (eds.), *The Biology of Human Intelligence* (Nafferton: Nafferton Books, 1984), pp. 99–127.

Mjøen, Jon Alfred. 'Harmonic and Disharmonic Racecrossings' in *Eugenics in Race and State*. Volume II. *Scientific Papers of the Second International Congress of Eugenics* (Baltimore: Williams & Wilkins, 1923), pp. 53–58.

Teper, Sue and E. M. Symonds. 'Artificial Insemination by Donor' in C. O. Carter (ed.), *Developments in Human Reproduction and their Eugenic, Ethical Implications* (London: Academic Press, 1983), pp. 19–52.

Articles

Aliens Act 1905

'Artificial Insemination (Donor)', *Eugenics Review* 48:4 (1957), pp. 203–211.

'Eugenics, Socialism and Capitalism. Debate at Members' Meeting', *Eugenics Review* 27:2 (1935), pp. 109–119.

'International Eugenics Federation: Report of the Annual Meeting', *Eugenics Review* 20:3 (1928), pp. 185–187.

'The Reith Lectures 1959. The Future of Man', *Eugenics Review* 52:3 (1960), pp. 145–150.

'The Sterilization Bill', *Eugenics Review* 23:2 (1931), pp. 153–154.

'Voluntary Sterilization. Report of the Conference at the Caxton Hall on May 23rd 1932', *Eugenics Review* 24:2 (1932), pp. 133–136.

Bartholomew, G. W. 'The Development of Artificial Insemination', *Eugenics Review* 49:4 (1958), pp. 187–195.

William Beveridge, 'Eugenic Aspects of Children's Allowances', *Eugenics Review* 34:4 (1943), pp. 117–123.

Binney, Cecil. 'Legal and Social Implications of Artificial Insemination', *Eugenics Review* 40:4 (1949), pp. 199–204.

Binney, Cecil. 'Legal Problems Raised by Modern Discoveries about Sex', *The Medical-Legal Journal* 21:3 (1953), pp. 90–98.

Blacker, C. P. 'Artificial Insemination: The Society's Position', *Eugenics Review* 50:1 (1958), pp. 51–54.

Blacker, C. P. '"Eugenic" Experiments Conducted by the Nazis on Human Subjects', *Eugenics Review* 44:1 (1952), pp. 9–19.

Blacker, C. P. 'Eugenics in Germany', *Eugenics Review* 25:3 (1933), pp. 157–159.

Blacker, C. P. 'J. B. S. Haldane on Eugenics', *Eugenics Review* 44:3 (1952), pp. 146–151.

Brewer, Herbert. 'Eutelegenesis', *Eugenics Review* 27:2 (1935), pp. 121–126.

Brewer, Herbert. 'Eutelegenesis and Sterility', *Eugenics Review* 28:4 (1937), pp. 343–345.

Burt, Cyril. 'Experimental Tests of General Intelligence', *British Journal of Psychology* 3 (1909), pp. 94–177.

Burt, Cyril. 'The Measurement of Intelligence by the Binet Tests, Part I', *Eugenics Review* 6:1 (1914), pp. 36-50.

Burt, Cyril. 'The Measurement of Intelligence by the Binet Tests, Part II', *Eugenics Review* 6:2 (1914), pp. 140–152.

Burt, Cyril. 'Problems of Subnormality', *Eugenics Review* 10:4 (1919), pp. 224–231.

Carter, C. O. 'Promising Families: Some Conclusions', *Eugenics Review* 52:4 (1961), pp. 197–200.

Cattell, Raymond B. 'Some Further Considerations between Intelligence, Fertility and Socio-Economic Factors', *Eugenics Review* 29:3 (1937), pp. 171–179.

Cox, Harold. 'The Reduction of the Birth Rate as a Necessary Instrument for the Improvement of the Race', *Eugenics Review* 14:2 (1922), pp. 83–92.

Crew, F. A. E. et al. 'Social Biology and Population Improvement', *Nature* 144 (1939), pp. 521–522.

Darwin, Leonard. 'Heredity and Environment. A Warning to Eugenists', *Eugenics Review* 8:2 (1916), pp. 93–122.

Darwin, Leonard. 'Some Birth Rate Problems', *Eugenics Review* 12:3 (1920), pp. 147–157.

Darwin, Leonard. 'The Society's Coming of Age. The Growth of the Eugenics Movement', *Eugenics Review* 21:1 (1929), pp. 9–20.

Drysdale, C. V. 'Scientific Bases of Birth Control', *Eugenics Review* 20:3 (1928), pp. 173–178.

Fraser Roberts, J. A. 'Population Problems in Light of Differential Fertility', *Eugenics Review* 36:1 (1944), pp. 9–16.

Galton, Francis. 'Foreword', *Eugenics Review* 1:1 (1909), pp. 1–2.

Galton, Francis. 'Hereditary Improvement', *Fraser's Magazine* 7:37 (1873), pp. 116–130.

Galton, Francis. 'Hereditary Talent and Character. Part I', *Macmillan's Magazine* 12: 68 (1865), pp. 157–166.

Galton, Francis. 'Hereditary Talent and Character. Second Paper', *Macmillan's Magazine* 12: 70 (1865), pp. 318–327.

Galton, Francis. 'On the Anthropometric Laboratory at the Late International Health Exhibition', *Journal of the Anthropological Institute* 14 (1885), pp. 205–221.

Galton, Francis. 'The Part of Religion in Human Evolution', *The National Review* 23 (1894), pp. 755–765.

Galton, Francis. 'Statistical Inquiries into the Efficacy of Prayer', *Fortnightly Review* 12 (1872), pp. 125–135.

Goodhart, C. B. 'Estimation of Illegal Abortions', *Journal of Biosocial Science* 1:3 (1969), pp. 235–245.

Goodhart, C. B. 'The Frequency of Illegal Abortion', *Eugenics Review* 55:4 (1964), pp. 197–200.

Herbert, S. 'Eugenics and Socialism', *Eugenics Review* 2:2 (1910), pp. 116–123.

Herbert, S. 'Socialism and the New Science', *The Socialist Review. A Monthly Review of Modern Thought* I March-August 1908 (London: Independent Labour Party, n. d.), pp. 305–309.

Hogben, Lancelot T. 'Modern Heredity and Social Science', *The Socialist Review: A Quarterly Review of Modern Thought* Vol. 16 No. 89 (April–June 1919), pp. 147–156.

Hutton, Kenneth. 'Intelligence Quotients and Differential Fertility. Some Observations from Winchester College', *Eugenics Review* 44:4 (1953), pp. 205–215.

Huxley, Julian. 'Eugenics and Society', *Eugenics Review* 28:1 (1936), pp. 11–31.

Inge, W. R. 'Some Moral Aspects of Eugenics', *Eugenics Review* 1:1 (1909), pp. 26–36.

Jones, L. T. E. 'Psychology and Vocational Fitness: The Inherent Danger of Intelligence Tests', *Socialist Review* 41 (June 1929), pp. 37–41.

Kemp, Tage. 'Danish Experiences in Negative Eugenics, 1929-45', *Eugenics Review* 38:4 (1947), pp. 181–186.

Kemp, Tage. 'Genetic-Hygienic Experiences in Denmark in Recent Years', *Eugenics Review* 49:1 (1957), pp. 11–18.

Simms, Madeleine, 'People Count. A History of the General Register Office', *Journal of Biosocial Science* 20:3 (1988), pp. 377–378.

Marshall, Wm. C. 'The Education Bill and Eugenics', *Eugenics Review* 10:1 (1918), pp. 18–23.

Mjøen, Jon Alfred. 'Harmonic and Disharmonic Racecrossings', *Eugenics Review* 14:1 (1922), pp. 35–40.

Mügge, Maximilian A. 'Eugenics and the Superman: A Racial Science, and a Racial Religion', *Eugenics Review* 1:3 (1909), pp. 184–193.

Pearson, Karl and Margaret Moul, 'The Mathematics of Intelligence. The Sampling Errors in the Theory of a Generalized Factor', *Biometrika* 19:3-4 (1927), pp. 246–291.

Pearson, Karl and Margaret Moul. 'The Problem of Alien Immigration into Great Britain, Illustrated by an Examination of Russian and Polish Jewish Children. Part I', *Annals of Eugenics* 1:1 (1925), pp. 5–54.

Pearson, Karl and Margaret Moul. 'The Problem of Alien Immigration into Great Britain, Part II', *Annals of Eugenics* 2:1–2 (1925), pp. 56–127.

Pearson, Karl and Margaret Moul. 'The Problem of Alien Immigration into Great Britain, Illustrated by an Examination of Russian and Polish Jewish Children. Part III', *Annals of Eugenics* 2:3–4 (1927), pp. 290–317.

Pearson, Karl and Margaret Moul. 'The Problem of Alien Immigration into Great Britain, Illustrated by an Examination of Russian and Polish Jewish Children. Part III', *Annals of Eugenics* 3:1–2 (1928), pp. 1–76.

Pearson, Karl and Margaret Moul. 'The Problem of Alien Immigration into Great Britain, Illustrated by an Examination of Russian and Polish Jewish Children. Part III (cont.)', *Annals of Eugenics* 3:3–4 (1928), pp. 201–264.

Penrose, L. S. 'The Galton Laboratory: Its Work and Aims', *Eugenics Review* 41:1 (1949), pp. 17–27.

Perrins, Cullwick. 'Eugenics without the Churches', *The Socialist Review* New Series Vol III no. 3 (January 1931), pp. 140–146.

Piddington, Marion. 'For Whom are we Building the Bridge? Task of Eugenics', *The Land*, 18 October 1929.

Race, R. R. 'Some Notes on Fisher's Contributions to Human Blood Groups', *Biometrics* 20:2 (1964), pp. 361–367.

Review by H. W. H., 'The Races of Britain', *Science* 7:155 (1886), pp. 84–86.

Reznichenko, A. S., C. Huyser and M. S. Pepper. 'Mitochondrial Transfer: Implications for Assisted Reproductive Technologies', *Applied and Translational Genomics* 11 (2016), pp. 40–47.

Schenk, Faith and A. S. Parkes, 'The Activities of the Eugenics Society', *Eugenics Review* 60:3 (1968), pp. 142–161.

Spearman, Charles. '"General Intelligence" Objectively Determined and Measured', *American Journal of Psychology* 15:2 (1904), pp. 201–292.

Spearman, C. 'The Measurement of Intelligence', *Eugenics Review* 6:4 (1915), pp. 312–313.

Spriggs, M. 'Lesbian Couple Create a Child Who is Deaf Like Them', *Journal of Medical Ethics* 28:5 (2002), p. 283.

Trevor, J. C. 'Some Anthropological Characteristics of Hybrid Populations', *Eugenics Review* 30:1 (1938), pp. 21–31.

Webb, Sidney. 'The Majority Report', *Eugenics Review* 2:3 (1910), pp. 233–241.

Whetham, C. D. 'Eugenics and National Baby Week', *Eugenics Review* 9:3 (1917), pp. 234–235.

Wythenshawe, Lord Simon of. 'Some Aspects of World Population and Food Resources', *Eugenics Review* 46:2 (1954), pp. 91–100.

Secondary Sources

Books

Asgeirsson, Hrafn and Salvör Nordal (eds.), *Reproductive Technology and Surrogacy. A Global Perspective* (Copenhagen: Nordic Council of Ministers, 2015)

Bangham, Jenny. *Blood Relations. Transfusion and the Making of Human Genetics* (Chicago: University of Chicago Press, 2020)

Bannister, Robert C. *Social Darwinism. Science and Myth in Anglo-American Thought* (Philadelphia: Temple University Press, 1979)

Barkan, Elazar. *The Retreat of Scientific Racism. Changing Concepts of Race in Britan and the United States between the World Wars* (Cambridge: Cambridge University Press, 1992)

Baron, Jeremy Hugh. *The Anglo-American Biomedical Antecedents of Nazi Crimes. An Historical Analysis of Racism, Nationalism, Eugenics and Genocide* (Lewiston NY: Edward Mellon Press, 2007)

Bland, Lucy. *Banishing the Beast. Sexuality and the Early Feminists* (New York: The New Press, 1995)

Bowler, Peter J. *Monkey Trials and Gorilla Sermons. Evolution and Christianity from Darwin to Intelligent Design* (Cambridge MA: Harvard University Press, 2007)

Bowler, Peter J. *Science for All. The Popularization of Science in Early Twentieth-century Britain* (Chicago: University of Chicago Press, 2009)

Brooke, Stephen. *Sexual Politics. Sexuality, Family Planning, and the British Left from the 1880s to the Present Day* (Oxford: Oxford University Press, 2011)

Brookes, Martin. *Extreme Measures. The Dark Visions and Bright Ideas of Francis Galton* (London: Bloomsbury, 2004)

Buchanan, Roderick D. *Playing with Fire. The Controversial Career of Hans J. Eysenck* (Oxford: Oxford University Press, 2010)

Carlson, Elof Axel, *Genes, Radiation, and Society: The Life of H. J. Muller* (Ithaca NY: Cornell University Press, 1981)

Challis, Debbie. *The Archaeology of Race. The Eugenic Ideas of Francis Galton and Flinders Petrie* (London: Bloomsbury Academic, 2013)

Chitty, Clyde. *Eugenics, Race and Intelligence in Education* (London: Continuum, 2007)

Claeys, Gregory (ed.). *Late Victorian Utopias: A Prospectus.* Volume 1 (London: Pickering & Chatto, 2009)

Crook, Paul. *Darwinism, War and History. The Debate over the Biology of War from the 'Origin of Species' to the First World War* (Cambridge: Cambridge University Press, 1994)

Crook, Paul. *Darwin's Coat-tails. Essays on Social Darwinism* (New York: Peter Lang, 2007)

Daar, Judith. *The New Eugenics. Selective Breeding in an Era of Reproductive Technologies* (New Haven CT: Yale University Press, 2017)

Deary, Ian J., Lawrence J. Whalley and John M. Starr, *A Lifetime of Intelligence. Follow-up Studies of the Scottish Mental Surveys of 1932 and 1947* (Washington DC: American Psychological Association, 2009)

Desmond, Adrian and James Moore, *Darwin's Sacred Cause. Race, Slavery and the Quest for Human Origins* (London: Allen Lane, 2009)

Dickens, Peter. *Social Darwinism. Linking Evolutionary Thought to Social Theory* (Buckingham: Open University Press, 2000)
Dubow, Saul. *Scientific Racism in Modern South Africa* (Cambridge: Cambridge University Press, 1995)
Engs, Ruth Clifford. *The Eugenics Movement. An Encyclopedia* (Westport CT: Greenwood Press, 2005)
Farrall, Lyndsay Andrew. *The Origins and Growth of the English Eugenics Movement 1865–1925* (New York: Garland Publishing, 1985)
Finegold, William J. *Artificial Insemination* (Springfield IL: Charles C. Thomas, 1976)
Gillham, Nicholas Wright. *A Life of Sir Francis Galton. From African Exploration to the Birth of Eugenics* (Oxford: Oxford University Press, 2001)
Gould, Stephen Jay. *The Mismeasure of Man* (London: Penguin, 1992)
Griffith, Gareth. *Socialism and Superior Brains. The Political Thought of Bernard Shaw* (London: Routledge, 1993)
Hall, Lesley A. *Sex, Gender and Social Change in Britain since 1880* (Basingstoke: Macmillan, 2000)
Hanson, Clare. *Eugenics, Literature and Culture in Post-war Britain* (New York: Routledge, 2013)
Hart, Bradley W. *George Pitt-Rivers and the Nazis* (London: Bloomsbury Academic, 2015)
Hasian jr., Marouf Arif. *The Rhetoric of Eugenics in Anglo-American Thought* (Athens GA: University of Georgia Press, 1996)
Hawkins, Mike. *Social Darwinism in European and American Thought, 1860–1945. Nature as Model and Nature as Threat* (Cambridge: Cambridge University Press, 1997)
Jones, Greta. *Social Darwinism and English Thought. The Interaction between Biological and Social Theory* (Sussex: The Harvester Press, 1980)
Jones, Greta. *Social Hygiene in Twentieth-century Britain* (London: Croom Helm 1986)
Kevles, Daniel J. *In the Name of Eugenics. Genetics and the Uses of Human Heredity* (Berkeley CA: University of California Press, 1986)
Kühl, Stefan. *For the Betterment of the Race. The Rise and Fall of the International Movement for Eugenics and Racial Hygiene* (Basingstoke: Palgrave Macmillan, 2013)
Lewis, D. S. *Illusions of Grandeur. Mosley, Fascism and British Society 1931–81* (Manchester: Manchester University Press, 1987)
Macintosh, Kerry Lynn. *Human Cloning. Four Fallacies and their Legal Consequences* (Cambridge: Cambridge University Press, 2013)
MacKellar, Calum and Christopher Bechtel (eds.). *The Ethics of the New Eugenics* (New York: Berghahn, 2014)

Mazumdar, Pauline M. H. *Eugenics, Human Genetics and Human Failings. The Eugenics Society, its Sources and its Critics in Britain*, (London: Routledge, 1992)

McGrath, Alastair E. *Darwinism and the Divine. Evolutionary Thought and Natural Theology* (Chichester: Wiley-Blackwell, 2011)

McKibbin, Ross. *Classes and Cultures. England 1918–1951* (Oxford: Oxford University Press, 2000)

McLaren, Angus. *Reproduction by Design. Sex, Robots, Trees, and Test-tube Babies in Interwar Britain* (Chicago: University of Chicago Press, 2012)

Merricks, Patrick T. *Religion and Racial Progress in Twentieth-century Britain. Bishop Barnes of Birmingham* (Cham: Palgrave Macmillan, 2017)

Paul, Diane B. *Controlling Human Heredity. 1865 to the Present* (New Jersey: Humanities Press, 1995)

Pearson, Karl. *The Life, Letters and Labours of Francis Galton.* Volume 1 (Cambridge: Cambridge University Press, 1914)

Pearson, Karl. *The Life, Letters and Labours of Francis Galton.* Volume 2 (Cambridge: Cambridge University Press, 1924)

Pearson, Karl. *The Life, Letters and Labours of Francis Galton.* Volume 3a (Cambridge: Cambridge University Press, 1930)

Pfeffer, Naomi. *The Stork and the Syringe. A Political History of Reproductive Medicine* (Cambridge: Polity Press, 1993)

Pichot, André. *The Pure Society. From Darwin to Hitler* (London: Verso, 2009)

Pickens, Donald K. *Eugenics and the Progressives* (Nashville TN: Vanderbilt University Press, 1968)

Plotz, David. *The Genius Factory. Unravelling the Mysteries of the Nobel Prize Sperm Bank* (London: Simon & Schuster, 2005)

Porter, Roy. *The Greatest Benefit to Mankind. A Medical History of Humanity from Antiquity to the Present* (London: HarperCollins, 1997)

Porter, Theodore M. *Karl Pearson. The Scientific Life in a Statistical Age* (Princeton NJ: Princeton University Press, 2004)

Pugh, Martin. *Women and the Women's Movement in Britain 1914–1959* (Basingstoke: Macmillan, 1992)

Richardson, Angelique. *Love and Eugenics in the Late Nineteenth Century. Rational Reproduction and the New Woman* (Oxford: Oxford University Press, 2003)

Rose, Jonathan. *Proletarian Philosophers. Problems in Socialist Culture in Britain, 1900–1940* (Oxford: Clarendon Press, 1984)

Rutherford, Adam. *Control. The Dark History and Troubling Present of Eugenics* (London: Weidenfeld & Nicolson, 2022)

Schaffer, Gavin. *Racial Science and British Society, 1930–62* (Basingstoke: Palgrave Macmillan, 2008)

Scott, Gillian. *Feminism and the Politics of Working Women. The Women's Co-operative Guild, 1880s to the Second World War* (London: Routledge, 2001)
Searle, G. R. *Eugenics and Politics in Britain, 1900–1914* (Leyden: Noordhoff International Publishing, 1976)
Searle, G. R. *The Quest for National Efficiency. A Study in British Politics and Political Thought, 1899–1914* (Oxford: Blackwell, 1971)
Soloway, Richard A. *Demography and Degeneration. Eugenics and the Declining Birthrate in Twentieth-century Britain* (Chapel Hill NC: University of North Carolina Press, 1995)
Squier, Susan Merrill. *Babies in Bottles. Twentieth-century Visions of Reproductive Technology* (New Brunswick NJ: Rutgers University Press, 1994)
Stack, David. *The First Darwinian Left. Socialism and Darwinism 1859–1914* (Cheltenham: New Clarion Press, 2003)
Stone, Dan. *Breeding Superman. Nietzsche, Race and Eugenics in Edwardian and Interwar Britain* (Liverpool: Liverpool University Press, 2002)
Szreter, Simon. *Fertility, Class and Gender in Britain, 1860–1940* (Cambridge: Cambridge University Press, 2002)
Thomson, Mathew. *The Problem of Mental Deficiency: Eugenics, Democracy, and Social Policy in Britain c. 1870–1959* (Oxford: Clarendon Press, 1998)
Timmins, Nicholas. *The Five Giants. A Biography of the Welfare State* (London: Fontana, 1996)
Trombley, Stephen. *The Right to Reproduce. A History of Coercive Sterilization* (London: Weidenfeld & Nicolson, 1988)
Turda, Marius. *Modernism and Eugenics* (Basingstoke: Palgrave Macmillan, 2010)
Turda, Marius and Aaron Gillette, *Latin Eugenics in Comparative Perspective* (London: Bloomsbury Academic, 2014)
Turda, Marius and Maria Sophia Quine, *Historicizing Race* (London: Bloomsbury Academic, 2018)
Turney, Jon. *Frankenstein's Footsteps. Science, Genetics and Popular Culture* (New Haven CT: Yale University Press, 1998)
Van Camp, Nathan. *Redesigning Life. Eugenics, Biopolitics, and the Challenge of the Techno-human Condition* (Brussels: Peter Lang, 2015)
Vickers, Rhiannon. *The Labour Party and the World*. Volume 1 (Manchester: Manchester University Press, 2003)
Waddington, Keir. *An Introduction to the Social History of Medicine: Europe since 1500* (Basingstoke: Palgrave Macmillan, 2011).
Waites, Bernard. *A Class Society at War. England 1914–1918* (Leamington Spa: Berg Publishers, 1987)
Weeks, Jeffrey. *Sex, Politics and Society. The Regulation of Sexuality since 1800* (London: Longman, 1981)

Weindling, Paul. *Health, Race and German Politics between National Unification and Nazism, 1870–1945* (Cambridge: Cambridge University Press, 1993)

Werskey, Gary. *The Visible College* (London: Penguin, 1978)

Winter, J. M. *The Great War and the British People* (Basingstoke: Palgrave Macmillan, 2003)

Young, Hugo. *One of Us. A Biography of Margaret Thatcher* (London: Pan Books, 1989)

Yuval-Davis, Nira. *Gender and Nation* (London: Sage Publications, 1997)

Zweiniger-Bargielowska, Ina. *Managing the Body. Beauty, Health, and Fitness in Britain, 1880–1939* (Oxford: Oxford University Press, 2010)

Chapters in volumes

Adams, Mark B. 'Eugenics in Russia 1900–1940' in *idem, The Wellborn Science. Eugenics in Germany, France, Brazil and Russia* (New York: Oxford University Press, 1990), pp. 153–216.

Alexander, Denis R. 'Creation and Evolution' in J. B. Stump and Alan G. Padgett (eds.), *The Blackwell Companion to Science and Christianity* (Chichester: Wiley-Blackwell, 2012), pp. 233–245.

Bashford, Alison. 'Julian Huxley's Transhumanism' in Marius Turda (ed.), *Crafting Humans. From Genesis to Eugenics and Beyond* (Goetingen and Taipei: V & R Unipress and National Taiwan University Press, 2013), pp. 153–167.

Bland, Lucy and Lesley A. Hall, 'Eugenics in Britain: The View from the Metropole' in Alison Bashford and Philippa Levine (eds.), *The Oxford Handbook of the History of Eugenics* (Oxford: Oxford University Press, 2010), pp. 213–227.

Cottingham, Jane. 'Sexual and Reproductive Connections and Disconnections in Public Health' in Ignatz Sanger (ed.), *Advances in Reproductive Technology Research* (New York: Nova Science Publishers, 2013), pp. 145–152.

Ekberg, Mervyn. 'Eugenics: Past, Present and Future' in Marius Turda (ed.), *Crafting Humans. From Genesis to Eugenics and Beyond* (Goetingen and Taipei: V & R Unipress and National Taiwan University Press, 2013), pp. 89–107.

Farquhar, Dion. 'Gamete Traffic/Pedestrian Crossings' in E. Ann Kaplan and Susan Squier (eds.), *Playing Dolly. Technocultural Formations, Fantasies and Fictions of Assisted Reproduction* (New Brunswick NJ: Rutgers University Press, 1999), pp. 17–36.

Flinn, Andrew. '"Mothers for Peace", Co-operation, Feminism and Peace: The Women's Co-operative Guild and the Antiwar Movement between the Wars' in Lawrence Black and Nicole Robertson (eds.), *Consumerism and the Co-operative Movement in Modern British History. Taking Stock* (Manchester: Manchester University Press, 2009), pp. 138–154.

Gaudillière, Jean-Paul. 'Le Syndrome nataliste: Hérédité, médecine et eugénisme en France et en Grande-Bretagne, 1920–1965' in Jean Gayon, Daniel Jacobi and Marie-Claude Lorne (eds.), *L'Eternel retour de l'eugénisme* (Paris: Presses Universitaires de France, 2006), pp. 177–199.

Haines, Erica. 'Egg Donation' in Annette Burfoot (ed.), *Encyclopedia of Reproductive Technologies* (London: Routledge, 2018), pp. 269–271.

Hall, Richard. T. 'Prenatal Diagnosis Technologies' in Annette Burfoot (ed.), *Encyclopedia of Reproductive Technologies* (London: Routledge, 2018), pp. 193–197.

Jasanoff, Sheila. 'Imagined and Invented Worlds' in Sheila Jasanoff and Sang-Hyun Kim (eds.), *Dreamscapes of Modernity. Sociotechnical Imaginaries and the Fabrication of Power* (Chicago: University of Chicago Press, 2015), pp. 321–341.

Kevles, Daniel J. 'Le Spectre de Galton et les nouvelles technologies de la reproduction: Genre et eugénisme, enjeux présents et passés' in Jean Gayon, Daniel Jacobi and Marie-Claude Lorne (eds.), *L'Eternel retour de l'eugénisme* (Paris: Presses Universitaires de France, 2006), pp. 201–229.

Levine, Philippa and Alison Bashford. 'Introduction: Eugenics and the Modern World' in Alison Bashford and Philippa Levine (eds.), *The Oxford Handbook of the History of Eugenics* (Oxford: Oxford University Press, 2010), pp. 3–24.

Mittwoch, Ursula. 'From Eugenics to Human Genetics' in *Penrose: Pioneer in Human Genetics. Report of a Symposium Held to Celebrate the Centenary of the Birth of Lionel Penrose Held 12th and 13th March 1998* (no place of publication: Centre for Human Genetics at UCL, no year of publication), pp. 40–42.

Muir, Cindy. 'Introduction' in *idem*, *Reproductive Technology* (Detroit: Thomson Gale, 2005), pp. 9–14.

Neuhaus, Richard John. 'The Return of Eugenics' in *idem*, *Guaranteeing the Good Life: Medicine and the Return of Eugenics* (Grand Rapids MI: William Eerdmans, 1990), pp. 1–28.

Pick, Daniel. 'The Politics of Nature. Science and Religion in the Age of Darwin' in Gareth Stedman Jones and Gregory Claeys (eds.), *The Cambridge History of Nineteenth-century Political Thought* (Cambridge: Cambridge University Press, 2001), pp. 649–690.

Richardson, Angelique. '"I differ widely from you": Darwin, Galton and the Culture of Eugenics' in Eckart Voigts, Barbara Schaff and Monika Pietrzak-Franger (eds.), *Reflecting on Darwin* (Farnham: Ashgate, 2014), pp. 17–40.

Roll-Hansen, Nils. 'Eugenics and the Science of Genetics' in Alison Bashford and Philippa Levine (eds.), *The Oxford Handbook of the History of Eugenics* (Oxford: Oxford University Press, 2010), pp. 80–97.

Searle, Geoffrey R. 'Eugenics: The Early Years' in Robert A. Peel (ed.), *Essays in the History of Eugenics* (London: Galton Institute, 1998), pp. 20–35.

Searle, G. R. 'Eugenics and Class' in Charles Webster (ed.), *Biology, Medicine and Society 1840–1940* (Cambridge: Cambridge University Press, 1981), pp. 217–242.

Shakespeare, Tom. 'Sex Selection Should be Outlawed' in Cindy Muir (ed.), *Reproductive Technology* (Detroit: Thomson Gale, 2005), pp. 66–68.

Soloway, Richard A. 'From Mainline to Reform Eugenics—Leonard Darwin and C. P. Blacker' in Robert A. Peel (ed.), *Essays in the History of Eugenics* (London: Galton Institute, 1998), pp. 52–80.

Squier, Susan. 'Negotiating Boundaries. From Assisted Reproduction to Assisted Replication' in E. Ann Kaplan and Susan Squier (eds.), *Playing Dolly. Technocultural Formations, Fantasies and Fictions of Assisted Reproduction* (New Brunswick NJ: Rutgers University Press, 1999), pp. 101–133.

Waldschmidt, Anne. 'Genetic Screening and German-based Eugenics—Old and New' in Annette Burfoot (ed.), *Encyclopedia of Reproductive Techniques* (London: Routledge, 2018), pp. 373–379.

Weldon, Estela V. 'Why Do You Want to Have a Child?' in Mariam Alizade (ed.), *Motherhood in the Twenty-first Century* (London: Karnac, 2006), pp. 59–71.

Articles

Becquemont, Daniel. 'Eugénisme et socialisme en Grande Bretagne 1890–1900', *Mil neuf cent* 18:1 (2000), pp. 53–79.

Bellomy, Donald C. '"Social Darwinism" Revisited', *Perspectives in American History* new series 1 (1984).

Brown, Ian. 'Who were the Eugenicists? A Study of the Formation of an Early Twentieth-century Pressure Group', *History of Education* 17:4 (1988), pp. 295–307.

Cain, Joe. 'Publications Produced by the Francis Galton Laboratory for National Eugenics', *The Library* 22:4 (2021), pp. 523–548.

Claeys, Gregory. 'The "Survival of the Fittest" and the Origins of Social Darwinism', *Journal of the History of Ideas* 61:2 (2000), pp. 223–240.

Clever, Iris. 'Miriam Tildesley and the Anthropological Politics of Standardizing Racial Measurements', *Perspectives on Science* 30:1 (2022), pp. 13–47.

Curthoys, Ann. 'Eugenics, Feminism and Birth Control: The Case of Marion Piddington', *Hecate* 15:1 (1989), pp. 73–99.

D'Arcy, F. 'The Malthusian League and the Resistance to Birth Control Propaganda in Late Victorian Britain', *Population Studies* 31:3 (1977), pp. 429–448.

Davin, Anna. 'Imperialism and Motherhood', *History Workshop* 5 (1978), pp. 9–65.

Farber, Steven A. 'US Scientists' Role in the Eugenics Movement' (1907–1939). A Contemporary Biologist's Perspective', *Zebrafish* 5:4 (2008), pp. 243–245.

Fraser Roberts, J. A. 'Reginald Ruggles Gates 1882–1962', *Biographical Memoirs of Fellows of the Royal Society* 10:10 (1964), pp. 83–106.

Freeden, Michael. 'Eugenics and Progressive Thought: A Study in Ideological Affinity', *Historical Journal* 22:3 (1979), pp. 645-671.

Harper, Peter S. 'Julia Bell and the Treasury of Human Inheritance', *Human Genetics* 116 (2005), pp. 422–432.

Hart, Bradley W. 'Watching the "Eugenic Experiment" Unfold: The Mixed Views of British Eugenicists toward Nazi Germany in the Early 1930s', *Journal of the History of Biology* 45:1 (2012), pp. 33–63.

Hart, Bradley W. and Richard Carr, 'Sterilization and the British Conservative Party: Rethinking the Failure of the Eugenics Society's Strategy in the Nineteen Thirties', *Historical Research* 88: 242 (2015), pp. 716-739.

Hopkinson, D. A. 'Harry Harris. 30 September 1919-17 July 1994', *Biographical Memoirs of Fellows of the Royal Society* 42 (1996), pp. 152-170.

Jones, Greta. 'Eugenics and Social Policy between the Wars', *Historical Journal* 25:3 (1982), pp. 717–728.

Karp, Laurence. 'Eutelegenesis: A Historical Perspective', *American Journal of Medical Genetics* 5:4 (1980), pp. 327–329.

Lucassen, Leo. 'A Brave New World: The Left, Social Engineering, and Eugenics in Twentieth-century Europe', *International Review of Social History* 55:2 (2010), pp. 265–296.

MacKenzie, D. 'Karl Pearson and the Professional Middle Class', *Annals of Science* 36:2 (1979), pp. 125–143.

MacKenzie, Donald. 'The Zero-Sum Assumption. Reply to Sullivan', *Social Studies of Science* 29:2 (2009), pp. 223–234.

Macnicol, John. 'Eugenics and the Campaign for Voluntary Sterilization in Britain between the Wars', *Social History of Medicine* 2:2 (1989), pp. 147–169.

MacNicol, John. 'The Voluntary Sterilization Campaign in Britain, 1918–1939', *Journal of the History of Sexuality* 2:3 (1992), pp. 422–438.

Magnello, M. Eileen. 'The Non-Correlation of Biometrics and Eugenics: Rival Forms of Laboratory Works in Karl Pearson's Career at University College London, Part 1', *History of Science* 37:1 (1999), pp. 79–106.

Magnello, M. Eileen. 'The Non-Correlation of Biometrics and Eugenics: Rival Forms of Laboratory Works in Karl Pearson's Career at University College London, Part 2', *History of Science* 37:2 (1999), pp. 123–150.

Mandler, Peter. 'Educating the Nation II. Universities', *Transactions of the Royal Historical Society* 25 (2015), pp. 1–26.

McMillan, John. 'The Return of the Inseminator: Eutelegenesis in Past and Contemporary Reproductive Ethics', *Studies in the History and Philosophy of Biological and Biomedical Sciences* 38:2 (2007), pp. 393–410.

Notestein, Frank W. 'The Report of the Royal Commission on Population: A Review', *Population Studies* 3:3 (1949), pp. 232–240.

Pearson, E. S. 'David Heron, 1881–1969', *Journal of the Royal Statistical Society. Series A (General)* 133:2 (1970), pp. 287–291.

Paul, Diane. 'Eugenics and the Left', *Journal of the History of Ideas* 45:4 (1984), pp. 567–590.

Pilgrim, David. 'The Eugenics Legacy in Psychology and Psychiatry', *International Journal of Social Psychiatry* 54:3 (2008), pp. 272–284.

Ray, L. J. 'Eugenics, Mental Deficiency and Fabian Socialism between the Wars', *Oxford Review of Education* 9:3 (1983), pp. 213–222.

Redvaldsen, David. 'Eugenics as a Science and as a Social Movement. The Cases of Denmark and Norway 1900–1950', *Moving the Social. Journal of Social History and the History of Social Movements* 48 (2012), pp. 133–155.

Renwick, Chris. 'Eugenics, Population Research, and Social Mobility Studies in Early and Mid-twentieth Century Britain', *Historical Journal* 59:3 (2016), pp. 845–867.

Renwick, Chris. 'From Political Economy to Sociology: Francis Galton and the Social-Scientific Origins of Eugenics', *British Journal for the History of Science* 44:3 (2011), pp. 343–369.

Richards, Martin. 'Artificial Insemination and Eugenics: Celibate Motherhood, Eutelegenesis and Germinal Choice', *Studies in the History and Philosophy of Biological and Biomedical Sciences* 39:2 (2008), pp. 211–221.

Roelcke, Volker. 'Eugenic Concerns, Scientific Practices: International Relations in the Establishment of Psychiatric Genetics in Germany, Britain, the USA and Scandinavia, c. 1910–60', *History of Psychiatry* 30:1 (2019), pp. 19–37.

Schneider, William. 'Towards the Improvement of the Human Race: The History of Eugenics in France', *Journal of Modern History* 54 (1982) pp. 268–291.

Searle, G. R. 'Eugenics and Politics in Britain in the 1930s', *Annals of Science* 36:2 (1979), pp. 159–170.

Semmel, Bernard. 'Karl Pearson: Socialist and Darwinist', *British Journal of Sociology* 9:2 (1958), pp. 111–125.

Shaw, Christopher. 'Eliminating the Yahoo. Eugenics, Social Darwinism and Five Fabians', *History of Political Thought* 8:3 (1987), pp. 521–544.

Shearmur, Jeremy. 'Beveridge and the Brief Life of "Social Biology" at the LSE', *Agenda: A Journal of Policy Analysis and Reform* 20:1 (2013), pp. 79–94.

Spencer, Hamish G. and Diane Paul, 'The Failure of a Scientific Critique: David Heron, Karl Pearson and Mendelian Eugenics', *British Journal for the History of Science* 31:4 (1998), pp. 441–452.

Stack, D. 'Charles Darwin's Liberalism in "Natural Selection as Affecting Civilised Nations"', *History of Political Thought* 33:3 (2012), pp. 525–554.

Tabery, James. 'R. A. Fisher, Lancelot Hogben, and the Origin(s) of the Genotype-Environment Interaction', *Journal of the History of Biology* 41:4 (2008), pp. 717–761.

Taylor, Barbara. 'The Demise of the Asylum in Late Twentieth-century Britain: A Personal History', *Transactions of the Royal Historical Society* 21 (2011), pp. 193–215.

Waller, J. C. 'Ideas of Heredity, Reproduction and Eugenics in Britain, 1800–1875', *Studies in the History and Philosophy of the Biological and Biomedical Sciences* 32:3 (2001), pp. 457–489.

Watt, David C. 'Lionel Penrose, F. R. S. (1898–1972) and Eugenics. Part Two', *Notes and Records of the Royal Society of London* 52:2 (1998), pp. 339–354.

Weindling, Paul. 'International Eugenics. Swedish Sterilization in Context', *Scandinavian Journal of History* 24:2 (1999), pp. 179–197.

Weindling, Paul. 'Julian Huxley and the Continuity of Eugenics in Twentieth Century Britain', *Journal of Modern European History* 10:4 (2012), pp. 480–499.

Weiss, Sheila Faith. 'After the Fall, Political Whitewashing, Professional Posturing, and Personal Refashioning in the Postwar Career of Otmar Freiherr von Verschuer', *Isis* 101: 4 (2010), pp. 722–758.

Wisscher, Peter M. and Michael E. Goddard, 'From R. A. Fisher's 1918 Paper to GWAS a Century Later', *Genetics* 211:4 (2019), pp. 1125–1130.

Woodhouse, Jayne. 'Eugenics and the Feeble-minded: The Parliamentary Debates of 1912-14', *History of Education* 11:2 (1982), pp. 117–137.

Theses

Clever, Iris Isabelle. 'The Lives and Afterlives of Skulls. The Development of Biometric Methods of Measuring Race (1880–1950)', Ph. D. thesis, University of California Los Angeles, 2020.

Merricks, Patrick T. 'Should Such a Faith Offend? Bishop Barnes and the British Eugenics Movement, c. 1924–1953', Ph. D. thesis, Oxford Brookes University, 2014.

Index

D. Redvaldsen, *A History of British Eugenics since 1865*, Medicine and Biomedical Sciences in Modern History,
https://doi.org/10.1007/978-3-031-72290-5

GPSR Compliance
The European Union's (EU) General Product Safety Regulation (GPSR) is a set of rules that requires consumer products to be safe and our obligations to ensure this.

If you have any concerns about our products, you can contact us on

ProductSafety@springernature.com

In case Publisher is established outside the EU, the EU authorized representative is:

Springer Nature Customer Service Center GmbH
Europaplatz 3
69115 Heidelberg, Germany

www.ingramcontent.com/pod-product-compliance
Lightning Source LLC
Chambersburg PA
CBHW062142200726
48454CB00005BB/351
* 9 7 8 3 0 3 1 7 2 2 8 9 9 *